Remodelling Geography

Luc Anselin

Remodelling Geography

Edited by
Bill Macmillan

Basil Blackwell

First published 1989

Basil Blackwell Ltd
108 Cowley Road, Oxford, OX4 1JF, UK

Basil Blackwell Inc.
3 Cambridge Center
Cambridge, MA 02142, USA

British Library Cataloguing in Publication Data

Remodelling geography.
1. Geography. Mathematical models
I. Macmillan, Bill
910′.0724
ISBN 0–631–16099–X

Library of Congress Cataloging in Publication Data

Remodelling geography.
Bibliography: p.
Includes index.
1. Geography—Mathematical models.
I. Macmillan, Bill.
G70.23.R46 1989 910′.01′51 88–7547
ISBN 0–631–16099–X

Typeset in 10 on 11 pt Sabon
by Vera-Reyes Inc., Philippines
Printed in Great Britain by
Camelot Press Ltd, Southampton

Contents

Preface

The genesis of this book lies in a meeting of the Institute of British Geographers Quantitative Methods Study Group. The meeting, entitled 'Models in Geography: 20 Years On', was held in the School of Geography at Oxford and formed part of the School's centenary celebrations. As the title of the meeting indicates, it was also the twentieth anniversary of the publication of Chorley and Haggett's seminal work *Models in Geography*. One of the purposes of the meeting was to reflect on the development of modelling since the publication of that book, to assess current modelling efforts and to speculate about the future. The other was to confront model-builders with some of their critics in the hope that some fruitful exchanges would ensue. This volume has an additional purpose: to counteract the notion – widespread among students who have a serial view of the development of the discipline – that modelling is something that people did in the sixties and seventies but have now largely abandoned.

The fact that the meeting was held in Oxford – a place barely touched by the upheavals of the quantitative revolution – was more or less accidental. However, the coincidence of the meeting with the School's centenary and, more importantly, the twentieth anniversary of the publication of *Models*, gave us the opportunity to do something out of the ordinary. The enterprise was made possible by the willingness of Richard Chorley and Peter Haggett to participate: without them, we would have been deprived of our theme and a splendid contribution; with them, we were able to attract a powerful set of speakers to examine a set of issues that were ripe for discussion.

In a sense, calling the meeting 'Models in Geography: 20 Years On' was misleading. We did not want to have a 'Where are they now?' review of modelling and modellers: it is worth noting that apart from Chorley and Haggett only one of the participants (David Harvey) contributed to the earlier volume. Nor did we want a simple stocktaking exercise of the past two decades – we were more interested in the future than the past. Nor even did we want to have a 'New Models in Geography'. The title *Remodelling Geography* (which was suggested by John Davey of Basil Blackwell) captures the agenda we had in mind very much better, so it has been adopted for the book.

The meeting lasted two days and was divided into six sessions. Each session consisted of two main papers and a discussion. The book follows the same format, its six parts having the same titles and contents as the sessions, though the contents have been supplemented and Haggett and Chorley's contribution has been promoted from last to first. Commissioning extra work obviously slowed the publication process but it also gave the contributors the opportunity to interrelate their papers. This has resulted in a finished product that is more coherent than a standard conference volume.

The temptation to move some of the other contributions around has been resisted even though many of them would fit equally well in more than one part of the book: the distinction between the applied chapters and the others is somewhat arbitrary; the changing technical environment is a constant theme rather than the particular preserve of the chapters in the section of that name; the critical chapters are concerned as much with making new proposals as with attacking old ones; many of the substantive papers are every bit as critical as those in the critical section; and everyone is concerned with the question of research priorities.

The six sessions were intended to be divided equally between human and physical geography and some effort was made, at least on the physical side, to obtain representative coverage of the discipline. Happily, the contributors were not so easily constrained. A number of papers neatly bridge, or pay little attention to, the notional human–physical divide and others roam beyond immediate subdisciplinary concerns: Haines-Young's paper, for example, considers problems of modelling geographical knowledge in a general sense and would not look out of place in the human or applied sections. On the whole, though, the physical contributions can be divided up by their subdisciplinary concerns: Thornes with geomorphology, Harvey and Godfray with biogeography; Henderson-Sellers with climate; Kirkby (at least for purposes of illustration) with hydrology. Unwin's discussion paper in the physical section adds a further climatological twist.

The human geography contributions, on the other hand, split along methodological or philosophical lines: broadly speaking, Openshaw is concerned with inductive modelling, Wilson with deductive modelling, my own chapter looks at the nature of models, and both Batty and Bennett consider the social context, or 'culture of the times'. Macgill's discussion at the end of the human section reflects the methodological tenor of the earlier papers. Substantive issues in economic, urban and regional geography do get an airing, however. Social and political issues also run through several contributions, but the problems of modelling in social and political geography in a narrow sense are not discussed.

The papers by Henderson-Sellers and Batty appear in the applied section along with a discussion by Beaumont. The difficulties posed by the demands of practice in climate work and urban modelling are highlighted in the main papers, while the discussion illustrates the general problems and possibilities of modelling from a practitioner's perspective.

In the section on the changing technical environment Rhind and Haining look at the new possibilities for modelling brought about by developments in computing and geographical statistics respectively. Cox assesses these contributions in the context of a discussion on the general relationship between modelling and data analysis.

Harvey, Harvey and Scott, and Cosgrove provide the critical view of modelling. Harvey's critique takes the form of an account of his own intellectual odyssey from models to Marx, while Cosgrove reflects on what he sees as the failings of modelling from a humanist perspective. Harvey and Scott, in their joint paper, are less concerned with criticism than with proposing a new project involving a return to grand theory. Flowerdew's discussion paper begins by classifying modelling criticisms then reviews the other contributions in terms of his classification system.

The final section looks at research priorities in physical and human geography through the eyes of Kirkby and Bennett. I have contrived to get the very last word in by making the editorial commentary an afterword. The first word goes, as it must, to Haggett and Chorley.

BILL MACMILLAN
Oxford

* * *

I would like to thank Professors Chorley and Haggett for their support, all contributors for their chapters – and their patience – and Moira Wise for her assistance with both the references and the organization of the Oxford conference.

B.M.

The Contributors

Michael Batty is Professor of Town Planning at the University of Wales College of Cardiff

John R. Beaumont is ICL Professor of Applied Management Information Systems at the University of Stirling.

Robert J. Bennett is Professor of Geography at the London School of Economics.

Richard Chorley is Professor of Geography at the University of Cambridge.

Denis Cosgrove is Reader in Cultural Geography at the Loughborough University of Technology.

Nicholas J. Cox is a Lecturer in the Department of Geography, University of Durham.

Robin Flowerdew is Lecturer at the University of Lancaster.

H. Charles J. Godfray is Lecturer in Pure and Applied Biology, Imperial College at Silwood Park, University of London.

Peter Haggett is Professor of Urban and Regional Geography at the University of Bristol.

Roy Haines-Young is Lecturer at the University of Nottingham.

Bob Haining is Reader in Geography at the University of Sheffield.

David Harvey is Halford Mackinder Professor of Geography at the University of Oxford.

Paul H. Harvey is Lecturer in Zoology and Fellow of Merton College at the University of Oxford.

Ann Henderson-Sellers is Professor of Physical Geography at Macquarie University.

Mike Kirkby is Professor of Physical Geography at the University of Leeds.

Sally Macgill is Lecturer at the University of Leeds.

Bill Macmillan is Lecturer in Human Geography and Fellow of Hertford College, University of Oxford.

Stan Openshaw is Senior Lecturer in Geography at the University of Newcastle.

David Rhind is Professor of Geography at Birkbeck College, University of London.

Allen Scott is Professor of Geography, University of California at Los Angeles.

John Thornes is Professor of Physical Geography at the University of Bristol.

David Unwin is Senior Lecturer in Geography at the University of Leicester.

Alan Wilson is Professor of Urban and Regional Geography at the University of Leeds.

From Madingley to Oxford: A Foreword to *Remodelling Geography*

Peter Haggett and Richard Chorley

The Road to Madingley

In one sense the road to Madingley began at Bodie. On the Labour Day weekend of September 1962, we set out with another Cambridge colleague, Roger Barnett, to drive from Berkeley on a circular loop. Our route was to be Yosemite, on across the Sierra Nevada and down into the Owens Valley, then cutting back to the Pacific at San Luis Obispo and north along the California coast road back to Berkeley.

One of us had been working with the U.S. Geological Survey at Denver, the other teaching summer school at the University of California at Berkeley attending (perhaps gatecrashing would be a more accurate term) the Regional Science Association meeting held there. Through the hospitality of Jay and Jean Vance our paths from Cambridge had met in the Bay Area. James Parsons, then Chairman of the Berkeley geography department, had kindly allowed us the use for the long weekend of one of the University of California's station wagons which, with its 'diamond E' mark, allowed access to state parks.

The late morning of the Monday found us in the northern part of Owens Valley at the abandoned mining town of Bodie. It was scorchingly hot in the car; hotter still outside. So we sat in the shade of the saloon bar (sadly, deserted many decades before) and turned to look forward to plans for the following summer when we would both be back again in England.

The Purpose of Madingley

School geography in Britain was in something of the doldrums. Regional geography had become somewhat routinized and we were both concerned that some of what we saw then as exciting new developments in the universities on both sides of the Atlantic should be passed on to young geographers struggling through their sixth-form courses. A letter from Ray Pahl, then geography tutor at the Cambridge University Extra-Mural Board (now Research Professor of Sociology at Kent) presented the possibility that a residential course for teachers might be the answer. John Andrew and Geoffrey Hickson, as senior members of the Board, supported the idea and suggested Madingley might be an ideal location; July 1963 an ideal time.

Madingley Hall is a rambling Elizabethan mansion which had been purchased by the university shortly after the end of World War II. It lies in a small village four miles from Cambridge, surrounded by formal gardens (with a croquet lawn, to be much used by the geographers) on one side; walled kitchen gardens on the other. Altogether very rural, very English, very un-revolutionary.

Madingley was to be an important venue in bringing together groups of university geographers and school teachers in high summer. The first meeting was in late July 1963, and the lectures given during this and the next summer were brought together in a book which we edited together entitled *Frontiers in Geographical Teaching*. This was published in 1965, and consisted of eighteen contributions from geographers drawn from university, college, schools and government. Its dustjacket was marked by an Isardian landscape superimposed on an early Elizabethan print of Madingley Hall on the front, while the back announced a considerably more ambitious volume, to be called *Models in Geography*.

The Idea of *Models*

The idea for *Models* had been launched in the Spring of 1964. An initial meeting was held at Cambridge in the Department of Geography's Clark Collection room, a handsome room with its walls lined with classical geographical and exploration volumes. It is not clear that their authors would have approved either of the 'robustly anti-idiographic' aim of the proposed volume, or of the distinctly junior status of its proposed authors. Certainly, they would have considered a veil should be drawn over events later in the day when the anti-idiographic discussions between contributors became so robust as to lead to fisticuffs in a pasture field (located somewhere along the Cambridge–Bristol axis) before a bemused circle of Friesian heifers.

The initial aim of the 'Models' movement was to influence the teaching of geography at the point where it seemed most in need of change and potentially most susceptible to it – namely the sixth-form level. Unlike the United States, where the more broadly-targeted High School Geography Project appeared in the mid-sixties to be foundering due to the then low status and lack of professionalism associated with the subject in many American schools, Britain possessed in school geography a clearly defined discipline with an established role and status and which was orchestrated by an organized group of well-trained, dedicated and highly professional geography graduates. A substantial proportion of these was clearly dissatisfied with existing syllabuses, recognizing them to be antique, demanding of memory rather than intellect, innocent of any real philosophical overviews, and increasingly divorced from the condition of the real world which was changing so fast after World War II. Of course, the Madingley group were not the first or only geographers to recognize this but, like a safe with a combination time lock, massive and apparently invulnerable structures may yield to a very small impulse if it is applied at just the right place and time. Thus it was with *Models*.

In those heady, rather confused months during which the book took shape it would have been difficult to identify our aims clearly, which is perhaps why it did not occur to any of us to write a preface, or to lay down hard-and-fast rules for contributors. However, we were all aware that the preceding fifteen years or so had witnessed a growing body of intellectually challenging geographical research which was capable of turning the fulcrum of the discipline if the right pressure was applied. In short, our most clearly definable aim was to make geography at all levels a more intellectually attractive and relevant subject. Associated with this was our desire to begin an active review of the goals of geographical scholarship, its relations with other disciplines, the changing dialogue between human and physical geography, and not least, to try and explore the unifying bonds in the subject at a time when conventional regional geography appeared to have run out of steam.

The Legacy of Madingley

The impact of Madingley is for others to judge. We should only note that those who took part seemed to have enjoyed it; notably the young and not-so-young group of teachers and college lecturers who attended, made it financially possible, and joined so enthusiastically in the discussions and classroom implementation.

As to the longer-term legacy, both *Frontiers* and *Models* seemed to diffuse widely, were reprinted more than once, appeared in paperback, and were translated into many languages. They were to lead directly to a continuing series of follow-up publications. Two years later Edward Arnold published the first volume of *Progress in Geography*, edited by four of the original *Models* contributors (including Christopher Board and David Stoddart), all of whom had been Demonstrators in the Cambridge department of geography. Nine volumes were to be published between 1969 and 1976. In the latter year the pressure of material led to the decision to replace the hardback series with a quarterly journal. While we had hoped and pressed for a single journal, a survey of readers suggested that two were more appropriate, and so the first issues of *Progress in Physical Geography* and *Progress in Human Geography* were published in that year. The original 'gang of four' was first supplemented by two further editors, Bruce Atkinson of Queen Mary College London and David Lowenthal of University College London, and later by Ron Johnston of Sheffield University and Andrew Goudie of Oxford. The profession owes a considerable debt to both Chris Board and Bruce

Atkinson for editing both journals in the decade since their inception.

Like the Madingley volumes, both journals strove for catholicity and to widen the range of geographic scholarship rather than to narrow it. Very deliberately their subtitles referred to international reviews of geographic work in the 'natural and environmental sciences' and the 'social sciences and humanities' respectively, and a greater flow of ideas across the traditional geographical boundaries has been worked for, even if not fully achieved.

But a second aspect of the Madingley conferences is more important, even if less definable. We refer to a change of mood in British geography which began in the early sixties. We are still far too close to judge why this take-off occurred and why it did so at this time. Perhaps the best explanation is that the Madingley conferences were launched at the right time. Universities were accelerating their growth; new departments were being founded; new geographers recruited. Second, the volumes were catholic in approach and laid no specific theological foundations on which geographic models should be built. A third reason was the relative simplicity of the analyses proposed. Models necessarily avoided most of the difficult technical and conceptual problems which have since emerged. What was happening in geography appeared also to be happening in other subjects. The late David Clarke's *Analytical Archaeology*, published in 1968, paralleled in many ways the theme and structure of the models volume.

In *Frontiers* we wrote rather naively about our preference that geography should 'explode in an excess of reform' rather than bask in 'the watery sunset of its former glories'. Not surprisingly, neither has happened. However, the past twenty years have been marked by a succession of innovations in geography, sometimes perplexing and conflicting but always exciting. Today minorities of scholars continue to advocate, on the one hand, the disintegration of the subject or, on the other, the need to return the 'solid virtues' of yesteryear. Thankfully, once again, neither will happen.

It would be presumptuous to assert that *Models* precipitated these changes: clearly it did not. But those involved in its production recognized that changes were both necessary and inevitable. At the very least, its publication gave impetus to a natural process. At the time the general geographical public could have been forgiven for thinking that *Models* was merely to do with spatial generalization, with mathematics, and with Kuhn. It is now clear that it was much more than this, being instrumental in throwing open the windows of a hitherto introverted discipline, and the fact that we have discovered as a result that our house has many mansions should not displease us. The rise of social and radical geography owes as much to the appearance of volumes such as *Models* as do quantitative and modern economic geography. Today at all scholarly levels geography is a vital, relevant, experimental, intellectually demanding and thoroughly fascinating subject in a way it was not more than twenty years ago.

From Madingley to Oxford

It is not our purpose to trace the history of geographic modelling between the sixties and the eighties, that is ably done in the substantive chapters of this volume, most of which were presented as papers at the Oxford Models Conference. It would also be presumptuous of us to try and comment on achievements which in some ways have moved so much faster and further than we had foreseen. However it may be of interest of record our impressions of the two modelling environments.

One immediate contrast which strikes us is in the levels of both optimism and of sophistication: Madingley was optimistic but naive, Oxford sophisticated but realistic. The early sixties was an optimistic period for geographical innovators. Those who were working at Madingley had grown up in a period in which young men in their twenties were cracking the genetic code for life itself, and in which men in their forties were leading the then most powerful nation on Earth. Universities in Britain and around the world were expanding at an unprecedented rate, and geography too 'had never had it so good'. For a brief period in the sixties geography PhDs were doubling in number every six years. Scientific methods which had shown such unprecedented success in solving technical problems in the physical and biological world showed some prospect of equal promise in the social world. Physical and human geography had rarely been so united, the bridges between the different parts of the subject rarely so strong. It seemed, at least for a brief window of time, that all that was needed was to find the language which converted the man-made environment into the same terms as the natural environment, and that spatial structures would provide one such language, systems analysis another. The naivety of this view is easy to see in retrospect. In geography it was like trying to replace the 140,000 characters of Chinese writing (each region to be learnt slowly and by heart) by a Roman alphabet of less than thirty. Dacey's spatial semantics and syntax showed ways in which a general-purpose geographic language might be built. If too much was tried too quickly, then we were in good company at the time and learning fast.

A second contrast between Madingley and Oxford relates to the closing of the frontier. Twenty years ago it was harder to spot the smoke of academic neighbours over the horizon; in the steady-state world of the late eighties, we can see clearly the fires burning in neighbouring camps. This greater packing of subjects into academic space has made for a greater consciousness of territorial boundaries.

A third contrast is in the intellectual environment within which geographic modelling now occurs. There is now a greater sensitivity to the assumptions and methodological issues inherent in modelling. Both the data environment and the data handling environment have been revolutionized in ways outlined in the papers by David Rhind and Stan Openshaw for computing, by Mike Kirkby for geomorphology, and by Ann Henderson-Sellers for climatology. The user environment has changed: 'Models I' was essentially pedagogic, aimed at an expanding but protected home market; 'Models II' is essentially concerned with R and D, aimed dominantly at stable and highly competitive export (i.e. non-geographic) markets. The communality has decreased and, as Alan Wilston's paper shows, geography is going through a period of parallel and plural developments with the scholarly rivers braiding and anastomosing.

A fourth contrast between Madingley and Oxford is demographic. The oldest geographic contributor to *Models* was in his mid-thirties, almost all were still on temporary posts or had newly gained lectureships, none was a professor, none was well-known. The Oxford gathering appears senior and professorial; its attitudes defensive and careful, in striking contrast to the aggressive and carefree climate twenty years ago. This is, of course, a sign of academic acceptance but, as David Harvey's contribution shows, the young Turks of Madingley are not now too busy polishing their medals to face the problems of the nineties. More important, a new generation of naive, aggressive and carefree young geographers are also holding meetings.

Postscript

The fact that this seminar on 'Remodelling Geography' was held in the 'British Isles' room of the Oxford School of Geography brought back especial memories for one of us, reinforcing the magnitude and necessity of recent changes in our discipline. Any account of the events which led up to this meeting, twenty years after the publication of *Models*, is bound to be a personal one. Several interesting accounts of the major changes in geography over that time have been written, and we are aware that other participants at Madingley, almost all of whom are active and well, will have seen things in different ways and their accounts may differ from our own. History is, after all, only a 'fable agreed upon', and we give here reflections which are essentially one-sided and personal. We are fortified only by the knowledge that by the next conference (presumably in 2007) our recollections may be even fainter and even more imaginative.

Madingley is now in academic terms a long time ago. We were tempted to try to write together for this volume the kind of agenda for geography, footnote-buttressed and serious in mien and purpose, that we might have done then. But that would be wholly inappropriate. The baton has passed to others and a pompous essay of that kind would now come ill from us, and we have other canvasses (smaller but equally demanding) still to paint. We hope that Madingley and *Models* played some small part in opening up geography to a wider range of scholarship so that it stands today deeply suspicious of too much orthodoxy, not least our own.

PART I
Modelling in Physical Geography: Retrospect and Prospect

1 Geomorphology and Grass Roots Models

John Thornes

Introduction

The production of *Models in Geomorphology* by R. J. Chorley in 1967 signalled the beginning of the end of a decade of effort to shift geomorphology from a largely qualitative, idiographic and historically based subject to one which was to become quantitative, nomothetic and process-oriented. There had previously been a preoccupation with regional denudation, river and glacial chronologies and climatic reconstructions based on speculative connections between landforms and sediments on the one hand and past climates on the other. This shifted to the intensive investigations of field processes, to statistical analysis of general morphometric data and to the acceptance of mathematical modelling. This revolution has been well documented recently, for example by K. J. Gregory (1986) and need not be repeated here except to note that the groundwork for this transformation had already been established in the previous decade. The important precursors included Strahler's advocacy of systems analysis (1950) and the dynamic basis of geomorphology (1952), Culling's (1957) introduction of the relevance of systems theory and its relationship to fluvial and hillslope processes and Scheidegger's (1961) contribution to the analytical modelling of hillslope processes and form. Chorley himself had already made major contributions to breaking the grip which denudation chronology held, by his contribution to the role of general systems theory in geomorphology (1962), his strong attack on W. M. Davis and his followers (1965) and his review of statistics in geomorphology (1967b).

At the same time there were strong awakenings in climatology. In Britain Hare published a widely read paper on 'The Westerlies' (Hare, 1962) and his book *The Restless Atmosphere* (1953) did much to shift climatology from an excessive concern with regional climatic statistics and climatic classification to an interest in global circulation and energy budgets. About the same time the first indications of the potential for satellite climatology made their appearance. In biogeography the notions of succession had been long established and the idea of systems was already accepted.

Given the lively academic progress of the sixties and the prevailing sense of change, it is hardly surprising that the notions of models and modelling were so quickly espoused by the academic community in physical geography as in other areas of science. However, despite this remarkable shift in thinking, the singular lack of the intellectual training and wherewithal to develop these ideas might have led to their failure had it not been for their enormous pedagological utility coupled with those of systems analysis, which were popularized soon after (1971) by the publication of Chorley and Kennedy's *Physical Geography: A Systems Approach*, and the pressures from cognate sciences. Among the latter were, of course, the developments in climatology and synoptic meteorology, but particularly influential were the strong fluvial and hillslope geomorphology schools in the United States led notably by L. B. Leopold and S. A. Schumm. Within geomorphology in the United Kingdom since that time, the major developments have been mainly in hillslope hydrology and erosion and in glacial geomorphology and they rest on the shoulders of only a few individuals such as M. J. Kirkby, G. Boulton and W. E. H. Culling. Even in the subject at large there have been remarkably few with the ability to create lasting structures from basic model building activities. Some of these are briefly reviewed in the second section of this chapter.

Another factor which led to easy acceptance of models was the parallel adoption of statistics in geography. Since statistics, and especially regression model concepts, are much more readily grasped than differential calculus, for many practitioners regression analysis and modelling were regarded as essentially synonymous. They were and remain consonant with experimental design, especially for field problems, and emulated the behaviour of other field sciences.

Major Developments Since 1967

Three areas can be identified as providing the cutting edge for model acceptance and implementation since 1967 in a British context. The first of these is the development of theory in a formal modelling framework, the second is the deployment of important conceptual notions from science at large and the third is the adoption of the massive technical changes which have facilitated a more experimental approach to theoretical model building. These provide an important prerequisite to contemporary changes and therefore are briefly reviewed in the following paragraphs.

Theoretical Developments

In formal model building, geomorphology, like most other sciences, has been characterized by a small number of important contributions which have led to major developments in the field. Of these, the provision of a paradigmatic framework for hillslope hydrology and processes has been largely provided by M. J. Kirkby. In particular three important papers describing theoretical models have sparked off a great deal of subsequent field and laboratory investigation. The first paper was concerned with the theoretical development of the influence of topography on the nature and location of throughflow (Kirkby and Chorley, 1967) based on the earlier empirical evidence of Hewlett (1961). This has led, for example, to the investigation of topographically controlled partial contributing areas (Anderson and Burt, 1978), the improvement of understanding of subsurface water movement in relation to solute chemistry (Burt et al., 1983) and the development of a general hydrological model for ungauged catchments (Beven and Kirkby, 1978). The second paper (Kirkby, 1971) was concerned with the development of characteristic profiles for hillslopes given a continuity-based process model. The significance of this work is that it offered a testable model for the relationship between processes and form for hillslopes which had hitherto existed only in the very important but relatively inaccessible work of Young (1963), which was based on graphically constructed iterative solutions and which Young himself had already tested to some extent in the field. The third paper (Kirkby, 1980) provided a theoretical development of Smith and Bretherton's (1972) concept of the stability of hillslope hollows to furnish a testable model of the controls of climate, lithology, topography and basal slope dynamics on drainage density as a function of drainage network growth. This paper also introduces the wider concept of dynamic stability discussed below.

In quite a different vein, the analytical work of W. E. H. Culling (1960, 1963) has also made a major contribution to theoretical modelling, though it is perhaps generally less well known. Culling's later work has focused on the development of the application of rate process theory, first to the theory of soil creep (Culling, 1983) and later to a more general theory of particulate sediment transfer (Culling, 1986). This work is unique in mobilizing the blurred crystal model for the description of the soil matrix and then providing stochastic models for the solution to the net diffusion of soil particles as a result of independent random shocks to the system. The application of this theory to slow diffusive movement around fixed objects of known geometry (e.g. telegraph poles) has recently been tested by Flavell (1985).

In glacial geomorphology the work of G. S. Boulton (1972, 1974) provided the first breakthrough for many years by modelling the relationship of ice-sheet dynamics as controlled by thermal regimes to landform development. Although glaciologists had made enormous strides in modelling the mechanics of glacier response to budget changes (for example in the outstanding work of Nye, 1960) geomorphologists were still largely concerned with descriptive inferential attempts to explain glacial landforms and deposits. The regime work of Boulton and his many subsequent contributions established a scientific basis for glacial geomorphology which has still to be fully realized, though there have been some notable large-scale successes in this field (Sugden, 1978).

The third area in which formal theoretical modelling has played a decisive role in geomorphology has been in the investigation of fluvial processes, notably in the U.K. through the work of R. A. Bagnold and J. R. L. Allen. Bagnold's (1966) work on the general physics of sediment transport led to the provision of a measurable and hence testable law of transport based on the concept of stream power, and although some of the earlier transport models are physically based, most are empirical approximations which have little utility or relevance in geomorphological research. Allen (1974) not only provided the stimulus for the development of models of meander behaviour, but also developed more general theory of bedform generation, migration and death which provides an interesting comparison with the notions of persistence and sensitivity of dynamic geomorphic systems (Brunsden and Thornes, 1979).

Conceptual Developments

In addition to the developments in formal modelling there have been a number of important conceptual advances since 1967 which helped to strengthen the modelling approach to geomorphology when it might otherwise have failed. The most important of these was the utilization of systems analysis or more general systems theory. As Huggett (1985) points out, the terms model and system are not synonymous and Strahler (1980), too, takes pains to distinguish between them. Whereas systems actually exist (though can never be completely described) a model is an attempt to describe, analyse, simplify or display a system. The systems approach codified for physical geographers by Chorley and Kennedy (1971) and thoroughly reviewed by Bennett and Chorley (1978) provided a readily assimilated and easily reproducible set of ideas which have proved to be especially attractive in a teaching context. In practice, however, the number of significant papers in this field in British geomorphology is really quite small. Of these Bennett's (1976) adaptive filtering of river channel slope series, and Lai's (1979) investigation of the Box–Jenkins structure of spatial responses of soil water in two dimensions represent the transfer-function modelling approach and Kirkby (1980) and Thornes (1980) have used dynamic systems theory to describe the non-linear behaviour of hillslopes and channels respectively. In geomorphology, systems analysis is a popular concept enjoyed by many, but whose deeper ramifications are understood by relatively few. This is in strong comparison with hydrology in which both the theory and practice of systems analysis are well embedded.

Another important modelling concept to develop over the period, and directly related to the theory of dynamic systems, has been the notion of thresholds and instability. Initially conceived and formalized in a geomorphic context by Smith and Bretherton (1972) it was subsequently illustrated and popularized by Schumm (1973) and will be further discussed in the penultimate section of the chapter.

Technical Developments

The third group of changes which are responsible for the position reached today are essentially technical and technological. In the sixties the range of problems that could be addressed were limited by the ability to analytically solve the governing differential equations in a problem, the difficulty in obtaining numerical solutions even where analytical methods were available, or the Herculean task involved in statistical modelling where more than a few variables were involved. This appears to have led to some quite specific modelling outcomes. First there was a natural tendency to posit the problems such that known available solutions could be applied. Witness the number of geomorphic problems that were regarded as analogous to the heat diffusion problem, for which a range of solutions for well-defined boundary conditions were well known from the classical work of Carslaw and Jaeger (1959). Secondly a series of approximations were devised to deal with complexities that were otherwise incapable of solution. A good example of this has been the widespread adoption of the kinematic approximation first to channel (Wooding, 1965), later to hillslope (Kibler and Woolhiser, 1970) and finally to subsurface flow (Beven, 1977). Thirdly problems were limited in their intellectual scope by what could be solved within the technology available.

Three developments have removed many of these constraints. The first is the incredible spread of cheap, massive and user-friendly computing power. The second has been the increase in available software for tasks ranging from simulation, through numerical solution to advanced statistical problems. The third has been the development of much-improved numerical techniques for handling spatially complex distributed models which are intrinsically the type which geomorphologists should be developing. This revolution has come about in less than a decade and the strategies adopted for computer modelling of geomorphic systems in the mid-seventies (e.g. Ahnert, 1976; Armstrong, 1976) have now been more or less completely displaced. What would then have taken four hours to run, now may take as little as four minutes at a fraction of the cost.

An important consequence of the lifting of these restrictions has been the development of very large-scale system simulation models. These have been most conspicuous in engineering (especially chemical engineering) and have become known as 'engineering models', but they are most familiar to geomorphologists in the guise of distributed catchment models, ecological systems models or models of soil erosion. Of these the ecological models have been the first, the best-developed and the most rigorously tested. At a time when hydrologists are already well down the path (e.g. Anderson and Howes, 1986) and geomorphologists appear to be about to make major investments in this area (e.g. Chisci and Morgan, 1988) it seems appropriate to take stock of the achievements of the mega-modellers and learn from their experience. In particular, since the introduction of vegetation cover into geomorphological problems and models in a serious way will increase their complexity by an order of magnitude, it seems to invite the direct deployment of some or part of the many ecological models already in existence. It is to these grass root models we now turn.

The Ecological First-generation Mega-models

Although ecologists had been using mathematical models for over 100 years, it was only after the development of the energy flows and trophic–dynamic concepts by Lindemann in 1942 that interest focused on the flow of energy and matter through ecosystems. This led in the early sixties to the field of tracer kinetics and in particular to the earliest analogue computer models of the flow of radioactive substances through biological systems. Specifically Olsen (1960) created an analogue model by setting up linear differential equations of the rate of change in each compartment of the system and solving for the distribution of radioactive elements throughout. At about the same time Garfinkel investigated the Lotka–Volterra models in the context of grass–herbivore competition on the Univac I computer (Garfinkel 1963) and systems ecology as we know it today was born. Throughout the sixties modelling proceeded haltingly as the traditional mathematical ecologists expressed doubts about the value of digital simulations presuming to handle the entire system (Holling, 1964). By contrast Davidson and Clymer (1966) fully explored the potential of computer modelling for the spread of epidemics, the study of the dynamics of disease syndrome, for public health aspects of water pollution and for modelling the growth and reproduction of parasites. Meanwhile Forrester (1961) had produced the first major systems model in social science which proved to be very influential.

The International Biological Programme

The major stimulus to total system models, however, came with the International Biological Programme (IBP) in which, from the beginning, modelling was envisaged as playing a major role. It was believed (Shugart and O'Neil, 1979) that the model would enable synthesis of the total problem and that sensitivity analysis could be used to specify and adjust the programme goals so that the final model would represent in some way the entire dynamics of the system under study. When the fully tested model was available it would be used for the management of resources and for the assessment of environmental impacts. Van Dyne (1966) expressed the view that in applying ecosystems concepts there would be no limit to the size and complexity of the models, but as actually implemented the Biome programmes were more conservative and less optimistic about the possibilities for sweeping breakthroughs. Nonetheless the models were large by any standards and typically had 20–50 variables and several hundred parameters.

This international programme (IBP) was established to promote integrated studies of the productivity of the major biomes of the world, to identify potential uses of both new and existing natural resources and to study human adaptability to changing environments worldwide (Patten, 1975). The 'Analysis of Ecosystems' part of the United States' contribution was organized for a comprehensive, coordinated and integrated investigation of five major biomes at the ecosystems level: grassland, deciduous forest, desert, tundra and coniferous forest.

Each of these major programmes was approached in rather a different fashion, though all had modelling of the ecosystem as the underlying philosophy. Indeed in the original Biome proposal to the National Science Foundation (in 1969), the word 'model' appeared 348 times! Whereas the grassland group, under Van Dyne took a relatively mechanistic approach, the Desert Biome group, under D. W. Goodall emphasized the need to model transient system dynamics at a variety of different scales (Goodall, 1975). At the end of the scheduled programme there were few formal publications on which a judgement could be made about the IBP's success (see for example the views of Watt, 1975) but between 1978 and 1983 all the groups published major works dealing with the individual programmes. Of these programmes the grassland ecosystems group come closest to achieving large-scale simulation of an entire (but not comprehensive) ecosystem.

The IBP Grassland Model

Work on the grassland model began in 1969 and culminated in the publication by Innis of the *Grassland Simulation Model* in 1978 after about 70 professional person-years of effort. The original objectives were, 'to develop a total systems model of the biomass dynamics for a grassland which is representative of the sites in the U.S.–I.B.P. Grassland Biome network and with which there can be relatively easy interaction' (Innis 1978).

Early in the programme it was agreed that this should be attempted through deterministic rather than stochastic modelling, difference rather than differential equations, non-linear rather than linear mathematics and mechanistic rather than holistic concepts. Some of the earlier work on the PWNEE and LINEAR models was curtailed and the main effort was put into the ELM model which became the major grassland model of the research programme. This model has five parts – abiotic, producers, consumers, decomposers and nutrients. The abiotic sections deal with flows of heat and water and utilize hydrological and soil-physical sub-models. The producers submodel represents biomass dynamics for warm season and cool season grasses, forbs, shrubs and cacti. These flows are influenced by nitrogen and phosphorus from the nutrient submodel and governed by heat and water from the abiotic sector. The consumer submodel is for cows, sheep and grasshoppers and the decomposer submodel is divided into hard and soft components distributed across several soil layers. A full description of the essential features of each submodel is contained in Innis (1978). Whilst accepting that sooner or later process geomorphologists will have to come to grips with all aspects of the processes dealt with in the models, only the producer section will be illustrated here.

Sauer (1978) used a novel approach in developing the producer component by discretizing the simulated years into seven phenophases, the onset of each being determined by temperature and soil moisture conditions. Thus for example the transition to senescence may be determined by drought conditions and to the reproductive phase by passing a temperature threshold and thence to fruiting by soil drying. This general model seems appealing to the geomorphologist interested in erosion because the litter and cover conditions significantly interact with the incidence of rainfall to determine the available erosive energy. Within the phenophases the magnitude of carbon flows is controlled by soil water potential, temperature, nitrogen, intercepted insolation, quantum efficiency, light saturation and phenology. This model, based on Loomis and Williams (1964), essentially sets gross productivity and then constrains this to a minimum as determined by the limiting factors.

As a result of comparison with the PWNEE site data for two years, Sauer concludes that a significant inadequacy of the submodel is its failure to properly incorporate the controls of litter fall. This is critical because the shift from standing dead to litter is a major control in the progress of the carbon cycle and, from a soil conservation and productivity point of view, it is a major source of cover and soil organic matter, especially in areas of sparse vegetation, as in semi-arid environments. The breakdown of litter is provided for in the decomposer submodel. Here the litter is not differentiated into species, but into 'hard' and 'soft' litter which is distributed across the surface and three subsurface layers and the rates of breakdown judged in terms of CO_2 released. The decomposers were also considered as a group. Since the decomposer models were available 'off the shelf' (Innis himself was already a major contributor to this field) it is hardly surprising that this submodel appeared more successful and was able to provide some of the new results to come from the exercise.

The Benefits and Lessons of the IBP Programme

It is impossible in the space available to do justice to the richness and ingenuity of ideas deployed in these major model-building exercises and all deserve careful scrutiny rather than a superficial judgement. Already by 1975, at the end of the formal modelling period, some of the short-comings of the approach were becoming apparent (Innis, 1975). A detailed analysis of the strengths and weaknesses of the strategy followed the validation and sensitivity analyses of the model and are outlined in Woodmansee (1978), who was a member of the team.

The major benefits to emerge from this programme appear to have been the following:

1 There was a significant increase in the understanding of the major biomes involved as a result of the modelling effort. In fact modelling was firmly established as a method of research into complex (as opposed to simple) systems.
2 The programme illustrated that it was actually possible to build models of such complex systems despite the doubts expressed at the outset of the programme (e.g. by De Wit, 1970).
3 In particular areas, simulation experiments were possible which would otherwise have been impossible over significant time and space scales and some properties emerged which had been suspected but hitherto untested. In this respect the process of modelling highlighted inadequacies and provided a multitude of fresh hypotheses on which subsequent work has been developed.

These educational benefits are easily underestimated. SWECON, the Swedish Coniferous Forest Project, which started in 1972 and ended in 1981, produced a fund of trained ecologists, including 19 Ph.D.s whose worth was of immediate relevance to the problems of forest clear-cutting and acid rain which emerged at the end of the 'seventies. In addition to the major publication *The structure and function of Northern Coniferous Forest* (Persson, 1980) some 200 papers were published on forest ecology arising directly from the programme (Lundholm, 1984).

The lessons learned from the IBP models fall into three general groups: (1) problems of specific inadequacies of model formulation; (2) problems relating to the magnitude of the modelling enterprise; (3) problems relating to objectives.

Lesson 1: Model Formulation

All models are by definition incomplete representations of reality and all represent the limitations of extant knowledge at the time of construction. It is therefore easy to criticize models with hindsight. The more inventive the modeller, the more likely this is to be the case. Thus, for example, the core weakness of the producer component of the grassland model, through which many other effects flow, is in the failure to embody an explanatory model of the transition from standing dead to litter. A major question could also be posed concerning the assumption of Liebig's 'Law of the minimum' in determining carbon production, a feature which has been carried over into the erosion models discussed below. In addition to such errors of formulation, there are many errors of aggregation – spatial aggregation, aggregation of species, aggregation of age groups. These errors are very difficult to detect through conventional validation and sensitivity analyses, especially since they affect different parameters and state variables differently. There are, too, errors which arise from inbuilt, and sometimes hidden, constraints deployed to ensure that the model runs mechanically, e.g. by incorporating capacity limits on populations to ensure stability of the model. The actual provision of parameter values itself represents a major unknown in many of the mega-models. Sometimes parameters are based on field data, sometimes culled from literature and often, obtained from the best-guesses of experts. Finally errors arise from failure to incorporate feedbacks. Thus for example in the grazing component of the grassland model the relation between grazing, growth and primary production fails to accommodate the effects of grazing on vegetative reproduction itself though this has for long been a lively debate among ecologists.

Lesson 2: Model Magnitude

Some of the most critical difficulties arise from the sheer magnitude of the models. In the grassland model there are over 1000 parameters to be estimated for the complete model. The complexity is such that the number of complete runs that can be achieved is limited. For ELM by 1978 there had been some seventy runs including six-year sequences. The core problems of such large scale models are: (1) limitations on data for parameterisation and testing; (2) inability to evaluate the problem by conventional methods of sensitivity analysis; (3) the impossibility of both incorporating and appreciating the interactions between the separate components and submodels. The demand made by these models quickly outstrip the capacity to supply them with data, even where national or international agencies are involved. For example the weather data alone were found to be inadequate to operate even the abiotic component of the ELM model. Even where extensive data are available, the noise levels involved in collection are such that it may be impossible to determine when it is the model, and when the data, which is at fault. Model evaluation is very sensitive to field sampling techniques. In the grassland models the performance was evaluated by the criterion that the model results should be within one standard deviation of the real field data. However this criterion proved hopelessly inadequate, mainly because of the poor quality of the available test data. With such large models evaluation of the entire model becomes impossible, at least by the conventional methods so far developed. In sensitivity analysis by parameter perturbation, for example, for n parameters there are 2^n ways of testing. The method resorted to in the ELM model was to invite the submodel designers to select ten macroparameters which were then subjected to a fractional factorial design allowing thirty-two treatments of combinations of the ten parameters. In an ideal world one would like also to evaluate the models against structural and driving variable changes, as well as parameter changes. The first was constrained for fear of a never-ending revision, and failure to accommodate more realistic driving variables for weather inputs to the abiotic section has been a major source of criticism. Nevertheless, the ELM model moved further towards testing than any geomorphological model yet produced, and important lessons are to be learned from the results.

It is fair to ask whether validation in the conventional sense has any real meaning for these large scale models. Mankin, O'Neil, Shugart and Rust (1975) differentiated between validation and utility. If a model is to predict the unmeasurable then by definition it cannot be validated in the usual sense, and such models can only provide the best-possible simulations. A model can then only be invalid if we can devise an experiment in which the model's outputs disagree with systems measurements within the specified area of interest. A model is useful if it accurately represents some of the systems behaviour under consideration. Then a critical question becomes not whether a given model is valid but whether its development is a more efficient methodology than experimentation alone. In the view of Mankin et al., too much time had already been spent on the validation problem. This raises some important points of principle, but it is difficult to accept that abandonment of all forms of validation would be either desirable or acceptable.

The third major problem arising from the sheer size of the exercise is the failure actually to build in some of the most important interactions of the processes. The magnitude of the task of bringing together the various subcomponents leads inexorably to treating the problem as one of computability. Processes become subroutines and maintain a strict partition when in nature they operate side by side or interactively. Where interactions are involved, for example in constraining photosynthesis, they are mainly confined to multiplicative submodels to yield combined rather than interacting effects.

Lesson 3: Model Objectives

The last group of problems associated with these mega-models are encompassed in the general term objectives. It should be clear that the intellectual and training benefits that flow from these models are enormous, and that they offer endless opportunities for experimentation which could not be achieved in the real world. However the models failed in two other very important general respects. First they failed to achieve portability and hence could only be used by the modellers who constructed them. Second, they appear never to have been used to simulate the biomes in a way that could be used seriously to predict the impact of environmental changes resulting from human activity or shifts in the abiotic driving variables. The objectives for the biome models were clearly spelled out but too often with large-scale models objectives, especially relating to their relative contribution in research and application, become seriously blurred and the construction of the model itself becomes the major driving force behind the operation. There is a great deal of political, emotional and organizational capital and prestige bound up in the largest of these models, and original objectives have a tendency to be lost in the momentum developed in the operation and the established wisdom resulting from it. The question raised by Watt (1975) appears with hindsight to have been singularly relevant. Is there any conceivable advantage to be gained from constructing one super-model that could deal with all (these) questions? Watt himself argues that models should be developed to address specific problem-oriented questions. The final word on the grassland models must go to Woodmansee (1978): 'the cycle of modelling seems only to end when the modeller expires or moves'.

The Soil Erosion Second-generation Mega-models

At the very point in time when the first generation ecological models were being evaluated, U.S. Department of Agriculture (USDA) mobilized its Soil Conservation Service (SCS) and Agricultural Research Service (ARS) to produce a set of major models for the investigation of chemical and biological processes, soil–water–plant relations, hydrology and erosion. Modelling was called for, in the view of the USDA, because it provides for an efficient time–space scale collapse, allows an unlimited set of management strategies to be considered and, with its emphasis on agricultural processes, should enable gaps in knowledge to be identified. As a result of a consultative report from the Stanford Research Institute emphasizing the role of off-site rather than on-site effects (which were considered economically negligible) the SCS redefined its research priorities (Theurer, 1985). It was also accepted that few universities or research groups have the people or resources to mount and sustain long-term efforts. In the development of the suite of models which followed much of the experience of the ecological modelling effort was carried over, including some of the personnel. The major difference appears to have been the clear remit to develop practically usable models.

As with ecological modelling, soil erosion modelling was already far advanced and the personnel involved included the best research scientists in the erosion field. The ARS–SCS has developed a family of models (indicated in table 1.1) which have different objectives but many common underlying subroutines. Two models are of particular interest in identifying the advantages and difficulties that accompany large-scale simulation efforts in the field of soil erosion. These are EPIC (Erosion–

Table 1.1 Examples of USDA models used for erosion, vegetation and soil interactions

Acronym	*Model*	*Comment*
CREAMS (1 and 2)	Chemical, runoff and erosion from agricultural management systems.	Field-scale model for runoff erosion and chemical output from homogeneous units with wide range of options.
SPUR	Simulation of production and utilization of rangelands.	CREAMS with additional components for grazing, insect consumers and herding management.
EPIC	Erosion–productivity impact calculator.	Small area, multilayered soil, crop yield in relation to erosion and management practices. Includes snow melt and channel flow.
NTRM	Nitrogen–tillage–residue management.	Detailed analysis of soil–plant–water continuum. Research model for soil chemistry under cropping. Practical and simpler equivalent is COFARM.
SWAM	Small watershed model.	Essentially CREAMS plus off-site effects.
SWRBB	Simulator for water resources in rural basins.	Predicts effects of management decisions on water and sediment yields from ungauged rural catchments, including snowmelt.
SPAW	Soil–plant–air–water model.	A comprehensive model to compute daily actual ET incorporating plant phenology and variable root distributions.

Source: based on United States Department of Agriculture, *ARS–30*, 1985

Productivity Impact Calculator) and CREAMS (Chemical, Runoff and Erosion from Agricultural Management Systems).

The EPIC model has been designed specifically to evaluate the effects of soil erosion on long-term soil productivity and a complete description of the physical component is found in Williams (1985). The surface runoff model is determined by the SCS curve number method, as with most of the members of this family of models and erosion through a modified Universal Soil Loss Equation (Onstad and Foster, 1975). For crop growth a single model is used for all crops, though each crop has its own model parameters and the conversion of intercepted energy to biomass is very similar to that deployed in the ecological models, i.e. there is a conversion factor which is constrained by a multiplicative 'stress factor' ranging between 0 and 1 and incorporating soil moisture, temperature and nitrogen and phosphorus limitations. Unlike the earlier models different cropping and tillage processes are simulated. To convert the cover into an agent affecting erosion, the crop management factor of the MUSLE (Modified Universal Soil Loss Equation) is invoked. This takes account of above-ground biomass, crop residues on the surface and a 'minimum factor' for the crop. Williams (1985) reports that simulations have been performed on 150 test sites in the continental U.S. and thirteen sites in Hawaii. Unfortunately the data presented consist only of fifteen of the test sites and are for crop yields for thirteen of the sites and erosion amounts for three of the sites. It is not clear how far these sites are typical of the data set as a whole, and whether or not the results reflect calibration or optimization procedures. Nevertheless the results presented are quite remarkable given the crudity of some sections of the model, the difficulty of measuring the variables in the field (even in controlled experiments) and the variability of the weather at the sites. It seems highly likely that EPIC or even its more sophisticated version ALMANAC, will be used for the next Resource Conservation Act cycle (Grossman, 1985).

CREAMS2 represents an even higher level of complexity and sophistication, and is derived from the earlier CREAMS model published in 1980 and used by Federal and State agencies, consulting firms and private industry nationally and internationally (Smith and Knisel, 1985). It has as its objective the prediction, through simulation, of runoff and pollutant transport in and from agricultural fields, especially as a result of agricultural management. In addition to and in conjunction with water movement, the model includes the processes of snow melt, erosion and sediment transport, nutrient cycling, adsorbed and dissolved chemical transport, soil heat flow, crop growth and residue decay. The scope of the model is limited to areas which can be described by a single soil profile regime, first- and second-order catchments and a single cropping system and single management practice. It can model perennial grass, forest and grazed and harvested cover as well as annual crops. The hydrological submodels are particularly well developed with a series of more complex optional procedures (up to fully distributed water and sediment routing) depending on available data.

The erosion submodel of CREAMS2 (Foster, 1982) is based on the now conventional distinction between interrill and rill flow. In interrill flow the vegetation operates through an effective intensity factor and a soil loss ratio which represents ground cover by mulch, crop residue and ponded water. Its effect on rill flow is considered minimal, so that although the model may be well adapted for mid-western row crops, its utility over much of the heavily eroded western rangelands is in some doubt. In this respect it is interesting to compare the workshop papers for the Natural Resources Modelling Symposium (United States Department of Agriculture, 1985) with those for the workshop on estimating erosion on five rangelands (United States Department of Agriculture, 1982). The latter were almost totally dominated by the USLE (Universal Soil Loss Equation), the former almost totally by CREAMS, EPIC and SPUR. Although the disengagement from the USLE is proceeding only slowly it seems likely to be achieved in the next decade (Foster, Laflen and Alonso, 1985). In particular the very poor handling of vegetation cover and the absence of adequate modelling of gully growth are two outstanding weaknesses. If sections of the model lack credibility, then the whole edifice is undermined as far as application is concerned. This is especially true of the USLE and MUSLE when applied to shrublands in semi-arid environments (Trieste and Gifford, 1980; Pendleton, 1985).

These models are regarded as second-generation models because they involve more sophisticated modelling styles and technology, they address much more clearly defined problems and because they are built on the experiences gained in the first generation, both in terms of team activity and in terms of specific models and model expertise. It is difficult to avoid the conclusion, however, that despite the successes in application so far reported, they suffer from many of the shortcomings of the earlier models. Specifically they are virtually impossible to parameterize except in special cases, such as experimental stations, or in high-technology environments. They are difficult, if not impossible, to validate in a meaningful manner and they are tools for research rather than application. Some of these difficulties might be expected to disappear in the near future if geographical data bases, remote sensing and geographical information systems and expert systems fulfil their widely advocated potential. On the other hand, paradoxically, the accumulation of more data is likely to reveal further the inadequacy of existing models to deal with the tremendous temporal and spatial variability of man-made as well as natural environments. The success of these second-generation models when applied in the arable fields of Iowa looks rather fragile when viewed in the context of the rangelands of Africa or even of the Mediterranean, with their crusted soils, gullied hillsides and complex vegetation covers. This is not a criticism of CREAMS or EPIC or SPUR, but simply a restatement of the need to tailor the models to the correct time and space scales and to seek solutions that can be applied in the environmental domains for which solutions are required. It would be a major blow to erosion modelling prestige and practice if the USLE was replaced by more sophisticated, more costly but equally irrelevant models.

Hydrological Mega-models

This discussion of mega-models would not be complete without some reference to modern large-scale simulations of hydrological systems. These are well reviewed in the literature. See for example Anderson and Howes (1986) for HYMO, Beven, (1986) for SHE and De Coursey (1982) for SWAM. With these models the history of development is much longer, the complexity of a lower order of magnitude and the prospect of calibration and verification much greater. The ungauged catchment problem comes closest to the kind of problems outlined in the first two sections. Here too there is serious doubt as to what can be achieved in terms of large-scale deterministic systems models. In 1972 Dooge asserted, with respect to time-series forecasting models and their application that: 'a model is something to be used rather than believed'. More recently Klemes (1986) has argued that: 'For a simulation model, on the other hand, it must be stressed that it should not be used unless it can be believed.' The point being made here by Klemes is that simulation models of many types of change cannot be tested except by implementation, and unless they are sufficiently credible they will never be implemented. The development of simulations of nuclear reactor fires and rocket failures brings home this message more forcibly than most examples of interest to earth sciences, though flood forecasting comes fairly close. Generally speaking the credibility of all models diminishes as the scientific base on which they are constructed is brought into question. Both surface and groundwater models have been brought into question, for example, by the progressive assertion of non-Darcian flow models. The recognition that distributed models are essential strains the credibility of the models yet further when the quantity and quality of the data required to drive such models is recognized. For scientific research and development the challenge is tremendous. In application the prospect of ever being able to meet the cost is too daunting (Robbins, 1985). There is a real prospect that, without a clear attempt to provide a way of handling this problem, models such as the SHE model will, like the ecological models, become archives of abandoned hypotheses (Beck, 1987). Again Klemes puts the point forcibly: 'Simulation is a safe game for the modeller, but a precarious game for the user.'

A New Generation of Models

Four major trends can be identified which have changed the style and subject matter of modelling in the eighties:

1 There has been a relaxation of subject boundaries within the earth sciences, with a consequent fruitful invasion of physical geography by other scientists anxious and well equipped to do the job.
2 There has been a consequent development of modelling at the regional and global scales.
3 The development of evolutionary models has revitalised flagging interest in longer-term behaviour and development, and is coupled to the ideas in (1) and (2).
4 Some of the areas identified as neglected in the mega-models have been taken up in the context of the later developments; in particular there has been a renewed and much stronger interest in vegetation cover.

In addition despite much arm-waving about man-environment interactions, our knowledge of such interactions is still quite feeble and our ability to model them even weaker. This too is likely to change in the next quinquenium as scientists in related fields realise the potentially rich pickings.

Cross-boundary Models

Perhaps the most important change has been the increased development of models which transcend traditional subject boundaries. This has largely come about as a result of: (1) the development of large-scale simulations of the global climate; (2) the greatly increased confidence which could be placed on information about Quaternary climatic changes; (3) the appreciation that man's activities continue to create major changes in the cover that may have significance for long-term climatic change. These developments are encapsulated in the work of Eagleson and his associates. Recognizing the need for a realistic treatment of the natural vegetation cover in general circulation models (GCM), Eagleson (1978) attempted to establish the relationships between soil hydrological properties, vegetation cover and climate, using an annual deterministic model. This first, lumped model produces a vegetation cover density which is optimum with respect to the annual soil moisture balance. Later models deal with this set of interactions in the context of specific biomes such as the savanna problem (e.g. Eagleson and Segarra, 1985). Although the principles embodied in these models are already very familiar (such as the plant water use coefficient) and sometimes grossly oversimplified (such as the assumption that water is the only limiting factor), the models make such an important intellectual leap that they are highly innovative and significant. Moreover, unlike the large-scale simulation models they reveal important properties which derive from the modelling exercise itself and provide a rich crop of stimulating hypotheses for testing. Although they too require parameters, Eagleson has been skilful in limiting the parameters to the most critical and in showing that the parameters are not impossible to evaluate for a variety of situations (Eagleson and Tellers, 1982). The models are as yet still rather primitive ecologically, for example in the key but questionable assumption that communities evolve in such a way that water-use coefficients provide for an optimal condition of aggregate growth (and consequently cover) under zero-moisture stress. However they also incorporate some of the sophisticated dynamic systems modelling strategies which are further discussed below.

Global and Regional Models

These models indicate the shift in scale brought about by interest in global climates and circulation models which are discussed at length in

Chapter 9 by Henderson-Sellers. As the oxygen isotope data on ice mass magnitudes became available in the late seventies, it prompted the development of an important set of global lumped models of ice-sheet dynamics which also spread across traditional boundaries. For example Pollard (1979) and Pollard, Ingersoll and Lockwood (1980) have developed models in which the relative role of external forcing, driven by Milankovitch dynamics, is qualified by ice-sheet response, tectonic rebound and ocean–atmosphere interactions. As with the Eagleson models it is the focus on interactions, so evidently lacking in the engineering simulation models, that is critically important, and that allows new theory and explanations to emerge from the activity. From Pollard's work, for example, the conditional stability of ice sheets with respect to latitude goes some way to accounting for the sawtooth behaviour of recession that is indicated by the palaeo-environmental data (see e.g. Barber and Coope, 1987). Ultimately the hope is that these GCM models will be able to incorporate not only the effects of intrinsic changes in ocean–ice–atmosphere interactions, but also the effects of human activity at the global scale. As yet the problems of matching global dynamics with regional scale changes, even for areas as large as the Amazon Basin, and even in the context of rain-forest removal, seem formidable (Henderson-Sellers, 1987). To geomorphologists, however, the message should be clear, the desired models of past climates and past vegetation covers for inputting into models of geomorphic change will be much closer by the end of the decade than they were at the beginning. Perhaps then the prospects for meaningful rather than speculative palaeohydrology will also be much better (cf. Thornes, 1987).

Evolutionary Models

The third major trend has been in the development of models of evolutionary behaviour, as foreseen by Brunsden and Thornes (1979). These models, which have a long history in ecology, are concerned to explain and predict the probable trajectories of state variables on manifolds representing the systems of interest using the methods of dynamic systems analysis. The potential and implications of these non-linear modelling strategies are reviewed by Thornes (1985) and Culling (1987), and there have recently been some significant applications in geomorphology in hillslope studies (e.g. Trovimov and Moskovin, 1984), in channel modelling (Kirkby, 1980; Thornes, 1980) and in vegetation cover–soil erosion interactions (Thornes, 1987). They shift the interest from equilibrium conditions *per se* to stability or otherwise of the equilibria. Here again there is an important link with the global models, where these techniques have also been applied. For example Eagleson and Segarra (1985) argue that the only stable equilibrium in their savanna model is one that involves a mixture of woodland and grass, and they are able to show that this equilibrium is stable to perturbations relating to human activity but metastable with respect to climatic changes. Similarly, in evaluating a global ice-sheet model Saltzman and Sutera (1987) simulate the shift from warmer, low-amplitude ice-sheet oscillations to colder, high-amplitude oscillations in the middle-Quaternary by variations in only three free parameters and without recourse to external changes in the earth's orbit.

Modelling Neglected Processes

These major changes have been taking place against a background of continuing research developments in the areas identified as lacunae in the mega-models. In particular there has been a major effort to model the development of rill and gully systems without which the erosion models developed by the USDA are of very limited utility outside cultivated fields in the Middle West. This work is discussed by Kirkby in Chapter 19. Its significance lies in the provision of a more general theory of network behaviour as well as in its applicability to soil erosion modelling. Likewise, the modelling of geomorphology and vegetation cover interactions (De Ploey, 1982; Thornes, 1987; Kirkby and Neale, 1987) has been initiated. It is interesting to note that in fluvial studies large-scale efforts at simulating channel form and change have been essentially abortive. The problem seems to revolve around the fundamental difficulty of providing adequate bedload theory, even though deterministic models of channel bifurcation go back a long way in the literature. The recent dynamic systems models of multistable states with very simple dynamics appear to be especially well suited to model the conditions of stable one-, two- or multithread channels, just as the stable limit cycle models of ecology and medicine appear to provide an obvious lead into the pool and riffle problem. As yet there appear to have been no efforts in these directions.

Modelling the Human Impact?

We conclude this chapter by noting that although physical geographers write extensively about man's impact on the environment (e.g. Jones, 1983) very little real effort is being made to model these effects either for predictive purposes or in an effort to understand them better. It seems likely that, as with global-scale physical geography, other scientists will step in where geographers fear to tread. Innovative modelling is demanding both intellectually and technically. Technical accomplishment will not guarantee progress, as the section on mega-models has shown. It is, however, important that physical geographers tackle these problems if the subject is to survive into the next century.

2 Modelling Geographical Knowledge

Roy Haines-Young

The soul's dark cottage, battered and decay'd
lets in new light through chinks that time has made;
stronger by weakness wiser men become,
as they draw near to their eternal home,
leaving the old, both worlds at once they view,
that stand upon the threshold of the new.

Edmund Waller, 'On the foregoing divine poems'

The Chinks that Time has Made

Models belong to that class of objects which my son, aged two, calls 'slippery things'. It's just at the moment when you think you have them cornered that they're off. The difficulty with any review of models in geography is simply trying to pin down just what they are.

The rot seems to have set in from the early days of modelling in geography. Haggett and Chorley, for example, suggested that a model is:

> either a theory, a law, a hypothesis or a structured idea. It can be a role, a relation or an equation. It can be a synthesis of data. Most important, from the geographical viewpoint, it can also include reasoning about the real world by means of translation in space . . . or time (Haggett and Chorley, 1967: 21–22)

Elsewhere, models have also been widely conceived as idealized or simplified representations of reality. Despite such a range of ideas it is difficult to identify what is so distinctive about models that has generated so much excitement over the past two decades. Theories, laws and hypotheses have specific and separate roles in science, and it seems to be of little benefit to describe them all as 'models'. Moreover since all verbal, visual and numerical statements about the world are in some sense simplifications of it, to label them as models also seems to be of dubious value. By including everything the term model, as it is generally used, describes nothing in particular.

In an attempt to place some limit on what it is that the term model describes, and to give it some content, one working definition might be 'any device or mechanism which generates a prediction' (Haines-Young and Petch, 1986: 144). The definition could be strengthened by adding that the device or mechanism should be constructed on the basis of some theory, and that the prediction should be testable in the sense that it is potentially falsifiable. According to this view, modelling, like experimentation and observation, is simply an activity which enables theories to be examined critically.

Despite its lack of elegance, the definition suggested above has the advantage that it places the concept of a model firmly within the framework of the 'hypothetico-deductive' view of science. It also provides a criterion by which progress in geography can be judged. One might ask, for example, to what extent has the use of the range of modelling techniques presently available stimulated the development of a strong theoretical basis for the discipline? Or, are the trends in the way in which models are now being designed and used likely to deepen our understanding of the earth as the home of man? As we shall see, the restrictive definition of models suggested above may not be acceptable to all. Indeed it seems unlikely that many could be persuaded to rethink an idea which been used so casually for so long. In the end perhaps the only virtue of the definition suggested here is that, although we may reject it, it allows us to focus momentarily on the strengths and weakness of a 'battered' and 'decay'd' concept in something of a new light.

The essay has two parts. In the first, the scientific and engineering styles of modelling are considered together with some contemporary criticisms of the 'scientific' or 'positivist' paradigms with which such work is associated. It is suggested that modelling is an essentially design-based, deductive enterprise which does not preclude the analysis of the qualitative aspects of natural and human systems. The view that one should reject science because quantitative modelling has tended to dehumanize and narrow our view of the world is questioned. The second part of the essay examines the way in which techniques derived from the field of artificial intelligence and expert systems' engineering provides a richer language for modelling than has been available in the past. Models, it is suggested, are not representations of the world but representations of our knowledge and assumptions about it. Future work in this area of information technology is important because of the emphasis it places on modelling our knowledge about the world, rather than the simple collection, manipulation and display of facts about it. Quantitative and qualitative models remain indispensable tools for testing our understanding of the world, despite claims of the 'post-modernists' to the contrary.

Stronger by Weakness

A final judgement on the extent to which models and modelling have stimulated a deeper understanding of mankind and its environment is perhaps best left to the reader of this book as a whole. In coming to some conclusion the idea of models as devices for making predictions is helpful because, by attempting to apply it, one exposes several weaknesses in current thinking about models. The recognition of these weaknesses, it will be argued, is important since it may stimulate a more widespread and fruitful use of models in the future.

Science vs Technology

There is, in any account of modelling, the problem of how to treat the difference between the way in which models are used in a technological or engineering sense, and the way in which they are employed in 'pure' science. The engineer, like the pure scientist, uses a model to make a prediction, but while the aim of the scientist is usually to compare this prediction with the real world in order to test the theory on which the model was based, the engineer is more interested in using the prediction as the basis for some decision or action. To what extent are these two styles of modelling fundamentally different?

Although the term 'engineering model' is widely used such models are better described as technological ones, since the term engineering suggests a preoccupation with physical systems when the same contrast could be drawn over the use of models of the human world. Examples of these two styles of modelling in geography are manifest. Thus the models in island biogeography described by Harvey and Godfray in chapter 3 might be regarded as scientific because they have been used to test propositions relating to the role of competitive exclusion in controlling the distribution of island species. In contrast, many of the 'black box' models of hydrological systems described by Kirkby in chapter 19 would not. These are technological models in the sense that they are based on the recognition of a set of empirically derived, functional relationships between variables rather than on some underlying scientific rationale. As Batty's account in chapter 10 illustrates, such models have also found widespread use in urban geography. If, as Kirkby suggests, black-box models are the most widespread in the literature of physical geography, then this does not seem to augur well for the development of the theoretical basis of the discipline. Moreover, since such models are essentially 'data-based', or 'data-driven', preoccupation with them would seem to be a recipe for directing attention away from developing a theoretical framework.

If the distinction between the scientific and engineering styles of modelling can be maintained then the first issue posed by the working definition of models given above arises from the fact that it appears to exclude a whole class of activity normally subsumed under the term 'modelling'.

The contrast which is often drawn between the engineering and scientific styles of models is symptomatic of a more general distinction which is often drawn between science and technology. Curran (1987) has provided a recent review of the topic in the context of remote sensing. His comments extend to the other information-based technologies and are relevant to any view which one might form of models. Curran argues that the technological approach to research in geography has received relatively little attention but that it does, nevertheless, represent an important and distinct tradition within the discipline. He argues that the technological approach is mainly stimulated by human need and involves: 'the systematic application, of human ability to achieve greater control over nature and human processes' (Curran, 1987: 1266). Its major distinguishing characteristic, Curran claims, is the emphasis it places on design.

Although there is much to dispute in Curran's general thesis, his paper is of value in highlighting this aspect of design in technologically based work. He illustrates the process by means of the following quotation:

> all designs of successful remote sensing efforts involve at a minimum: (1) clear definition of the problem in hand, (2) evaluation of the potential for addressing the problems with remote sensing techniques, (3) identification of the remote sensing data acquisition procedures appropriate to the task, (4) determination of the data interpretation procedures to be employed and the reference data needed, and (5) identification of the criteria by which the quality of information can be judged. (Lillisand and Kiefer, 1979: 30)

If the term 'model' were substituted for 'remote sensing' then this proposition could almost equally well describe the technological use of models in geography. Moreover, Curran's account illustrates that rather than setting technological applications of techniques like remote sensing and modelling apart from mainstream science, there is in fact an essential unity between them. If Lillisand and Kiefer's view does encapsulate the key characteristics of the technological approach then it is clear that the methodology of technology is essentially a deductive one (cf. Haines-Young and Petch, 1980, 1986).

With access to the new information handling technologies the siren calls of induction are likely to remain strong. Such a tendency is illustrated by Openshaw's contribution to this volume (chapter 6), in which he examines the case for a return to the inductive approach, involving the evolution of models from theory-based and data-free mathematical tools towards data-based models. Such developments, he argues, are in part stimulated by the external demand for applied geographical modelling and are facilitated by the range of computer-based techniques of data handling which are now becoming available.

Black-box or engineering models may be based on empirical relationships whose theoretical basis is often obscure. Such models may therefore appear more as instruments for making a prediction or a forecast, than as devices for deepening our understanding of natural systems. However, the variables of which they are composed are not thrown together in a random and unstructured way. It seems absurd to suggest that access to a large data base and the range of modelling tools which the new information technologies provides eliminates the need to consider critically the way in which variables should be combined and processed. Indeed, in view of the speed with which modern computers can generate nonsense it is even more important to consider how output can be evaluated critically. Even the simplest kind of modelling which could be undertaken using a geographical information system (GIS) requires such consideration. For example, justification of a land capability map produced by the overlay of several data planes held in a GIS as a basis for rational decision making, is most likely to be found in the extent to which the prediction represented by the map can be corroborated independently, rather than on claims about the 'quality' and 'coverage' of the input data or the expertise involved in their manipulation.

The unity of the technological and scientific approaches is further emphasized by the fact that it has increasingly been recognized that applications-based work both depends on and stimulates advance in pure science. As Brunsden (1985) has argued, the application of existing theory in fluvial geomorphology is one of the best forms of hypothesis testing since: 'failure to solve a practical problem in the natural landscape laboratory will quickly reveal deficiencies in the theory and stimulate new discoveries which model more exactly the real situation' (Brunsden, 1985: 236–7). This seems to have been the case with the theory of island biogeography. Without the attempt to apply it to the design of nature reserves it is unlikely that the theory would have developed so rapidly and its limitations seen so quickly.

The deficiency of technologically based models does not lie in the inclusion of empirical relationships but in the failure of workers to make explicit what theoretical foundations exist and to expose them to critical evaluation. The increasing focus on 'physically based' models in physical geography is therefore to be welcomed. The need for a more explicit treatment of theory underlying a model is even more important given the trend, noted by Thornes and Kirkby in their chapters, towards large interactive models and the general developments in the computer-based processing of geographical data. Progress is most likely to lie with the inclusion of a wider range of theoretical concepts in such models in order to cope with their growing complexity.

The first important weakness which the idea of models as predictive devices exposes is the false dichotomy which has been drawn between the technological and scientific styles of modelling. By ignoring the theoretical basis which exists in technological models we fail to exploit them as fully as we might. In particular, we fail to use the opportunity created by applied research to test our theoretical understanding of systems in new and more challenging situations.

Science and the Retreat from Rationalism

The retreat from science or positivism takes many forms. Although science and positivism are not necessarily the same, for present purposes we may consider them so. There are three strands of the reaction against science in contemporary geography which are worth considering here. Each is thrown in to sharper focus by the idea of models as predictive devices. The first and second concern the limitations which the language of science seems to impose on the way in which the world is perceived. The third focuses on the supposed ideological credentials of science. The criticisms of science mounted in each of these areas are valuable in helping us to identify further weaknesses in present thinking about models and modelling.

A central theme of the humanistic critique of science is the failure of the 'scientific paradigm' to capture or deal with concepts like meaning, value and imagination. Above all, with its emphasis on generalization, science is criticized for its neglect of the individual:

> What happens when we unthinkingly borrow the mathematical forms of mechanism, forms used to describe purely unconscious and non-sentient *things*, to describe aspects of our human world? The mathematical language of mechanism is going to make our human world . . . mechanical! Why? Because in the language we have chosen to describe it, that's all it can possibly look like. The language doesn't allow us to *think* anything else. And smearing the problem with a bit of statistical probability doesn't solve it – it just sweeps it under the rug (Gould 1985b: 283).

The humanistic tradition seeks to emphasize the variety of human responses to the environment, space, place and people, and rejects the attempt to pigeon-hole it simply as a series of examples of general behavioral models (cf. Johnston, 1985: 330). The humanists would claim, moreover, that the recognition of universal relationships and the ability to predict does not provide a complete basis for understanding.

The problem of finding an appropriate language in which to describe the world is an important one. Rationalists such as Popper argue that common languages can be found in which to discuss and criticize alternative views (Popper, 1970: 56–7). At the other extreme, methodological anarchists such as Feyerabend argue that languages are incommensurable. The position of the humanists is closer to the latter:

> The post-modern perspective distrusts claims for a privileged path to truth or to accurate representations of reality. It takes intellectual stimulus from a playful celebration of difference, from the mirroring of multiple perspectives. (Cosgrove, chapter 11 of this volume)

Without wishing to endorse such a view entirely there are aspects of the criticisms levelled at the restricted nature of contemporary modelling techniques which are valid. Such models have focused excessively on quantitative relationships and there is a sense in which they merely serve to articulate what is already implicit in their premises. There is a sense in which ideas important to humanistic geographers, such as the notions of 'becoming' and 'creativity', cannot be captured by such techniques.

The problem of language and universality is also emphasized in the second strand of criticism of science, namely that offered by the realists. Sayer (1984, 1985), for example, asks us to consider two 'methodological injunctions' which serve as competing propositions about the way in which explanations might be developed. On the one hand there is the positivist, or nomothetic. This, he suggests, argues that if we are to explain a process then we must discover the regularities or universal laws which govern its behaviour. Thus the thrust of research must be towards the discovery of order and can be characterized as essentially 'extensive'. In contrast, Sayer offers the realist view. This suggests that if we are to seek to explain why things behave as they do then we must understand their structure and the properties which enable them to produce or suffer particular kinds of change. The realist position, Sayer argues, calls for a more intensive kind of research. One which, it seems, very largely eschews large-scale, quantitative modelling. Whether all attempts at formal modelling are to be avoided is not altogether clear. What is certain, however, is that in adopting the realist position one is demanding much more attention be given to the analysis of qualitative relationships than present quantitative modelling techniques seem to allow.

The third strand in the reaction against science and its models considered here involves an attack on its ideology. This critique often begins with an attack on the claims of science to be 'objective' and 'value-free'. The milder forms of the criticism are illustrated by Batty (chapter 10), who suggests that urban modelling never represented a science because its intellectual structure was determined by a volatile social environment. The humanists also make a similar point. Describing the style of modelling of the 1960s, Cosgrove writes: 'Moral questions were systematically excluded from geography by the simple act of denying scientific status to alternative value systems to that ordained by the liberal democratic ideology of the post-war western consensus'. (Cosgrove, chapter 17) The more extreme forms of this kind of criticism are perhaps offered by the radical geographers. Bennett (1985: 218) summarizes the position succinctly. He notes that it rests on a largely unstated assumption that quantitative geography is necessarily positivist. Its character detracts researchers from the central question of social distribution and creates a false sense of objectivity by separating the observer from the observed. Moreover, through its use of quantitative analyses it reduces people to an atomistic level which eliminates 'considerations of soul and humanistic concerns'. Moreover it tends to support the *status quo*, facilitates the manipulation and control of society, and allows no consideration of values and hence 'the norms by which society *should* be organized'.

For those committed to modelling as a respectable undertaking the criticisms represented by the three caricatures of very different philosophical positions presented above may seem irrelevant to the practical problems which are to be faced. The tensions which these criticisms represent are not merely of academic interest, however. They pose a dilemma for those who still see a unity in the study of the earth as the home of man. To leave such criticism unanswered is to accept that there are these fundamentally different methodologies for the study of physical and human systems. Unfortunately, many of the most interesting and pressing problems which confront mankind are at the human–environment interface. Faced with the loss of what few natural ecosystems remain, the wholesale loss of the genetic resources of the biosphere, and the disruption of many of the earth's essential life support systems, do we simply lapse into the intellectual lethargy of post-modernism and say . . . well, all views are valid?

The idea of models as 'predictive devices' is useful because not only does it invest the concept with some content, but it also throws into sharper focus some important contemporary issues concerning the nature of science. Indeed some might suggest that it plays into the hands of those who would criticize science, since it would seem to bolster all that they appear to despise. It is unlikely that much can be said to affect those already committed. For the more open-minded there may be some room for accommodation.

The criticism that the modelling techniques employed by geographers over the past two decades have permitted consideration of only a very narrow range of phenomena is a valid one. There has been an almost complete preoccupation with things which can be quantified. This is a weakness in the use of models in geography which practitioners should actively seek to redress. However, the task, though difficult, is not an impossible one because models as predictive devices need not be mathematical, any more than theories need be. Logical models can just as easily be developed to make predictions as can quantitative ones.

Why is prediction so important? The critics of science may be correct in pointing to the restricted range of concepts which present modelling techniques allow one to consider, but this is not a sufficient reason to reject the 'scientific method' completely. For without prediction they lack purchase on the problem of how to evaluate any claim to have described accurately some aspect of the world. To describe the values held by an individual, and the forces in society which may shape them, as the humanist might do, is to create a structure which is just as theoretical as any which might be found in science. And, pressed to justify that the description does capture the essence of what it is that is observed, the humanist cannot avoid pointing to the other aspects of the description which are consistent with the view, or which are inconsistent with some other. Descriptions of the world, including even the 'slippery dialectics' so valued by Cosgrove (chapter 17), are essentially logical models.

The problem with the reactions to science described above is that they are based on a naive view of science. As Daniels (1985: 154) notes in presenting a case for humanistic geography, while it is understandable that humanistic geographers might wish to reject theoretical concepts 'to be on principle anti-theoretical is to relinquish all explanatory powers'. Descriptions of the world, whether they be based on humanistic, radical, realist or scientific methodologies, must be internally consistent. They are, in fact, all realist in the sense that they presume a reality which exists independently of our minds. Thus, if we are to accept one account of reality rather than some other as a better description of the way things are, these descriptions must have testable consequences. It is precisely because the scientist cannot be objective and value-free that testing and falsifiability are so important:

> My answer to your question How do you know? What is the basis of your assertion? What observations have led you to it? would be 'I do not know: my assertion is only a guess. Never mind the source, or the sources, from which it might spring – there may be many possible sources and I may not be aware of half of them; and the origins and pedigrees have in any case little bearing on truth. But if you are interested in the problem which I have attempted to solve by my tentative assertion, you may help me by criticising it as severely as you can; and if you can design an experimental test which can refute my assertion, I shall gladly, and to the best of my power help you to refute it. (Popper, 1972: 27)

Models, whether qualitative or quantitative, are not simply models about the world, but models of *our knowledge* about the world. The act of modelling is to put that knowledge in a form in which it can be seen more easily for what it is.

The second weakness in thinking about modelling which the idea of models as predictive devices exposes concerns the narrowness of the view which is taken of them. It is clear that they are neither necessarily quantitative, nor are they associated with a particular ideological position. The use of models to make predictions as a way of testing our understanding of the world is, moreover, by no means confined to what is traditionally labelled as 'science', since testability should be a prerequisite of any attempted description or explanation of the way the world is.

The Threshold of the New

To summarize: the concept of a model in geography has become so inflated that the term carries very little meaning. A more restricted view of models as predictive devices has been suggested. The value of this idea is that it highlights two key weaknesses in the way in which models have been used. On the one hand preoccupation with engineering models has suggested that modelling is an inductive process, when in fact technological applications are essentially design-based, deductive enterprises. On the other, preoccupation with modelling phenomena which can be represented quantitatively has led to a very narrow and naive view of what science is. More attention should be given to the problem of modelling qualitative relationships. Models are not simply calculating devices but re-presentations of our knowledge about the world in a form which is more easily criticized. What ways are open to us for modelling our geographical knowledge?

In looking to the future it is all too easy to suggest a 'technological fix' for our problems. However, there are in the developments known collectively as the 'new information-based technologies', techniques which may allow us to carry the problem of modelling geographical knowledge to a deeper level than has been possible at present. The techniques are those associated with artificial intelligence (AI) and expert systems.

So What's New?

An expert system can be described as a computer program designed to help people make better decisions. The programs attempt to encapsulate the kind of knowledge possessed by an expert or experts in a given field and make it available to the user of the system in such a way that it may be brought to bear on a particular problem. Such systems have been used in a variety of situations including the provision of medical diagnoses, legal advice, and the identification of likely locations for mineral deposits.

On the basis of the rather conventional picture of an expert system presented above, it may not be easy to see how the technologies which enable such systems to be built can be of assistance in geography. The significance of developments in artificial intelligence in general, and with expert systems in particular, lies in the fact that they show how our notion of modelling can be expanded to capture the more qualitative aspects of the world and our knowledge about it.

In formal terms, expert systems attempt to solve problems by what Jackson (1986) describes as 'pattern-directed inference'. The process depends on the capability of the 'knowledge engineer' to represent knowledge about a particular problem area, or domain, in such a way that patterns of incoming data can trigger the actions necessary to achieve a goal. The systems operate using 'symbolic computation', a process in which symbols stand for concepts, ideas and facts, rather than the numerical entities of conventional programs. It is this property of symbolic computation and the potential it offers for modelling a wider range of entities and relationships between them where the novelty of AI techniques lies. In considering the wider implications of these systems one should not be misled by the adjective 'expert'. Expert systems are essentially devices which can be used to represent our fallible knowledge about a given domain in a dynamic and better testable way.

Reviews of the developments in the field of AI from the geographical perspective have been provided by Smith (1984) and Couclelis (1986). Although such techniques are now being widely considered as tools in the fields of remote sensing and geographic information systems, the ideas remain sufficiently new for it to be necessary to describe some of the methods and the way in which they can be used in more detail. Readers who are familiar with the concepts may skip the next section.

How do Expert Systems Work?

Despite the variety of expert system architectures which exist their operation can be described in terms of the interaction of three major components:

1 a knowledge base;
2 a data base or working memory;
3 a rule interpreter.

The knowledge base, as the name implies, holds the knowledge about the domain of interest. Various formalisms for knowledge representation are available for the construction of the knowledge base which actually holds the domain-level information. They include such techniques as production rules, structured objects and predicate logic (Jackson, 1986; Waterman, 1986). The details of each formalism cannot be reviewed here. It is important, however, to describe something of their character in order that some of the implications of expert systems technology might be better appreciated. Thus we will consider one knowledge representation technique in more detail in order to show how it can be used to solve problems.

Production rules are probably the most widely used method of knowledge representation for building expert systems. A production rule is simply a construction of the following form:

IF <condition> THEN <action>

The rule should be read such that if the condition(s) (or the left-hand side) is true then the action(s) (the right-hand side) is initiated. An operational production system is composed of many such rules. Depending on the data supplied to the program particular rules are triggered and conclusions drawn.

Examples of expert systems in the field of environmental sciences which use production rules include BURN (Starfield and Bleloch, 1983), GEOMYCIN (Davis and Nannings, 1985), RCS (O'Keeffe et al., 1987) and PROSPECTOR (Gaschnig, 1982; Waterman, 1986). Table 2.1 shows part of the rule base of the GEOMYCIN application described by Davis and Nannings. The system was developed to help predict fire behaviour in the Kakadu National Park, Australia. The goal set for the system is to reach some conclusion about the level of fire risk in an area. In order to achieve this goal the system uses information supplied both by the user and a geographical data base. If during the running of a session the premise of one of the rules is satisfied then the rule is triggered and the

Table 2.1 Knowledge base taken from an application of GEOMYCIN

AVAILABLEBIOMASS Rule Set		
Rule 1	If	the season is wet
	then	the available biomass is negligible
		Applicable Decision Units: All
Rule 2	If	the season is cold, cool or hot
	then	the available biomass equals the grass biomass
		Applicable Decision Units: All
FIREDANGERRATING Rule Set		
Rule 3	If	season is cool, and the temperature is less than 30°C
	then	the fire damage rating is moderate
		Applicable Decision Units: Mixed scrub. Forest
Rule 4	If	the season is hot, and the temperature is greater than 30°C, and the wind strength is high
	then	the fire danger rating is extreme
		Applicable Decision Units: All
Rule 5	If	the season is hot, and the temperature is less than 27°C
	then	the fire danger rating is moderate
		Applicable Decision Units: Forest, Woodland
FLAMEHEIGHT Rule Set		
Rule 6	If	the fire danger rating is high or greater, and the available biomass is high
	then	the flame height is 6 m
		Applicable Decision Units: All
Rule 7	If	the fire danger rating is moderate, and the available biomass is low
	then	the flame height is 1 m
		Applicable Decision Units: All
Rule 8	If	the fire danger rating is negligible
	then	the flame height is 0 m
		Applicable Decision Units: All
FIREDAMAGE Rule Set		
Rule 9	If	the flame height is greater than 5 m, and the vegetation includes *Livistona humilis*
	then	the fire damage is extreme
		Applicable Decision Units: Forest, Woodland
Rule 10	If	the flame height is less than 5 m, and the flame height is greater than 0 m
	then	the fire damage is moderate
		Applicable Decision Units: All

Table 2.1 continued

Rule 11	If then	the flame height is 0 m the fire damage is negligible
		Applicable Decision Units: All

Source: Davis and Nannings, 1985

condition described in the action part of the rule is activated.

The second major component of the system, the working memory, is a dynamic data structure used to store information and to record actions and conclusions as the various production rules are fired. Table 2.2 gives an extract of part of a session with GEOMYCIN, and shows the kind of interaction which might occur when using the system.

Although representation of knowledge in a particular formalism is an important step in the development of an expert system, such knowledge acquisition on its own is insufficient to construct a 'problem-solving machine'. Like the domain expert both knowledge *and* reasoning have to be used by the expert system to solve a problem. The final component of the expert system, the interpreter, provides this facility.

The interpreter, or 'inference engine', controls the selection and operation of the production rules and the pattern of reasoning which they support. Its actions can be described in terms of a 'recognize–act cycle', involving:

1 matching the calling patterns of the rules against the rules in working memory;
2 resolving any conflict if more than one rule is triggered;
3 modifying the working memory when the rules fire.

The GEOMYCIN interpreter uses a 'backward chaining' strategy in an attempt to reach a conclusion. At the start of the session those parameters whose value needs to be found are identified. The rules which establish these values when their premises are satisfied are selected. The establishment of each of these premises then becomes an interim sub-goal which has to be achieved before the main goal can be satisfied. If the values of these parameters cannot be found from other rules or the geographic data base then the system asks the user to supply the relevant information.

This brief account of the operation of an expert system conceals many of the important differences that exist between the wide range of systems that have been developed. In particular it ignores the variety of strategies which can be used by the interpreter.

Expert systems have usually been developed in the context of particular applications and so were designed with a style of reasoning suited to that particular domain. However, the success of these early systems has

Table 2.2 Example consultation with GEOMYCIN

The user queries the possible degree of fire damage in mixed scrub:
User supplied data are underlined
Comments in parentheses

Decision Unit: Mixed Scrub

RULE 9 fails (not applicable to this decision unit)
Trying RULE 10 (to determine FIREDAMAGE)
  Trying RULE 6 (to determine FLAMEHEIGHT)
    Trying RULE 3 (to determine FIREDANGERRATING)

Question 1	WHAT IS THE SEASON IN THE MIXED SCRUB? cold RULE 3 Fails Trying RULE 4 RULE 4 fails Trying RULE 5 RULE 5 fails (not applicable to decision unit)
Question 2	I WAS UNABLE TO INFER FIRE DANGER. WHAT IS THE FIRE DANGER RATING? moderate (FIREDANGERRATING=MODERATE transferred to FOREST and WOODLAND decision units) RULE 6 fails Trying RULE 7 (to determine flame height) Trying RULE 1 (to determine AVAILABLEBIOMASS) RULE 1 fails Trying RULE 2
Question 3	WHAT IS THE GRASS BIOMASS IN THE MIXED SCRUB? unknown (obtains GRASSBIOMASS=low as default in geographic data file) (AVAILABLEBIOMASS=low) RULE 2 succeeds RULE 7 succeeds (FLAMEHEIGHT=1 m)

RULE 10 succeeds
THE FIRE DAMAGE IN THE MIXED SCRUB IS MODERATE

Source: Davis and Nannings, 1985

led to the attempt to generalize the kinds of approach developed by the construction of 'shells'. A shell is simply a system stripped of its knowledge. In terms of the three components of an expert system noted above it is the inference engine and the working memory minus the rule base. Usually such shells provide facilities for users to enter their own knowledge so that a new application can be built. Thus GEOMYCIN has been developed using the shell called EMYCIN, an expert system shell derived from work in the field of medical diagnosis (Shortliffe, 1976; Mackenzie, 1984). PROSPECTOR, initially designed for geological applications, is available as a shell for other applications.

Methodological Issues

When presented with a 'knowledge base' such as that illustrated in table 2.1 it is easy to be misled into thinking that the construction of expert systems is a mechanical exercise, little different, the post-modernist might claim, from the creation of a quantitative model. The concept cannot be rejected so easily. The simplistic account of expert systems presented above camouflages a number of important issues. Three of them will be considered here. They are the problems of knowledge representation, knowledge elicitation and explanation. The importance of expert systems lies in the fact that in requiring us to address these issues in a formal way we may significantly enrich our understanding of geographical systems.

The *problem of knowledge representation* concerns finding an appropriate way of representing knowledge about a particular domain. The production system formalism is a technique best suited to those kinds of domain where our knowledge is of a *procedural* kind. Other formalisms have been developed to cope with *declarative* kinds of knowledge. Figure 2.1 shows an example of a simple semantic network. A number of expert systems have been built up using such alternative knowledge representation techniques. In building an expert system there is always a danger of forcing a structure upon a particular knowledge domain which is inappropriate. An important aspect of future developments in expert systems technology is to design more sympathetic and perceptive ways of representing our knowledge about the world.

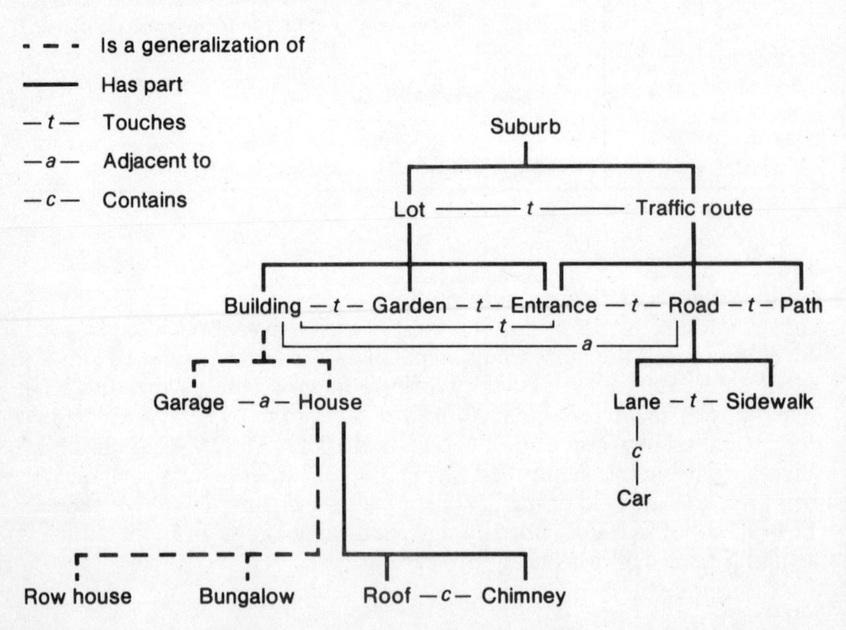

Figure 2.1 A semantic net (after Nicolin and Gabler, 1987)

The problem of knowledge representation impinges upon what has been called the 'cognitive analogy'. This concerns the extent to which expert systems model the way human minds work. The issue has been explored in some detail by both Smith (1984) and Couclelis (1986). Nystuen (1984) has argued that there is little potential in AI methods for addressing problems considered important in spatial decision-making, such as the decision to migrate and decisions involving shopping behaviour. He suggests that 'engineering' applications are most likely to hold promise for the future. As systems such as those built using RESHELL (Goodenough et al., 1987) and ECO (Muetzelfeldt et al., 1987) illustrate expert systems as 'user-friendly' interfaces to more complex technologies is likely to be one important application. But there seems little basis to Nystuen's general claims. His argument rests mainly on the expense and effort in developing an AI application. With the availability of versatile, low-cost shells for microcomputers it is unlikely that limitations really apply. There seems little to prevent the use of these technologies for modelling the qualitative aspects of our knowledge of both natural and human systems. These qualitative aspects might include, for example, opinions, beliefs and values and their supposed behavioral consequences. The availability of the expert systems' formalism may provide an opportunity to model human situations in more concrete and better-testable ways than is possible at present.

The search for ways of representing the diverse nature of geographical knowledge in forms which might enable expert systems to be developed will not be easy. It seems evident, however, that rather than narrowing our view of the world in the same way that the development of quantitative models seems to have done, the search for solutions to this first methodological problem is likely to enrich our understanding as we seek to apply expert systems' techniques to new situations. As Couclelis (1986: 6) notes, one of the most important opportunities which the availability of expert systems' technology offers may be the extent to which it enables us to greatly 'expand the realm of modelable phenomena'. Access to such techniques may therefore do much to overcome the criticism levelled at the narrowness of 'positivistic' approaches to geography.

The *problem of knowledge elicitation* is a second important issue which arises out of the use of expert systems. For some applications much of the knowledge available will exist in written form. For other problem areas most will reside in people's minds. As Burton and Shadbolt (1988) point out, the process of getting the appropriate information out of an expert's head and into some manageable form is one of the most difficult steps in developing an expert system. An appreciation of the process of knowledge elicitation, like the issues involved with the search for appropriate ways in which to represent knowledge, is important since it is likely to enrich our understanding of particular knowledge domains. An aspect which is highly problematic is the way in which experts deal with uncertainty. Although expert systems such as MYCIN and PROSPECTOR have been designed to take account of uncertainty the values assigned to conclusions and evidence are not simple probability factors. The development of ways of eliciting these qualitative aspects of knowledge represents a considerable challenge for those who seek to build any expert system.

In his review of some of the pitfalls of building expert systems Waterman (1986: 186) stresses the importance of 'choosing an appropriate problem'. Some problems, he argues, are more appropriate than others, and so one should not court failure by selecting one which is so difficult that it cannot be solved by the available resources. Well-structured knowledge domains and problems for which an agreed set of solutions can be perceived are those which are most likely to yield success compared to domains in which knowledge is unordered and goals vague. Unfortunately most problems lie somewhere between these two extremes, and only by attempting to elicit knowledge about a particular area can one begin to make such a judgement. Even if an expert system cannot be developed for a particular domain it is likely that the process of trying to develop a formal representation of its structure is likely to yield important insights into its character. Commercial application does not have to be the only goal of those concerned with building an expert system. The technology also provides a set of concepts which may enable us to expose knowledge about the world to a closer, more critical scrutiny.

The *problem of explanation* is the final issue considered here. It arises because it is unlikely that any advice-giving or 'reasoning system' would be taken seriously unless it could explain to the user the way it came to a particular conclusion. As Jackson (1986) notes, initial attempts to develop such facilities concentrated on providing a simple trace of the rules which were triggered in forming some conclusion. More recent attempts have concentrated on using the method of inference control and the structure of the knowledge base to construct such an explanation. Clancy (1983), for example, has considered the epistemology of the rule-based representation of knowledge in MYCIN. He argues that despite the wide use of the production systems' formalism it has limitations in the context of developing explanations. The analysis of the rules used to construct MYCIN showed them to have different roles, different justifications, and different rationales for ordering the choice of premise clauses. Clancy (1983) argues that the ability to make explicit this structural, strategic and support knowledge is essential if adequate explanations are to be developed.

The search for explanation is not confined to those concerned with building expert systems. It is a goal which is fundamental to most of the social and natural sciences. However, the discipline required to build explanations into expert systems could be an important one, for like the tasks of knowledge representation and elicitation, the process may cause us to look critically at what we suppose to be 'true'. We explain things in particular ways by virtue of the theories that we hold about the world.

The importance of expert systems lies in the fact that they are devices which enable us to represent that 'knowledge' in more formal and comprehensive ways than has been possible with the range of modelling techniques that have been available over the past two decades. An expert system represents both a model of some part of the world *and* of our knowledge about it. They are devices which may, therefore, enable both to be investigated critically.

Conclusion

The suggestion that we should broaden the range of phenomena included in our geographical models is hardly revolutionary, and yet the ability to model the qualitative aspects of systems in a dynamic way may turn out to be an important avenue of future research. While it is unlikely that AI and expert systems technologies can meet all of the criticisms of those who reject the scientific paradigm, it is clear that the wider availability of these techniques and concepts can serve to enrich the language we use to describe the world. Languages, as Gould (1985b) notes, always place limitations on us. But as we have seen, the recognition of such limitations is important in testing our understanding of the world. The importance of models lies in the fact that they allow such understanding to be evaluated critically. The significance of expert systems lies in the extent to which they allow this knowledge to be modelled in a dynamic way.

In their examination of the long-term benefits of implementing and using expert systems Starfield and Bleloch (1983) note: 'there is currently an awareness of the need for long-term ecological data; an expert system that is used, criticised and updated offers a mechanism for recording long-term experience'. (Starfield and Bleloch, 1983: 267) There is little to suggest that geography is any different in ecology in having a need to capture 'experience' in its models. Whatever else models are, there is a case for viewing them primarily as predictive devices, that is as tools which allow theories to be tested. Through the application of expert systems' techniques we may learn to represent the understanding which this process brings in more open and more tractable ways.

3 Models in Island Biogeography

Paul H. Harvey and H. Charles J. Godfray

Introduction

We have only a rather vague idea of how many species of plants and animals inhabit the earth: estimates range from about 3 million to 30 million. Every one of those species is limited in its geographical distribution. Biogeography is the study of the factors that determine animals' and plants' geographical distributions.

In the first half of this century, biogeographers focused chiefly on physiological and historical influences on distribution. If an organism is physiologically adapted for, say, living in deserts, or growing on alkaline soil, then its range will be strongly influenced by climatic or edaphic factors. A large number of studies have involved the correlation of the distribution of animals and especially plants with aspects of the physical environment. Historical factors that may exert an influence on current distributions include the pattern of past evolutionary events, ancient climates and the contingencies of chance colonization. In addition, the acceptance of the theory of continental drift revolutionized historical biogeography and led to the resolution of many distributional conundrums.

In this chapter we are more concerned with ecological factors influencing geographical range. Though it would be wrong to suggest that such factors were ignored prior to 1950, they gained increasing attention during the 1950s and 1960s, stimulated by such seminal papers as Brown and Wilson (1956) and Hutchinson (1959). Perhaps the most influential single work was MacArthur and Wilson's (1967) The Theory of Island Biogeography, which had ramifications far beyond the study of islands or biogeography. According to this school, the key to understanding the effect of ecological interactions on distribution was an appreciation of the force of interspecific competition – the testing ground for these ideas was the pattern of animal distributions among the islands of an archipelago. Species distributions were affected by competitive exclusion: species with populations limited by the same resource cannot coexist, so a checkerboard distribution is expected, with the species showing complementary distributions among the islands of an archipelago. Species morphologies were affected by character displacement: morphological similarity implies ecological similarity, so one way in which natural selection would result in reduced competition would be for similar species to diverge in morphology when they occurred together on an island. Buoyed by elegant theory and apparently consistent empirical studies, population biologists with an interest in biogeography were in the early 1970s reasonably content with their lot.

Two types of model have led to more sceptical attitudes. First, there have been statistical models: distributional data may accord with statistical null hypotheses which do not assume the action of competition. The second type of model has been genetic: if we are to interpret morphological change in evolutionary terms, then we must have an evolutionary model to work from. When an explicit genetic model is placed in an ecological context, the outcome may differ from expectation.

Our aim in this chapter is to outline one set of statistical models and one set of evolutionary models. The statistical models test for the presence of competitive exclusion, while the evolutionary models seek the ecological conditions under which character displacement is likely to be found.

Competitive Exclusion and Statistical Models

To what extent does interspecific competition influence the geographical distributions of species? This straightforward question is not easily answered. On first encounter, species occurrences on the islands of archipelagoes offer suitable material with which to work, particularly if the study organisms are plants or vertebrates with poor dispersal abilities. Habitats can be impoverished and islands may be isolated from each other, with the result that only a selection of species is found on each island. The effects of competitive exclusion might be particularly evident under such circumstances. However, the conclusions drawn by different authors – even those working on the same data set – are often at variance because they have used different assumptions when designing their statistical models. Even when distributional patterns are of the sort that could be caused by interspecific competition, other explanations are usually possible. Evidence of competition from species distributions will always be circumstantial though, of course, some patterns are stronger or more convincing than others. Manipulative field experiments are necessary before a case of competitive exclusion can be conclusively demonstrated.

This topic has been reviewed in depth elsewhere (Harvey et al., 1983; several chapters in Strong et al., 1984). Our aim here is to summarize the salient arguments together with illustrative examples. Parts of this section follow Harvey (1987) and, accordingly, we concentrate on the issues involved rather than providing a detailed historical bibliography. We start by comparing the different ways distributional data have been analysed. We then point to the problems inherent in such analyses, illustrating them with the results of a simulation model produced by

Colwell and Winkler (1984). The same results also provide guidelines for the design of more appropriate statistical null hypotheses against which to test the available data. Improved methods are evident in several recent papers and, by way of example, we shall summarize one of those, by Graves and Gotelli (1983).

Analyses of Species × Island Tables

Presented with a species × island table (denoting which species are present on which islands), how might we detect the effects of competitive exclusion? A common technique is randomly to reallocate the species among the islands. The production of many such 'null archipelagoes' allows a statistical comparison with the original data. It is important that the only differences between the real and artificial data are those due to the effects of competitive exclusion. In attempts to conserve in the null tables pattern that does not result from competition, three common restrictions are employed. These are listed below with three common statistical techniques used to detect the presumed effects of competition.

Restrictions on the analysis

1 *Retain number of species per island.* Some islands, by virtue of being larger, nearer source pools or containing more habitats, might be expected to contain more species. When creating null tables it may be useful to rearrange the cells, subject to the constraint that the number of species on each island is conserved.
2 *Retain relative abundance of species.* Some species are more common than others, and this need not be caused by competition. For example, generalist species may persist in a wider range of habitats than those with more specialized diets. We may wish to conserve the relative species abundances in our rearranged table.
3 *Retain incidence functions.* Some species are found only on islands of a particular size (Diamond, 1975). We conserve this aspect of the original table (known as species incidence functions), by constraining the rearranged table so that particular species are found only on islands of the sizes on which those species were actually recorded.

Statistical models used How might we detect differences between the original and the rearranged data? Three methods of analysis have commonly been used.

1 *The number of missing species combinations.* This is a commonly used but very weak statistic. For example, if a species pair was expected to occur on 100 islands of an archipelago but only occurs on one island, it would not be recorded as a missing species combination.

2 *Comparing species distributions.* If species pairs occur less often than expected, they have complementary species distributions, possibly as a result of competitive exclusion.
3 *Comparing islands.* Some islands may be dissimilar because particular combinations of species are able competitively to exclude others.

Combinations of restrictions and statistical models We have outlined three possible types of restrictions that may be used for constructing the null tables, which makes seven types of null tables (restrictions 1, 2, 3, 1&2, 1&3, 2&3, or 1&2&3). We have also mentioned three types of statistical tests used to compare the null tables with the observed. Any one of the three tests could be used in combination with any of the seven types of constraints, which gives up to twenty-one ways the data might be analysed. At least eleven of those combinations have been used in published studies (see Harvey et al., 1983). It is hardly surprising that different conclusions have been drawn by different investigators. Simberloff, Strong and others working from Florida State University have been influential both in promoting the above techniques and emphasizing differences among them (see Strong et al., 1984).

Problems with Analyses of Species × Island Tables:

Colwell and Winkler's Simulation

Colwell and Winkler (1984) illustrate a number of the problems associated with the types of analyses outlined above. They employ a series of computer simulations which begin by generating phylogenetic trees. As the hypothetical species evolve, their morphology (e.g. beak length or width) changes gradually so that, on average, closely related species are the most similar. This evolution takes place on a large land mass. The organisms then disperse to the islands of a nearby archipelago, their destinations being determined by additional computer routines.

One routine allows random invasion of the different islands followed by interspecific competition, resulting in the loss of some species from some islands. When morphologically similar species invade an island, some are more likely to become locally extinct as a result of competitive exclusion. An alternative routine allows random invasion of islands from the mainland without subsequent competition. As competition occurs only in the first routine, the second makes a suitable null model for comparison.

But a different type of comparison is often made with real data. Species found on different islands are randomly redistributed among the islands (by computer, of course), and the reassortment is compared with the original distribution pattern (see Strong et al., 1984). Colwell and Winkler incorporate a computer routine that mimics this procedure. They find that the species distributions differ very little between the reassorted archipelago and the one on which competition has occurred, thus confirming that the effects of competition are indeed partly incorporated into the null model derived from the reassortment routine. Consequently, a comparison that uses this procedure is less likely to detect the effects of competition than is Colwell and Winkler's first procedure.

Another factor, usually ignored, that leads to underestimating the effects of competition, is the varied abilities of different taxa to disperse. For example, ospreys are more mobile than ostriches, and so would be likely to be found on islands that ostriches cannot reach. In a second series of comparisons, Colwell and Winkler allow dispersal abilities as well as morphological characteristics to evolve in the phylogenies generated by their computer. The result is that closely related species tend to occur on the same islands. Some species are extirpated when competition occurs, thus producing a species assemblage on each island that is taxonomically less biased than that of the original invaders. Because differential dispersal abilities produce the opposite pattern from competition, if they are ignored when designing null models the effects of competition will be underestimated.

Colwell and Winkler's model makes the reasonable assumption that competition is more intense between more closely related species. Comparisons made among closely related taxa – for example, among species within a genus – are most effective for detecting the statistical consequences of competition, for they represent the level at which the process acts most strongly. Here, Colwell and Winkler have an advantage in knowing the likely relations between competitive intensity and phylogenetic relationships among species, whereas such relations may be far less clear in the real world. Colwell and Winkler could, if they wished, give objective and usable criteria for assigning subsets of species into 'guilds' (species with similar patterns of resource utilization), whereas such decisions about grouping real species usually involve some subjectivity.

Using Colwell and Winkler's example, we have introduced three of the problems associated with the interpretation of species × island tables. They are: (1) incorporating the effects of competition into the null model; (2) ignoring the differential dispersal abilities of different species; and (3) taking into account the likely importance of competition between species.

The Interpretation of Results

If a successful statistical analysis can be completed, taking into account all of the above problems, then groups of species with positively and negatively associated distribution patterns will be identified. One of the most thorough analyses that has been completed to date is by Gilpin and Diamond on the avifauna of the Bismarck Archipelago. Gilpin and Diamond (1982) found far more species with positively than negatively associated distribution patterns. The positive associations were considered to result from shared habitats, endemism (speciation tends to occur on large islands where extinction rates are low), shared geographical origins, and shared incidence functions. On the other hand, negative associations were thought to result from competitive exclusion, differing incidence functions, and differing geographical origins. We might add unequal distributions of habitats across islands as an additional factor contributing to negative species distributions. Diamond and Gilpin (1982) used other forms of evidence in order to distinguish among these explanations in particular cases. However, there is always the problem that a force selecting for positive association (e.g. shared geographical origins) may be balanced by one selecting for negative association (e.g. competitive exclusion). Unless all the constraints are built into the original statistical model, cases of competitive exclusion may be overlooked because the null hypothesis accords with the observed distribution.

Neotropical Land-bridge Avifaunas: a Case Study

The recent shift of emphasis towards a statistical approach to distributional data has lent improved rigor to ecology. A good example is provided by comparing Graves and Gotelli's (1983) study of the bird species occupying various neotropical land-bridge islands with previous investigations. They tackled two questions: '(i) At the family level, are island communities a non-random subset of adjacent mainland communities? (ii) Are species with restricted mainland distributions under-represented on land-bridge islands?'

Graves and Gotelli chose seven neotropical land-bridge islands off the north coast of South America and the south coast of the Panama peninsula. How did they overcome problems ignored or inadequately dealt with by previous investigators?

1 *Defining total source pools.* Graves and Gotelli's largest island was Trinidad. They found that a source pool including species found within 300 km of the island was appropriate on the grounds that the source pool contains the species found on Trinidad, but that going beyond that distance included many habitats not found on Trinidad. Thus the 'total source pool' for each of their islands was designated as all the bird species found within 300 km.
2 *Defining habitat source pools.* Different bird species occur in different habitats. Graves and Gotelli narrowed their total source pools for islands to 'habitat source pools' containing only those species occurring in the actual habitats found on particular islands.
3 *Defining colonization potential.* Species with wide geographical ranges were considered to be likely to have higher island colonization potentials than those with restricted ranges. Using various measures of geographical range, they were able to answer question (ii) above.

Using appropriate statistical techniques, Graves and Gotelli were able to show that: (1) certain families were consistently over-represented on the islands (pigeons, flycatchers and warblers), whereas no family was consistently under-represented; (2) the habitat pool is a significantly better predictor of species number in each family than the total pool; (3) the number of families on each island is consistent with the expected value; and (4) species with widespread mainland ranges are disproportionately over-represented on several islands.

Of course, Graves and Gotelli's study is not perfect. Even the casual reader could focus on methods that could be improved, but such is the nature of ecology. Nevertheless, their study brings our attention back towards species autecologies and this is a useful pointer. For example, whether their preferred habitat is present may be of paramount importance in determining whether vagile animals like birds *attempt* to breed in an area. The *success* of breeding attempts may be more dependent on the outcome of interspecific competition for resources.

Character Displacement and Evolutionary Models

In this section we discuss an application of population genetic modelling to a problem in evolutionary biogeography. We continue with the historical theme that species with similar niches will compete for limiting resources where their geographical ranges overlap. It has been suggested that this competition will result in otherwise similar species showing morphological divergence in regions where they are found together. If bill length is correlated with seed size in finches, two species may have the same length bill in areas of non-overlap (allopatry), but in areas of coexistence (sympatry) competition may act as a selection pressure leading to the evolution of a larger bill in one species and a smaller bill in the other. The phenomenon, known as ecological character displacement (Brown and Wilson, 1956), has been the subject of extensive empirical and theoretical research. Before proceeding to models of character displacement, we discuss briefly the field evidence.

The Field Evidence

Character displacement is an intuitively reasonable process, but there are few good field examples. In the early 1970s there were about twenty examples which were used as undergraduate textbook cases, but one by one they were whittled away as being either inadequate or incorrect representations of the data (in particular by Grant, 1972, 1975). What was wrong with them?

There were two problems. First, it is usually possible to *select* two areas of allopatry and one area of sympatry which, taken together, fulfil the required criteria of 'similar when separated and different when coexisting'. A clear example of this concerns the stoat (*Mustela erminea*) and weasel (*Mustela nivalis*). Samples selected from the correct parts of Ireland and the British mainland show the required pattern (Hutchinson, 1959; Williamson, 1971). But samples from elsewhere in Ireland, continental Europe, the U.S.S.R. and North America can be used to show either no pattern or the opposite pattern (Fairley, 1981; Ralls and Harvey, 1985). The second problem facing other cases of character displacement was that some patterns could be explained by geographical patterns in morphology which seemed to be independent of the presence of a second species. The nuthatches *Sitta tephronata* and *Sitta neumayer* were often cited as one of the best examples of character displacement in beak size until it was pointed out that the beak size differences found in sympatry were merely those expected from an extrapolation of longitudinally correlated changes in beak size which occurred where the species did not co-exist (Grant, 1975).

One case of ecological character displacement which was reported at the same time as Grant's critiques has fared better, and two others have become established in the more recent literature. Fenchel (1975) and his co-workers studied intertidal snails. Two species in the genus *Hydrobia* are normally the same size when they occur by themselves. However, where they co-occur, one species is consistently smaller than the other. How is size related to competition? Because of the mechanics of feeding in mud snails, the size of the animal is related to the size of the food items that can be consumed. Fenchel showed that in sympatry the two species had evolved (the size differences are inherited) to feed on different-sized food particles. Fenchel was also able to show in laboratory experiments that competition was more intense between similar-sized snails.

It has been suggested for many years that character displacement is found among Darwin's finches (*Geospiza*) in the Galapagos Archipelago. On some islands, particular pairs of species are found to coexist, but on other islands only one of the species is present. Differences in bill morphology had been claimed to be due to the evolutionary avoidance of competition, though this claim was disputed on the grounds of insufficient data. However, more recently Schluter et al. (1985) have provided convincing evidence of character displacement.

Fenchel and Schluter et al. obtained evidence for character displacement by the detailed study of two or a few competing species of animal. An alternative approach is to examine a far greater number of possible interactions in less detail. J. Fjeldså studied all twenty species of the world grebe fauna and found that, when species coexist, they tend to diverge in both morphology (bill size and shape) and diet; the original papers are published over a fourteen-year period in specialized journals, but the results are well summarized by Diamond (1987). Fjeldså's comparative studies are circumstantial evidence for evolutionary character displacement but cannot exclude an alternative, purely demographic explanation: if a species is very similar to another, it may not be able to invade its range. This demographic sifting of possible sympatric associations may ensure overlapping species differ more than allopatric pairs without any coevolutionary changes (Grant, 1972).

Evolutionary Models

Turning to theoretical models of character displacement, our intention is not to review this large and expanding field, but to sketch the essential structure of one influential model which illustrates how this approach can be useful in determining the conditions which can and cannot lead to character displacement. The modelling exercise is also useful in that it can reveal crucial components of a field system upon which research workers should concentrate their efforts. The final advantage of these genetic models, like so many others, is that the translation from a verbal to a mathematical argument forces assumptions to be made explicit.

Slatkin's (1980) model describes two species competing for a single type of resource that varies along a single dimension. The members of both species show morphological variation, again along a single dimension, which influences the resources they can use. For ease of explanation, we describe the model using a concrete example. We assume the two species are finches and that the resource they compete for is plant seeds which vary in size. Individual finches differ in bill size and the size of the bill influences the size of seeds they can eat. Slatkin (1980) constructs a model that incorporates three components:

1 the resource spectrum: the abundance of different sized seeds;
2 the population structure of each bird species: the number of individuals with different sized beaks;
3 a description of competition: how a bird of any particular species and bill size competes with another arbitrarily chosen individual.

Slatkin then sought conditions for reciprocal selection pressures on each species, caused by competition, leading to a divergence in beak size. The shape of the resource spectrum was assumed to remain the same over time. Biologically, we are assuming that the rain of large and small seeds is unaffected by the action of birds. Obviously, although constancy is a reasonable first approximation, there will be some environmental variation and selective feeding by birds may ultimately affect plant recruitment probabilities. Many different seed size distributions could be chosen, and Slatkin himself explored a number. We can begin by assuming that seed size is normally distributed with its variance providing a measure of the diversity of resources in the habitat.

The size of a bird's bill is determined both by genetic and environmental factors. The genetic component will, most likely, be under the control of many different genes, each exerting a small influence. Such characters tend to be normally distributed even (and this is important later) under moderate selection. The assumption that bill size is a quantitative character with a normal distribution makes it possible to describe the population structure of each bird species by three quantities: the total population size, the mean bill size and the variance in bill size. The distillation of the complexity of population structure into three parameters is one of the most elegant aspects of Slatkin's model.

In assessing how bill size will change over time, we need to know the relative reproductive success of birds with different-sized bills. Although many things will influence reproductive success, we are concerned only with factors involving resource competition. Since bill size affects the size of seeds an individual can eat, the shape of the resource spectrum will influence reproductive success: if there are very few big seeds, large-billed birds will fare badly. In addition, the number of competitors with different-sized beaks will influence fitness. While the large-billed bird in an environment with few large seeds has low fitness, his fitness will be lower still if there are many other large-billed birds with which he must compete.

Slatkin modelled these competitive effects by using an extension of the well-known Lotka–Volterra equation, which assumes that the reproductive success of an individual experiencing competition is reduced by a factor proportional to N/K where N is the number of competitors and K is a measure of the quality of the environment. Following Slatkin, instead of assuming all individuals suffer identically from competition, we calculate the effects of competition on a particular-sized individual of each species. First, in place of N, the total number of competitors, we use the number of competitors weighted by the difference in bill size. Thus an individual with a similar-sized bill exerts a greater competitive effect than an individual with a bill of a very different size. It is possible that even if individuals of the two species have the same-sized bill, they may for other reasons exert different competitive effects. A term to include difference between intra- and inter-specific competition can be included. Second, we replace K, the quality of the environment, by a term representing the quality of the environment for a bird with a particular-sized bill. This term is thus determined by the resource spectrum. We can now describe the fitness of an individual with a particular-sized bill, of either species, as a function of the resource spectrum and the number and bill sizes of its competitors.

We now need to describe the dynamics of the situation: how bill sizes change over time. For mathematical convenience we assume non-overlapping generations. The problem reduces to finding a set of equations that, given the system is in a particular state this generation, allows us to calculate the state in the next generation. As the resource spectrum and the description of competition are assumed to be constant, the only things that can change are the population structure of the two species, each of which is described by the three quantities: population size, mean bill size and variance in bill size. There are thus six (or three pairs of) state variables whose values we want to track over time. Consider population size: in the absence of resource limitation all the members of one species will have on average a fixed number of offspring each generation. Shortage of resources and competition reduce reproductive success, measured by the number of offspring, but we already have a measure of the reproductive success as a function of bill size. The population size in the next generation is simply the sum of the reproductive successes of all the individuals in the current generation.

To obtain the values of the other state variables in the next generation, we can make use of a result from theoretical population genetics which describes the change in the mean and variance of a quantitative character with frequency-dependent selection (Slatkin, 1979). To use this result we need to know the relative fitness of a bird with a particular-sized bill (which we determined above) and one other factor: the heritability of bill size. Bill size will be determined both by genetic and environmental factors, and heritability is a measure of the genetic fraction. If variance in bill size is chiefly genetic, selection will lead to a faster response than if it is chiefly environmental. Characters vary in their heritabilities and at the moment there is active debate about the size of the genetic component of different characters.

We now have a complete description of the system at any particular time and a set of equations that allow us to predict the system into the future. Unfortunately, the equations are too complex to solve analytically for any but the most restrictive (and unrealistic) assumptions, and we have to resort to computer simulation. There are (at least) four possible outcomes of the model: one species may go extinct or, if they both survive, mean bill lengths may diverge (character displacement), they may remain the same or they may converge. We now discuss some of Slatkin's chief conclusions about when character displacement will occur.

Consider a single species in isolation. The mean bill length will normally evolve so that it can most efficiently utilize the middle of the resource spectrum. What will happen to the variance in bill length? Competition between individuals near the centre of the resource spectrum will lead to selection to increase the variance in bill length, but as more individuals find themselves in parts of the resource spectrum with little food there will be opposing selection to reduce variance. Change in bill length mean and variance will only occur if there are intraspecific differences in fitness so the two opposing selection pressures will balance each other when all individuals have the same fitness. The benefit of having a bill length which allows a bird to utilize the centre of the resource distribution is offset by the large number of competitors. This result is common to many analyses of frequency-dependent selection.

Slatkin first examined what happened when individuals of the two species were identical in all the attributes dealt with by the model. He found that very quickly the numbers and bill sizes of the two species equilibrated such that the fitness of all individuals was approximately the same. This happened relatively quickly and without character displacement, though curiously the details of the equilibrium depended on the initial starting values. Slatkin concluded that character displacement did not always result from competition between two species on a single resource axis.

These results assumed the resource spectrum was described by a normal distribution and the next step was to explore the effect of relaxing this assumption. With symmetrical resource spectra, significant character displacement was only found when the distribution was approximately rectangular. The flat top of the rectangular distribution allowed the mean bill sizes to drift apart while the steep sides of the distribution acted as a strong selection pressure to reduce variance as the bill size distribution approached the edge of the resource spectrum. However, character displacement was found much more commonly if the resource spectrum was skewed. At equilibrium with a distribution skewed towards, say, small seeds, one species would evolve to have a mean bill length near the mode of the resource spectrum while the other was displaced towards larger seeds. Significantly, a skewed resource distribution led to the two species having different equilibrium abundances. If this difference was observed in the field it would give a clue to the mechanism leading to character displacement.

In the discussion above, there have been no constraints on the values of any of the state variables. However, it might be argued that there will be limits on the variance in bill size, perhaps through pleiotropic effects of the genes determining bill size. Restriction on variance affects the achievement of the equilibrium at which all individuals have constant fitness by concentrating all the members of one species within a fixed interval of the resource spectrum. This leads to reciprocal selection on both species to avoid the crowded areas, and thus to strong pressure for character displacement. Alternatively, an assumption that the variance in bill size could not fall below a certain threshold can lead to character convergence.

Finally, we have assumed moderate heritabilities. If the environmental component to bill size variance is very large, both species will come to occupy the centre of the resource spectrum and selection will act to reduce the genetic component of the variance to zero.

Relaxing the assumption that the two species are identical also affects the likelihood of character displacement. A well-known result of the Lotka–Volterra model is that coexistence of two species only occurs if intraspecific competition is stronger than interspecific. This condition is also necessary for coexistence in the present case, especially for symmetric resource spectra. A second difference between species is in the manner in which they exploit the resource spectra. Up to now we have assumed that a certain-sized seed is equally valuable to individuals of either species with the same bill size. However, suppose that one species has an inherent advantage in utilizing large seeds; perhaps a more efficient gizzard. It turns out that even quite small differences in the perception of the resource spectrum can lead to significant character displacement.

Thus, to conclude, character displacement with both species displaying equal abundances will result most frequently from differences in the ways both species perceive the resource spectrum or when there are constraints on the variance of the character in question. Skewed resource spectra can also lead to character displacement, but under these circumstances the species will display different equilibrium abundances.

Slatkin's model combines population dynamics and population genetics to predict the outcome of an interaction between two species. Our aim in describing it in some detail is to give an example of the type of coevolutionary model that is valuable in understanding problems in biogeography. We chose this model because of its elegance and because the original paper is well explained.

Subsequent papers have extended Slatkin's work. Milligan (1985) takes a structurally very similar model and argues that character displacement should be more frequently observed than Slatkin suggested. Virtually any difference between the species can lead to character displacement. The absence of character displacement observed by Slatkin when the two species had identical competitive effects on each other was a consequence of this unrealistic assumption. Indeed, Milligan argues, the equilibrium without character displacement was not truly stable and, given time, one species would go extinct. Only when the environmental component of bill size variance is very large is convergence predicted. Taper and Case (1985) also extend Slatkin's model and relax the assumption that the resource spectrum is constant, allowing it to be influenced by the competing species. They also allow the range of resources that a single individual is able to use (the within-phenotype niche breadth) to be under selection.

Other workers have approached the problem of character displacement in different ways. One class of models assumed non-quantitative genetics while another uses phenotypic models, a level of abstraction above the genetic. Abrams (1986) is a more recent example of the last class of model and we recommend both this paper and Taper and Case (1985) to readers wanting an entrée to this literature.

Summary

Twenty years ago, interspecific competition was considered to be important for influencing species distributions among islands of archipelagoes, and for influencing intraspecific morphological differences. The key ideas were competitive exclusion and character displacement. More recently, the evidence for competitive exclusion has been reassessed by comparing species' distribution patterns against statistical null models. At the same time, the evidence for character displacement has been criticized, and useful evolutionary models of the process have been developed.

4 Three Questions about Modelling in Physical Geography

David Unwin

Introduction

In his 1968 review of *Models in Geography* the late P. R. Crowe, a climatologist who had published numerous papers using statistical methods long before any 'quantitative revolution' in British geography, was evidently uneasy about the entire Cambridge enterprise (P.R.C., 1968). With the benefit of twenty years of progress in modelling in physical geography since he wrote, and with the advantage of that most exact of the sciences, hindsight, it is instructive to examine the reasons underlying his unease. In this discussion I shall concentrate on three questions that were posed in Crowe's review, all of which surface from time to time in the other contributions to this volume from physical geographers. Most of the examples will be drawn from climate modelling.

Question 1: What is Meant by the Term 'Model'?

> 'A model can be a theory or a law or an hypothesis or a structured idea. It can be a role, a relation or an equation. It can be a synthesis of data' (p. 21). Apparently it can also be a word ('peasant'), a map, a graph and finally, what most readers might have expected, some type of hardware arranged for experimental purposes. By opening the gate so wide the authors sow a confusion from which the work never recovers'. (P.R.C., page 424)

To raise this issue may seem churlish, and concern for it has a distinctly 1960s feel about it, but it is important. One suspects that the reason for the extremely catholic view adopted by the original *Models* volume was simply that, even in 1967, it was extremely difficult to find any work within geography that could clearly be called modelling in the more restricted sense in which the term is used elsewhere in science. In science, definitions of the term 'model' vary, but a reasonably succinct one is: 'any rule that generates outputs from inputs' (Haines-Young and Petch, 1986: 145). In other words, models generate new information that may be used to test whether or not the theories embedded in them are adequate, or to make useful predictions about the future states of the systems which they describe. As the contributions to this volume by Haines-Young, Thornes, Harvey and Godfray, Henderson-Sellers and Kirkby show, models in this more restricted sense are now commonplace in physical geography. Moreover, the requirement that a model predict data that can be used to test whether or not the real world behaves as we think places modelling activity firmly in a hypothetico-deductive framework of explanation (see Haines-Young's chapter in this volume).

In climatology, so-called 'climate models' of increasing complexity and dimensionality (Schneider and Dickinson, 1974; Shine and Henderson-Sellers, 1983) have been used to predict new information in at least three ways. First, the intention has been to determine whether or not the physics in them is adequate to allow predictions of the natural evolution of the climate system. Secondly, they have been used to investigate the sensitivity of that system to hypothetical external changes such as variations in the solar constant, or internal changes involving carbon dioxide, chlorofluoromethanes, tree clearance, even nuclear war. Thirdly, they have been used to reconstruct, or 'retrodict' climates consistent with known or suggested past boundary conditions, for example, during a full glacial.

At the lowest level in this hierarchy are zero-dimensional energy balance models that solve a simple energy conservation equation for a selected variable of state which is the model's 'climate'. Usually this is some equilibrium surface temperature, as, for example, in the well-known Myrup/Outcalt local model (Myrup, 1969; Outcalt, 1971). At the global scale similar models have been developed from initial formulations by Sellers (1969) and Budyko (1969). These examine the overall radiative balance of the earth and its atmosphere, and predict as their 'climate' a rather poorly understood effective radiation temperature that is not, and indeed could not be, measured, but which is a useful guide to thinking. It will be clear that these simple energy balance models are wildly oversimplified representations of the climate. They have no internal dynamics and no geography, and cannot address questions about the variation in temperature around their predicted state value. Yet, as more recent formulations by Ghil (1976) and Fraedrich (1978, 1979) show, they can be solved 'on the back of a cigarette packet' and make interesting and important predictions that address fundamental questions about the ambiguity and stability of today's global climate system. These simple 'scientific' models are analogous to those produced by geomorphologists to describe hillslope and glacial processes, and by ecologists such as Slatkin (1980; see Harvey and Godfray's chapter) and mathematical analysis of the model properties is yielding insights in these disciplines as rich as those that have been found in climatology (Degn et al., 1987).

At the other end of the hierarchy of climate models are general circulation models (GCMs) which are clearly 'engineering' or 'technological' models. They are the products of large institutionally based teams, are strongly compartmentalized, and are devised for application. GCMs developed from the weather forecasting models of the 1950s and 1960s (Meehl, 1987; Simmons and Bengtsson, 1984) and attempt to simulate the global atmospheric circulation with as much fidelity as possible. In particular they explicitly include the evolution of the motion systems that give day-to-day weather so that any 'climate' is obtained by integration of their output over some suitable time interval. Improvements over the past two decades have involved the successive incorporation of more and more processes together with better spatial resolution and improved methods of solution. Such is the amount of computation that modern GCMs are entirely contingent on access to the most powerful computers thus far devised. Even so, investigations of climatic change using them are restricted to establishing the model global climate given by changes in their external forcing or boundary conditions. The evolution of the system towards these new states is not studied. The climatology literature is nowadays dominated by studies involving the use of GCMs and, as Henderson-Sellers (this volume; see also Unwin, 1981a) makes quite clear, it is remarkable how little climatologists with a geography background have contributed to their development and use. Certainly, had he lived long enough to make them, Crowe's comments on modelling in climatology would have been well worth reading!

Question 2: What is the Role and Relevance of Statistical Models?

> it should be remarked that these young workers do indeed share a common faith though they do not openly declare it. It is an apparently unbounded confidence in the efficacy of 'regression' and related statistical procedures. The intellectual load now being placed upon this little bit of mathematical sleight of hand might well have appalled its inventor. (P.R.C., page 424)

In the twenty years since the publication of *Models*, the perceived role and relevance of statistical methods has changed considerably and it is suggested by Thornes (in his chapter) that, in geomorphology at least, general statistical modelling has lost much of its earlier significance. The reasons for this decline are superbly illustrated by the example of species models in island biogeography used by Harvey and Godfray in their chapter. They show that what is at first sight a clear-cut problem that can be addressed by standard techniques is actually rather more complex with a multiplicity of possible tests each evaluating a slightly different hypothesis, and with the predictable consequence that different conclusions have been drawn by different investigators. It is clear that classical statistical methods are often inappropriate, and I have argued elsewhere that their use tends to force investigations into the analysis and description of system morphology rather than process description

(Unwin, 1977). The temptation is to follow the tenor of Thornes' argument and reject all but the most elementary of statistical methods as inappropriate in physical geography.

I would argue that such a rejection is a mistake and that statistical methods have several important roles to play in physical modelling. With some notable exceptions in hydrology and related fields, the full potential of stochastic process modelling of time, space, and space–time series does not seem to have been realized. In 1977 I wrote:

> It may be that we are now on the threshold of a second wave of application of statistical models to physical geography based . . . upon methods developed originally for the analysis of fluctuations and noise in statistical physics where, like the environmental systems of interest in goegraphy, 'chance' mechanisms play an important role. (Unwin, 1977: 208)

A decade later I see no sign of this second wave! It would seem that physical geographers continue to believe that all 'real' science is deterministic. Such a belief may well align them with Einstein in a famous debate in the history of science, but physicists since the 1930s have sided with Bohr in accepting that causality is a matter of probability and thus of statistical analysis.

As almost all the examples given in this volume indicate, the distinction that is often made between probabilistic and deterministic models refers to the ends of a continuum with most models lying somewhere between them. There are several ways by which statistical concepts enter into modelling. First, and most traditionally, they enter into the calibration of *a priori* models that may well contain within them considerable insights into the underlying processes. Secondly, good statistical methods are needed in the efficient parameterization of engineering models, such as the GCMs discussed above, where processes operating at scales below that of the model's resolution are represented by empirically derived values or expressions involving parameter estimation from real-world data. Thirdly, as far as is possible models should be validated against the real world and this can often be surprisingly difficult. Spatial statisticians with a good working knowledge of autocorrelation effects in space–time series will, for example, be much amused by the procedures adopted by GCM modellers who frequently use Student's t to compare simulation results obtained at enormous cost in corporate effort and computer time with real-world data at the same grid points (Laurmann and Gates, 1977). Fourthly, there is an argument developed for GCMs by Chervin and Schneider (1976) that can be applied to almost any apparently 'deterministic' engineering model. At first sight every realization from such a model should be uniquely determined from the start, yet it is known that these models produce their own stochastic noise from a number of sources of uncertainty such as in the specification of their initial state and boundary conditions, their parameterization(s), the finite differences used in their solution, and even in undetected errors in their computer code. This would not matter if they were stable to small disturbances but, like the real atmosphere, they are capable of magnifying small differences as they run. It follows that it is not straightforward to decide whether or not a given model result indicates a real climatic change due to changes in its boundary conditions or is simply a reflection of the inherent system noise (Chervin, 1978). Again, the solution to this problem would seem to imply very careful use of methods from spatial analysis.

Question 3: How do we Teach the New Models?

What then is its object, and to whom is it addressed? (P.R.C., page 423)

My final question concerns education, and it leads me to a conclusion that some may find difficult to accept. The original Madingley lectures that gave rise to *Models* were delivered to school teachers and their aims were pedagogic. The enterprise might well have failed had it not been for the classroom and lecture hall utility of the ideas it introduced. True, *Models* contains one or two 'difficult' essays containing smatterings of formal logic, differential equations and linear algebra but, by and large, its contents are accessible to all. Looking at the variety of modelling in physical geography undertaken since 1967, it is evident that this work is now much less accessible. Our systems of interest are structurally complex and they behave in ways that can be awkward to model. To understand them the practitioner needs not only reasonable mathematical ability but also an effective training in a wide range of physical, chemical and biological principles. For example, full understanding of even the simple global energy balance models introduced above requires familiarity with differential calculus, gradient systems and eigenvalues (see Fraedrich, 1979).

This lack of accessibility to the average student I regard as a major challenge to education in physical geography. Even with my own third-year climatology option classes where the entire group has good school, sometimes university, qualifications in mathematics and physics, climate models are difficult to teach properly (for an attempt see Unwin, 1981b). I do not think that this teaching in any way equips them to participate in anything other than very low-level modelling research, as users of scientific models formulated and implemented elsewhere or as eclectic data-gatherers for input into other people's engineering models. The implications for the future of physical geography as we now know it seem obvious: either we become the last generation of geographers able to contribute to modelling research or we pay much more attention to the problems of training the next generation of modellers than we have of late. The recent texts by Hendersen-Sellers and McGuffie (1987), Kirkby et al. (1987) and Washington and Parkinson (1986) are clearly a start in this direction.

PART II
Modelling in Human Geography: Retrospect and Prospect

5 Classics, Modelling and Critical Theory: Human Geography as Structured Pluralism

Alan Wilson

Models and Fashions

In the late 1960s, and early 1970s, in the period following the publication of *Models in Geography*, modelling in geography had the status of high fashion. The tendency for a discipline to evolve through a sequence of fashions can be useful: new ideas are given an impetus and the resources to develop. But the less productive feature of such a sequence can be the abandoning of a fashion without its substance, its discoveries, being properly integrated into the body of the discipline. There is a danger that modelling has been rejected too quickly by too many geographers. Substantial discoveries have been made in recent years which considerably deepen the insights offered by modelling – even though it is still reasonable to see it as a new and immature subdiscipline.

The key argument of this chapter is that it is necessary to connect the ideas of modelling with both the earlier history of geographical theory – the 'classics' from von Thünen onwards – and later fashions which can now be summarized under the heading of 'critical theory'. It is then possible to argue that modelling skills have an enduring role and that it is wasteful to neglect them.

Ideally, some integration of concepts from different paradigms is required. Some would argue that this is impossible: the epistemological shifts have been too fundamental. However, I want to argue that what has been learned in the evolution of geographical epistemology does not invalidate modelling as such: it is a naive view to identify modelling with positivism and a narrow kind of 'science'. To make this case, however, it is necessary to spend some time on the analytical and philosophical underpinnings of different approaches to geography. The argument proceeds in six steps:

1 characterizing the basic dimensions of geographical analysis to generate a perspective which might be called 'structured pluralism';
2 looking for clues from the history of geography;

3 establishing a framework for a 'critical' geography;
4 reviewing modelling in the past, present and future of geography;
5 seeing what critical theory offers;
6 seeking to achieve a kind of unity through structured pluralism and a new regional geography.

The Basic Dimensions of Geographical Analysis: The Argument for Structured Pluralism

It is useful to begin by presenting a framework which demonstrates the immense variety of geographical subject matter, but which also provides a way of structuring the inevitable *pluralism* which this generates. If there is a position to be sought at the outset, it is one of *structured pluralism*; but within this, there is a kind of *unity* to be achieved by allowing the different approaches to interact fully and to influence each other.

The subject matter can be viewed in relation to six dimensions:

1 entitation
2 scale (level of resolution)
3 spatial representation
4 partial/comprehensive

(and these four between them constitute the activity of *system description*)

5 theory
6 method

(and the last two between them constitute the *approach*).

The number of combinations is immense. This way of looking at things is useful initially in that it also demonstrates that many of the decisions are made *unconsciously* and can be seen as surprisingly limiting when opened up and scrutinized. (For example, as we will see later, the classical theorists almost always use a continuous representation of space, almost as though this was 'natural', and this is very limiting in relation to the range of methods which can be applied; they run into insoluble problems very quickly).

Clues from the History of Geography

It can be argued (in the roughest of approximations) that there have been four main periods in the relatively recent history of the subject:

1 regional ('synthesis of knowledge about places');
2 substantive systems analysis ('systematic');

3 spatial analysis ('modelling');
4 radical.

Different styles of work within each of these headings can be interpreted (in some cases) in terms of different choices in relation to the first four dimensions.

If we think of the first innovators in geographical theory (von Thünen, Weber, Burgess and Hoyt and Christaller for example) as 'classical', then we can think of developments based on their ideas as (in a geographical sense) 'neo-classical'. The first two categories above largely fit this heading. What distinguishes the later developments is a more radical break from these traditions. However, there is no reason why the newer ideas should not be applied to the older approaches, and this will be an important part of the later argument (the terminology here should not be identified with that of 'classical' and 'neo-classical' economics, however, though there is, of course, a relationship).

It is also useful to bear in mind that distinct (and more occasionally overlapping) streams of work, progressing through some or all of these categories, can be associated with the alternative *disciplinary* inputs to geography. These include economics (neo-classical and Marxian), ecology, social physics and sociology, and certainly others. But the persistence of these over a long period is quite striking. One of the routes of progress to some kind of unity is obviously to be based on some more effective multi-disciplinary integration.

Structured pluralism, therefore, can be seen as based on: (1) the different ways of structuring the subject matter of geography; (2) the location of practitioners in historical periods – particularly in relation to the early part of their careers as career spans are typically much longer than periods; (3) the influence of different disciplines in formulating approaches to theory and method.

These three elements can be put together to form 'schools' – some obviously more important and influential than others. One of the main contentions of the present argument is that there is a tendency for members of schools not to see far beyond their own subject matter and approach; and even to argue quite strongly that theirs represents the best, even the only 'true', approach to the subject. Thus, one way towards unity of a kind is to attempt to break down the barriers, first by presenting a framework within which the contributions of the different schools can be understood – both for what they offer and in relation to each other – and secondly, by seeking out benefits of stronger connections between them. This is the objective – inevitably at best to be partly achieved – of the rest of the chapter.

At present we are far from a situation where the approaches of different schools are understood by practitioners of other schools. The position for students must be particularly difficult: the textbooks still contain a lot of uncritical presentations of classical and neo-classical theory; more modern approaches demand the knowledge of a technical jargon before it is possible to evaluate them effectively. So there are educational problems as well as research issues to be tackled.

A Framework for a Critical Geography

We are familiar with the idea of *literary* criticism: it involves the interpretation and understanding of texts, but it is also critical in the sense of attempting to deepen knowledge. In other disciplines closer to geography this style is adopted to a considerable extent – sociology for instance where (perhaps going too far in the other direction!) writings are often presented as critiques of the work of others. What follows can be seen as an attempt to develop the idea of geographical criticism by seeking to cross the boundaries of schools in a critical but constructive way.

It is useful to begin by amending the labels introduced earlier. It will be argued that there are three useful foci for discussing the future of geography:

1 geographical systems (or 'regional synthesis' and '(sub)-systems analysis');
2 modelling;
3 critical theory.

Geographical systems can be seen as absorbing the old 'regional' and 'systematic' and what is usefully left of the 'neo-classical' (though some of the latter obviously reappears under other headings). As a category, it will be the basis for an argument to be elaborated later that geography is failing in one of its main traditional functions: to synthesize (and contribute to the creation of) knowledge about regions (and about particular subsystems within regions). In other words, there will be a plea for a return to regional geography but in a reconstituted form: making the best of contemporary theories and insights.

Modelling is used as a term to characterize the most productive technical component of geographical analysis – essentially as an approach to complexity in conjunction with appropriate approaches to theory. It is not necessarily positivist and functionalist simply because it is (often) a mathematical approach.

Critical theory is seen, as the name implies, as providing the critical and theoretical base for the discipline. This is a broadening of what is often taken as radical geography: it is less explicitly Marxian in terms of substantive theory, but it is neo-Marxian in its approach. A key author is Jurgen Habermas.

These labels, therefore, provide the tools and perspectives which provide a basis for the development of substantive theory and, in particular, for a reconstituted regional geography. To make progress, therefore, we look in more detail at what modelling and critical theory have to offer in contemporary geography and then return to the issue of a new regional geography and a critical assessment of the discipline as a whole. These topics form the subject matter of the next three sections.

The Past, Present and Future of Modelling in Geography

The development of models in geography dates back at least to von Thünen. There are major streams, with some connections between them, from several disciplinary stances: economics; ecology; operational research; spatial interaction analysis ('social physics'); non-linear mathematics. (This list is related to the earlier one, of course, but differs slightly through the narrower focus on models.) As in the rest of the discipline, the practitioners have tended to form relatively distinct schools; but perhaps because of the common mathematical base, a smaller group has been involved for a number of years in setting out an understanding of the interrelationships – with quite striking results. It is also important to recall that geographers were relatively slow to pick up from other disciplines what have turned out to be very important ideas in modelling. It is also true that it has only been relatively recently that the relationship between 'model-based' and 'classical' geographical theory has been understood; and again the results are very striking. What remains, as a potentially exciting enterprise, is to exploit the skills of the modeller in the area of radical theory.

It is worth trying to give a broad indication of what has been achieved in modelling in recent years against a backcloth of the most common critiques. The criticisms come in two forms: first in relation to the inadequacies of the models available; second, as a comment on the inadequacy of the approach altogether.

One of the most common criticisms of models is that they are not dynamic; not process-oriented. Research carried out in the past ten years has largely removed the force of this criticism but this is still not widely known or understood. And it should be added that one of the products of the research – as often happens – is a realization and understanding that the nature of geographical dynamics is much more complex than might have been anticipated. But this itself offers new insights which are relevant beyond the bounds of modelling. This research has largely been carried out within the spatial interaction paradigm, though the ideas are so general that they are potentially applicable elsewhere, and the main insights gained are similarly quite general.

The main idea involves the addition of (process) hypotheses to previous model systems, and the use of some of the ideas of contemporary non-linear mathematics to solve the resulting problems. It is perhaps worth noting, however, that many of the problems are distinctively geographical and need non-conventional solutions; and that the hardest of the as yet unsolved problems are seen as 'difficult' within applied mathematics itself. No longer is geographical modelling the product of a simple-minded social physics, but itself generates interesting mathematical problems. It should also be emphasized that this paradigm is not dependent on neo-classical economics for its underpinnings, though the strength of that discipline within geography and regional science is such that its influence is undoubtedly strong.

It is worthwhile illustrating these developments with an example, whilst emphasizing the generality of the ideas involved by listing the range of actual and potential applications. As an example, it is simplest to take a trusty standby – retailing.

Retail modelling has a long history: back to the 1930s and Reilly and Converse; back to the mid-1960s to an archetypal model developed in similar forms by Huff, Harris, and Lakshmanan and Hansen. But there was then a gap to the late 1970s before Harris and the present author found a way of making the basic model dynamic; and many of the insights to be gained in a wide variety of fields from dynamic modelling can be seen in terms of this example. The dynamic retail model provided an 'explanation' of the spatial structure of retailing, introduced sudden change controlled by critical parameters and the associated phenomenon of multiple equilibria, and showed the potential complexities of dynamics.

The main idea of the new dynamic hypothesis is a representation of the process by which locational decisions are made. This idea can be applied, with appropriate modifications, beyond retailing. The complete list at present is: agriculture; industrial location; residential location/housing; public services; private services; transport; comprehensive modelling. In each of these cases it is possible to look back at classical theory and to understand the nature of its limitations – usually methodological in the light of the 'continuous space' tactic. We can therefore modify and generalize (for example): von Thünen; Weber; Burgess, Hoyt, Harris and Ullman; Reilly (though this work is based partly on non-continuous space); Christaller and Losch. These achievements are substantial in their own terms and should influence the teaching of geography by providing a constructive response to the failings of the theoretical bases which appear in most texts. But it may be that the most important steps are still to come. We next, therefore, briefly consider the main steps forward which remain to be taken:

1 hypothesis refinement (moving away, where appropriate from the neo-classical economic base – configurational analysis provides an example);
2 connections to radical theory (perhaps via the aggregation issue – the agency–structure problem and structuration);
3 better connections to applied work through performance indicators.

This last point can be illustrated directly in relation to the analysis and planning of health services, but generates concepts which can be applied much more widely.

Philosophy and Social Theory: Towards a Critical Geographical Theory

Radical geography has often been presented as substantive Marxist theory. This is often interesting, but it is important to take a step back and adopt a methodological stance to see what the radical perspective offers before the hypotheses of substantive theory are added. It is this argument which leads to the notion of radical geography as critical theory and provides a vital basis for the development of geographical criticism.

One of the features of critical theory is its combination of philosophy and social theory. This arises essentially from the recognition that 'knowledge' is socially produced; that there is a fundamental connection between epistemology and social theory. The argument is necessarily complex and the field develops its own ideosyncrasies and jargon. However, it is possible to make progress by summarizing the argument in as simple a way as possible to provide a basis for assessing its impact on geography and the future structure of the discipline.

There are five key linked propositions:

1 knowledge is a social process;
2 the evolution of language is a social process;
3 meaning evolves through intersubjective communication;
4 truth is a consensus;
5 much language has an ideological content.

This allows us to understand what knowledge is; it is always provisional. The ideas which result from this are more useful and constructive than those associated with what are often taken to be the main competing philosophical schools – positivism, phenomenology and structuralism – and much argument within geography is often related to these.

Positivism is presented, by its critics, as being associated with the view that the 'scientific' route is the only route to knowledge, and that this should be in some sense 'value-free'. It is not difficult to establish that the concerns of modern geography do not fit these criteria. However, it should not be forgotten that the original motivation of the Vienna school was the clarification of the idea of 'meaning', and particularly the establishment of criteria whereby 'nonsense' or 'metaphysics' could be identified. This remains a worthy aim. The problem was that the philosophers of the 1930s made the assumption that there was an unproblematic 'observation language' which could be used as the basis for the appropriate epistemological tests. In the light of even the above sketch of an alternative, this can be seen to be untenable. However, many of the objectives of positivism can be retained via the notion of 'intersubjective communication' and we return to this below.

Another criticism of positivism was its concern with an unachievable 'objectivity'. Again, elements of this can be recovered via intersubjective communication and a consensus theory of truth. But *phenomenology* can be seen as a response to this: phenomenologists argue that the subjective experiences of individual human beings are of the highest importance. It will be argued below that there is no need to disagree with this.

A part of the positivists' approach to objectivity via the observation language laid them open to what would now be called the *structuralist* critique: that that kind of 'science' was only concerned with 'surface' observations and phenomena whereas the events of the social (and geographical) world were determined by deeper structures. Again, there is now no need to disagree with this. 'We are all structuralists now!' (Anon).

To an extent, these 'schools', and the objectives of a critical theory, can all be better understood in terms of Habermas' three cognitive interests – technical, practical and emancipatory.

Habermas himself took technical interests to be associated with work and the empirical–analytical sciences; the practical as concerned with communication and interaction and the interpretive–hermeneutic disciplines; the emancipatory as concerned with power relations and constraints.

In the context of the present argument, the first can be seen as related to 'what can easily be agreed about' and in that sense begins to connect us back to the ideas of 'objectivity', and even 'positivism' – created out of intersubjective communication – without this being solely based on an idea of 'science'. The second is concerned with the process of language evolution and feeds both back to the technical and forward to the emancipatory. The third reminds us of the importance of power relations in society, and that these will be to an extent coded into the language. This therefore relates to our earlier assertion about 'ideology'. The concepts of what we took above as the three basic philosophical schools can easily be represented in this framework.

There is a sense in which all of the discussion in this section so far is *methodological*: it provides a framework for the epistemological scrutiny of any piece of analysis. In particular, it provides a framework for the critical analysis of any particular *theory* in such a piece of analysis. In this sense it can be used to review, for example, the inadequacies of either model-based research or neo-classical economics or Marxist theory within radical geography. It provides the tests through which the contributions of these theories can be assessed, modified or integrated. It is on this basis that we examine the possibilities of unification in the concluding section.

Unity via Structured Pluralism and a New Regional Geography

First, let us return to the idea that geography is now fundamentally failing to fulfil its traditional obligations and that these remain important. In particular, as a discipline, it does not take on board as a major task the provision of a synthesis of knowledge about places – say 'regions' for the sake of a term. There are some good reasons for this. If this had remained the raison d'etre of geographers, then they would have remained jacks-of-all-trades and failed to become involved in the exciting theoretical developments of the past three decades. But the obligations are not being met by anyone else (except perhaps by journalists, and to a limited extent by 'area studies' specialists – usually working at quite a coarse scale).

There is an obvious answer: to use all the methods, theories and insights of contemporary geography in all its varieties to write new *regional geographies* at various scales. (To this we could also add more specialized systematic geographies). This would have the added benefit of forcing the interaction of different approaches to the same region. The mere existence of such (at first) parallel studies would generate interaction and (it is assumed!) progress. This should include the application of model-building skills to theory building from radical perspectives. There is already some evidence that this is now happening.

The outcome of the argument can now be summarized:

1 a *critical* approach should be adopted at all times;
2 modelling approaches are important; their connection to social issues can be improved via a performance indicators' focus;
3 structuralist approaches tackle some areas of theory which cannot be investigated by modelling, though increasingly, there will be some connection;
4 *regional syntheses* are vital to progress and useful in themselves (and there will also be some connection here to the idea of *geographical information systems*).

6 Computer Modelling in Human Geography

Stan Openshaw

Introduction

Twenty-five years after the quantitative revolution was declared to be over (Burton, 1963) and some twenty-one years after the first real mathematical model-building methods appeared (Wilson, 1967), the dream of geographical generalization in the form of soundly based models and theory (Chorley and Haggett, 1967) is still alive. Progress, whilst at times technically spectacular, has been very slow in meeting the goal that was set. Chorley and Haggett (1967) quote Ackerman (1963) who specifies that this is 'nothing less than an understanding of the vast interacting system comprising all humanity and its natural environment on the surface of the earth' (p. 435). This is of course not attainable except in a highly generalized manner (see Harvey and Scott's chapter). Most of the subsequent modelling effort has been focused on the development of bundles of geographical theory related to selected urban and regional systems via mathematical methods (see Wilson's chapter). This contrasts with the more traditional spatial analysis route on which the first quantitative revolution in geography was based.

Wilson (1981b) is right to state that 'with the arrival of a range of techniques for dynamic modelling and the expansion of optimization techniques, it could be argued that most of the appropriate methods, again at least in a rudimentary form, are now available' (p. 276). Indeed as Wilson and Bennett (1985) subsequently note 'the statistical and mathematical approaches are fully absorbed into the discipline of human geography' (p. 3). It is an appropriate time therefore to consider what has been achieved and then to announce the beginnings of the computer modelling revolution as the logical next stage in the continued evolution of models in human geography.

The next section looks at the development of a paradigm based on mathematical modelling techniques in human geography at a time when the emerging information economy is reducing its relevance by focusing attention on a very different set of needs. The following section offers a detailed critique of some of the problems that have arisen. This leads to a presentation of the case for adopting an alternative perspective based on an inductive paradigm focused around computer data bases. The penultimate section develops this theme further and outlines some promising new developments and the last section offers some thoughts about the future.

The Hypothetico-deductive Approach to Theory Development by Mathematical Modelling

Data-free Models and Mathematical Tool Kits

The original 'models in geography paradigm' of Chorley and Haggett (1967) was formulated at a time when computer applications were the exception, computer hardware was considerably poorer than that offered by most early 1980s eight-bit microcomputers, and data sets were generally small. It is hardly surprising that mathematical models should within this framework offer an attractive route to theory formulation and knowledge generalization in geography. Quantitative techniques had previously been viewed as a means of developing theory by inductive means (Berry and Marble, 1968). The subsequent switch to deductive methods and normal science, aided by historical, technical and data limitations on spatial analysis, prepared the way for the emergence of a mathematical modelling paradigm. Here models were often viewed as data and 'computer free' expressions of 'pure' theory obtained by soundly based mathematical methods. It was an opportunity to put geography on to a scientific basis allowing it to compete with the 'real' sciences, and to a degree it has achieved this objective.

The purpose was the pursuit of knowledge and understanding by the application of a mathematical tool kit to urban and regional systems modelling within a hypothetico-deductive framework. As Wilson (1974) put it, 'The distinction between deductive and inductive approaches is very much the distinction between the approach of the (applied) mathematician and that of the statistician' (p. 4). The former was regarded as the more respectable route. Wilson (1981b) outlined the objectives as: handling complexity, identifying and understanding systemic effects, seeking methods which are applicable to a wide range of systems, and providing tools as an aid to planning and problem solving. He writes: 'At the heart of this programme lies the task of building models of systems and subsystems which provide an analytical basis of understanding and of planning' (p. 276). These are noble objectives and it is perfectly reasonable, whilst establishing these basic tools, that applications with real-world data would seem to be of secondary importance. Nevertheless, some empirical work was performed (Wilson, Rees and Leigh, 1977) although historically it has never been sufficient to complement the overriding interest in model development.

The problem is that this emphasis on mathematical modelling, as a largely data-free activity, has established a style of modelling which has continued despite massive changes in both computer hardware and data environments. Few of the texts on models or modelling methods in geography mention the word 'data' or 'computers'; see for example, Wilson (1974), Bennett (1979), and Thomas and Huggett (1980). Nor do they describe any of the models in sufficient algorithmic detail to allow for easy programming; the exception is perhaps Batty (1976). Studies containing details of empirical evaluations and applications are still the exception rather than the rule. Wilson (1981b) on the penultimate page of a brilliant book merely comments 'What is perhaps most lacking in many cases is an adequate level of empirical work which would lead to further refinement of theoretical models and to an ability to choose between alternatives' (p. 276). This neglect is greatly lamented. There is no disputing the great wealth of rich and varied models that are 'on the shelf in the shop window', awaiting application, or that a comprehensive tool kit of useful mathematical methods now exists. However, it is reasonable to question whether the real need now is for developing yet more, and ever more refined theoretical models, or whether more needs to be done with the models and methods that already exist. It is certainly a great disadvantage to have so much of the intellectual capital of quantitative human geography encapsulated in theoretical models that seemingly serve no useful purpose outside of geography itself. The whole enterprise has become narrowly focused, based on objectives which are relicts from the 1960s, and inward-looking. Seemingly few of its devotees have noticed the world changing about them in a most fundamental way.

The End of the Mathematical Modelling Era

The great age of mathematical geography put emphasis on the acquisition of formal mathematical skills as a prerequisite for understanding quantitative geography and as the basis for geographical theory. This perceived need still continues unabated. Wilson and Bennett (1985) justify their recent book by arguing the necessity for knowledge of a mix of mathematical methods in order to show how geographical theory can be assembled from a mathematical tool kit (p. 4). 'What geographical theory?' and 'Why an emphasis on purely mathematical methods?' are questions that, seemingly, people stopped asking about ten years ago. At that time, Sayer (1977) talked about 'mathematization beyond the call of duty' (p. 188) and provided a detailed critique of the theory-poor and restricted nature of the theoretical bases of many mathematical models of interest to geographers (Sayer, 1976, 1979).

The pro-mathematical modelling arguments were, on reflection, never actually won! They merely possessed a greater longevity due to a greater innate academic attractiveness. The 'faithless' merely opted out or were locked out by their failure to understand the language of mathematics (see Cosgrove's chapter) or changed their minds and tried to develop meaningless holistic theories of everything (see Harvey and Scott). Likewise, the 'faithful' were largely locked in due to their investment in intellectual capital. Academic life cycle effects also allow a maturity and persuasiveness to establish a degree of gloss that sometimes bears little resemblance to reality. Batty (1979) summarizes this process as follows:

> its proponents became more articulate and well versed in its foibles and special character. Its leaders tend to acquire more power relative to other disciplines with which they might compete for resources, and the outlets for publication become polarised in some senses. These changes generate a momentum which has been hard to resist'. (p. 874)

There is no dispute about either the quality of the work or of the utility of mathematical methods; rather, criticism is focused on the lack of applied relevancy, the lack of attention to empirical study, and the absence of explicit geography. Can human geographers really expect to study the world in a data-free environment? If yes, then it implies a degree of sophistication that is inappropriate and where are the theories? If no, then it begs the question, where are the data-related studies?

The criticism is made then that the deductive route to theory formulation has been taken too far and has not been particularly successful. Far too many so-called 'theories' have never been tested and many more are untestable! Yet prediction and empirical testing are not merely by-products of theory, nor are they the logical next step that can for ever be postponed; they are an integral part of science. The ability to make correct predictions is the only relevant scientific test of the depth and quality of understanding afforded by a mathematical model. Currently there seems to be a gross lack of balance.

Data-driven Computer Modelling in an Information Economy

In this paper the adjective 'computer' when applied to 'models' is deliberately symbolic. It is there to try to focus attention on the fact that, historically, mathematical modelling has not been regarded as a computer-based activity. It is also there to emphasize the importance of building operational, data-based models of human systems not as an exercise in pure theory formulation or in mathematics but for real practical and applied purposes. Finally, it is used to emphasize that the computer revolution in geography has yet to happen!

The world has changed considerably in the past 20 years. The early roots of mathematical models in planning applications in the mid-1960s has allowed the belief that there are still planning applications waiting in the wings, whenever we want them. The idea of urban models preserves this link with application even though the evidence of linkages with policy and practice are now extremely tenuous; indeed they were always vague (see Batty's chapter). Wilson (1981b) suggests that 'urban models have much potential for planning now' (p. 275). This ignores the fact that not only is there no available software for urban models of the sort he is referring to, but there are few (if any) planners wanting to use such models! A great gulf exists between theory and practice which renders most hopes of use and application of many theory-generated mathematical models in the real world an impossible dream (Openshaw, 1978).

Instead of traditional social science focused on cities and the operation of urban and regional systems, the emerging information revolution with its emphasis on the commodification of information is producing a very different perspective (Openshaw and Goddard, 1987). Attention is now on data and data bases. The new perspective is data- rather than theory-driven. The thirst for explanation is being replaced by demands for spatial analysis. The potential importance of models is no longer primarily as a basis for understanding systems behaviour but increasingly as a means of adding value to data. At first sight this is not so different from the mid-1960s planner looking to an urban model to help plan a city. However, there are fundamental differences. Planners only ever used models to inform decision-making in some vague fashion; model-using methodologies have only recently started to evolve and even then there are still no real ways of linking models directly to plan decision-making. The new opportunities are different in that models are and will increasingly be used in a directly coupled fashion.

The fundamental opportunity in both private and public spheres is that geographical data processing allied to computer models embedded in an increasingly data-rich but theory-poor environment open up major new opportunities for operational models that can assist both business and services to operate in a more efficient manner. This change of emphasis also brings with it moral and ideological issues that have lain dormant because previously hardly anyone used geographic models for anything important. The need for theory still exists but only to the extent that it will be used if it can be shown to deliver 'better' working models or is otherwise useful. Yet if geographers are not prepared to meet these emerging needs by re-orientating their modelling work then no doubt other computer sciences will be only too happy to oblige. The whole edifice known as 'Models in Geography' will be asset-stripped, the most relevant technology re-invented by people who last studied geography at school, and geographers will be left with no other course of action other than to become social theorists in the style of Harvey and Scott.

So What is Wrong with Mathematical Modelling?

A New Requiem!

It is easy to poke fun at current attitudes (Openshaw, 1985). A more serious attempt to start a debate about the need for modelling relevancy was provided later (see Openshaw, 1986). The basic cause of the problems appears to be that the external environment in which geography operates has now changed, resulting in the need to modify the prevailing paradigm by seeking a complementary inductive component. Whether geographers are prepared to adapt in order to survive probably depends on the extent to which they are content with the *status quo* and see themselves as traditional-style 'scholars of geography' rather than having to respond to external demands and stimuli. What follows then is a brief attempt to cast some doubts as to the past successes of models in human geography as a means of emphasizing the need for change in attitudes and orientation. The criticisms can be summarized under four headings: wrong objectives, insoluble problems in model construction, lack of empirical validation, and little evidence of success or of markets for the traditional produce.

The Wrong Objectives!

Some of these criticisms have already been dealt with, so attention here is restricted to three aspects.

Massively theoretical rather than applicable Many models can only be described as massively theoretical. They may constitute advances in terms of theory but they are limited because there is virtually no prospect of application or testing. They seem to be the result of a great international competition to develop ever more rigorous theory as a basis for developing a better understanding of the workings of urban and regional systems. But for what purpose? Research which is largely data-free and thus data-independent is no longer so relevant in the era of computer data bases. As Openshaw (1986) puts it:

> What a shame so much of this work is useless, in that there can never be an adequate empirical component to it and the questions being addressed relate to the pursuit of knowledge rather than more pressing applied matters of contemporary relevance and public concern. (p. 143)

Doubts about relevancy There is the basic question of model relevancy. It is taken for granted that the task of developing models is worthwhile. The basic question, 'Why are you doing this?' is hardly ever asked, and if asked seldom answered because it is assumed to be self-evident! Well what is it? How can it be justified compared with spending research resources on other alternatives? Openshaw (1986) put it as follows: 'It seems that the best people are studying completely the wrong problems and by so doing are denying society the full benefits of their unique skills and talents' (p. 143). This of course raises the question 'What are the correct problems?' and who, if anyone or anybody, should be permitted to decide what they are? Beaumont (1986) adds: 'Why are the best people unable to determine the most important problems? Do they want to address them?' (p. 419).

Outstanding Problems in Model Construction

The task of building models has never been easy. There are model building cook-books but little indication of how successful they have been. A number of issues can be identified which are problematical.

Systems thinking is flawed and ineffective Doubts can be expressed about the utility of systems thinking as a model design tool. Systems identification is a matter of faith and involves wish-fulfilment. It offers a conceptual tool but the resulting system map may well be only one of several different, equally plausible alternatives. How do you choose between them? If theory is this uncertain then can it be regarded as theory at all? The problem is that the systems being investigated tend to be too complex for satisfactory unambivalent process models to be built, relying only on human understanding as the principal design tool.

Easier to revive old models than build new ones Many of the models existing today are not of recent construction but are often no more than sophisticated versions of older models. How the first-generation models were actually 'created' in the first place is often not known. Ironically, the first-generation models tended to be simple enough to be capable of operational use; most of the subsequent attempts to 'improve' them tended to destroy all hope of new operational models emerging. The task of inventing new models has been historically far more difficult than that of reviving existing ones. The problem here is that seemingly no-one really understands how to create new models, from scratch. These difficulties place limits on the likely achievements of human being based analytical model building; indeed, these limits might have already been reached.

Aggregate patterns are most easily modelled but they are theory-poor The best system analytic tools available today do not go far enough to establish a good basis for proper understanding of the systems they are designed to represent. Aggregate models based on entropy-maximizing model building methods offer consistency, and they are appropriate tools for exploratory studies with an emphasis on description, but as the basis for explanation they are hopelessly inadequate. Mathematical modelling often starts to look fundamentally flawed when pushed too far, either by extreme disaggregation or by an overemphasis on the role of theory. Sadly also, it seems that attempts to reach the mechanisms responsible for generating aggregate spatial patterns usually result in disaster. The geography that is capable of model representation is lost, and with it all hope of achieving any practically relevant result.

Some important social processes will always be missing There are also severe doubts about the degree to which existing mathematical models are even based on sound theory. Sayer (1976) argued that it is not sensible to have models of spatial phenomena which are separate and abstracted from social processes. He writes, 'This search for determining forces and relations is not an infinite one, for we quickly arrive at the motor force of the capitalist economy – capital accumulation, and hence the manner of its production and reproduction' (1977: 188). The problem here is that widening the scope of models to incorporate relevant social processes, however desirable it may be, is also impossible! Even if the underlying theoretical structure could be mapped out, there is no reasonable prospect that any worthwhile operational model could be built; this criticism also applies to Harvey and Scott's suggestions.

How do you build theory-based models within existing data constraints? The task of building models that both represent theory and stay within contemporary data supply constraints cannot be solved in any satisfactory manner. It is also contrary to the objective of having models as representations of theory. Yet unless data feasibility is retained then the model cannot be tested. The easiest fix is to specify stripped down data-feasible versions of more complex models but this inevitably results in the criticism that the theoretical justification has been weakened to an intolerable degree.

The ultimate dilemma! The converse situation may also apply, and forms what has been termed 'the ultimate dilemma for theory-based models' (Openshaw, 1986: 145). The development of better theory usually leads to models with increased demands for data either by increasing their scope to capture more subsystems or through temporal and spatial disaggregation. In the limit (not likely to be reached until the third millennium!), this process will almost certainly result in purely mathematical models that cannot be operationalized, because of their impossible-to-meet data needs, but at the same time the models themselves will still probably be inadequate conceptual representations of complex open systems. Perhaps the objective that has been set is unachievable! The complexity and dynamics of urban and regional systems, together with an intrinsic level of randomness, suggest that there are probably real limits to the process of understanding that can be gained by deductive approaches in areas that are relevant to geography.

How do you build good performance into models explicitly? It is not obvious how to build, from scratch, well-performing models. The usual assumption that good theory automatically aids performance has yet to be empirically verified! Model goodness of fit is a characteristic that sometimes emerges later, if at all, in the model-building process. The use of entropy-maximizing methods offers no guarantees about performance as distinct from model design; even then the resulting model form is conditional upon the macrostate constraint equations that were used to represent existing knowledge. For example, the performances of the entropy-maximizing spatial interaction models are no better than the

Newtonian gravity models they replaced, despite their far more respectable scientific pedigree.

A Lack of both Empirical Validation and Application

At one time it was thought (wrongly as it turns out) that the development of models with data demands that could not be met, would stimulate the development of the necessary data bases. In the event it merely resulted in models that remain largely untestable academic curiosities. But even when data do exist, there has been no systematic and comprehensive attempt to validate existing theories, unlike physics where theories are constantly being tested experimentally. A related difficulty is the lack of any clear idea as to how a model should be evaluated. What criteria can be used to define acceptable levels of performance? What levels of descriptive, predictive, and forecasting performance are likely to be indicative of a well-performing model?

Absence of empirical testing Many model builders see no need to test or evaluate their models and theories against data. The criterion of success has been the number of academic publications rather than a goodness-of-fit statistic. There are no standard data sets useful for benchmarking models, no generally agreed methodology for model evaluation, and no history or experience of these activities.

Absence of software An emphasis on mathematical methods does not necessarily preclude computer modelling, and computers were available twenty years ago. However, there is nevertheless a paradox that many mathematical modellers have never made much use of computers in their work. It is hardly surprising, then, that the availability of software for working models is extremely limited and bears no relationship whatsoever, either to the range of theoretical models that exist in the literature or to the efforts that have been expended in this area of geography. There may be a large number of models outlined in excellent books by Wilson (1974, 1981b) and Batty (1976), but there is virtually no software readily available for any of them. The same sad situation also applies to many geographical analysis techniques.

No Real Evidence of either Success or a Market

Increasingly academic criteria are not by themselves sufficient to maintain a particular specialism. Hard questions may well be asked about performance and the degree to which the mathematical modelling paradigm is showing signs of triumph.

Lack of visible signs of success Partly because of the lack of attention to application, there is a major problem in providing evidence of success. Planners in general do not use models whilst those on offer by management consultants never fail! Can success simply be assumed because of the absence of any documentary proof of either failed theories or 'fallen over' models? Do they all really work equally well? Due to the absence of empirical application and testing, the principle of caveat emptor rules! There is accordingly no satisfactory way of discriminating between 'good' and 'bad' models, and documenting success is virtually impossible.

No external markets Finally, there is the problem of not having any external market or pool of users wanting and willing to use many of the models that have been produced. Traditionally, the assumed planning applications have turned out to be figments of the imagination. Planners are no longer interested in models (see Batty's chapter); indeed most never were! It was merely taken for granted, by geographers, that a model-based understanding of the workings of urban and regional systems would be of some value to planners and administrators. There are some exceptions. For example, transportation modelling is well established outside of geography, but again there is no tradition of hardcore empirical work. Indeed the continuing use of these models is predicated on the assumption that vital questions of performance and uncertainty are explicitly ignored!

No-one has seemingly ever asked who are the potential users for geographical models. Previously it never mattered; today it does! Marketing geography in the world outside of academia is now becoming essential to its future survival within. The world is seemingly littered with opportunities if only those geographers who had something to offer would start to look.

Data-driven Computer Models

Towards a New Perspective of Models in Geography

This discussion of the problems with mathematical models sets the scene for the argument that maybe the present moment is opportune for a substitution of the term 'mathematical' by 'computer' when applied to models. The need for a real change in emphasis of this sort is now overdue and fairly urgent. According to Batty (1979), 'changes in view only occur when the existing generation of scientists die out' (p. 873). By then it may be too late for models in geography. Beaumont (1986) neatly summarizes many of the previous criticisms by writing 'Are modellers asking the correct questions?' (p. 419). He identified four issues that are involved: the identification of problems, the development of appropriate modelling technology, the way forward, and the marketing of professional modellers; sadly he is unable to answer any of them! The view expressed here is that whether there is any future for models in geography depends on how geographers react to these issues and on the perspectives they hold regarding their own models.

Wilson (1972) complained that:

> slowness of progress in the past has probably resulted from a voluntary imprisonment of geographers within a restricted range of concepts and techniques, and progress in the future probably depends on our collective ability to branch out and be more ambitious. (p. 41)

At the time he was referring to the slow rate of development of theoretical geography and the need to use a wider range of mathematical techniques. More than fifteen years later the problems that now exist have been largely created as a result of the success of the approach Wilson favoured! Today the need is not solely for mathematical skills but for a much greater branching out in the direction of computer science and data bases.

The arguments in favour of application are themselves not new. Wilson's geography should be related to real geographical problems, with a concept of importance based on their relevance to urban or environmental planning issues. It's a pity that this part at least of his prescription was seemingly not more widely adopted. Somehow the objective in modelling has been too limited, too academic, and too inward-looking. The wrong perspective automatically generates many of the other problems that have been discussed. The way forward is fairly obvious, change perspective to favour applications with data. This will automatically focus attention on more relevant modelling methodologies, it will open up new external markets for models, and provide a paradigm more in touch with contemporary needs. It might not be such 'good' science, there were always doubts about this aspect of geography, but it will at least provide a firm basis for the future.

A Return to an Inductive Way of Thinking

Not many quantitative geographers would today welcome a shift back to the inductive approach of the early statistical modellers. Nevertheless, the arguments here would indicate that these pioneers almost got it right! Their crude models were somehow in balance with the available data environment and the results were often useful in relation to the needs of both academic and real worlds. The long search for better theory, better models, and mathematically rigorous and conceptually sound approaches has not delivered many of the promises that it once held. It is now massively out of step with the needs of real-world application, it has no relationship whatsoever to the available data supply, and it largely ignores the tremendous advances that have been made in computers and the new opportunities for adopting wholly computational rather than analytical forms of model technology.

It is possible that the systems and process orientations which dominate mathematical modelling, have in fact failed; except no-one has really looked hard enough at the visible signs! The search for understanding through process knowledge and casuality has never really worked. Explanations of systems and soundly based conceptual frameworks may be desirable, but are not necessarily useful. It is beginning to look as if many socio-economic processes are simply too complex for tractable and well-performing mathematical models to be built via a hypothetico-deductive approach under conditions of data restraint. New computer technology and vastly improved data bases may one day revive this basis for modelling in human geography, but at present it looks like a very unequal struggle. Meanwhile it would seem opportune to consider complementing these dreams with a more pragmatic form of inductive modelling as a basis for ensuring short-term survival, a re-orientation of priorities, and of the perspectives held of the existing heap of models.

Impacts of the Commodification of Information on Modelling Styles

The arguments for a return to a more inductive way of thinking are most strongly supported by what is happening in the emerging information economy. The fundamental technical change that is underpinning the development of the new post-industrial society is the transformation of knowledge into information which can be exchanged, owned, manipulated and traded. This commodification of information in the form of computer data bases both has economic implications and creates all manner of new opportunities for models as a means of adding value to data. The central importance of the traditional academic concerns such as the search for understanding of whole systems is reduced, whilst that of modelling relationships locked inside data bases for predictive and forecasting purposes is greatly increased.

The external force bringing about these changes is the ubiquity of the computer as a basis for administration and management in both private and public sectors, on both national and international scales. The convergence of computing and telecommunications into an integrated, network-based system focused around large data bases is opening immense new opportunities and markets for modelling technology related to data analysis, and many of these new needs are implicitly geographical. Yet currently few of these needs are being met by professional geographers (see chapter 12).

Models are urgently needed to make sense of the resulting massive data explosion. It should be noted that the data are mostly not in the public domain but are largely confidential to Governmental or commercially sensitive data about client trading behaviour. Yet never before in human history has so much information about so many people and their spatial behaviour patterns been stored in computers and therefore, theoretically speaking, accessible for analysis. Yet it appears that if geographers want access to such rich data sources as now exist they will in the future either have to pay or join the data-keepers in providing information services.

It is important not to underestimate the potential significance to human geography of these alternative and new sources of data coming on stream due to information technology. Nearly all the data are implicitly locationally referenced, much exists at the micro-level, and update channels are an integral part of the supporting information systems. Additionally, the costs of developing these systems are so large that the diversion of even a fraction of one per cent for the development of geographical models would amount to more resources than historically have ever been expended on models. If geographers want to find alternative funding for their work then it is here they will have to look. If they want access to some of the data to test out their theories, it is here they will have to work. Indeed, these new data sources can be used to breathe fresh life into hypothetico-deductive approaches suffering because of data starvation, but this is probably only going to be possible by riding on the back of inductive models, the main purpose of which to outsiders is data analysis, but to geographers is partly to open up the new data sources for secondary analysis.

On the Philosophy of Data Base Modelling

Modelling Data Without Any Great Depth of Understanding

The emerging new demand for models is not strongly related to what has been developed during the past twenty years. Many traditionally attractive model forms are destined to remain academically interesting but for most practical purposes utterly useless! The principal instantly transferable technologies are not initially the mathematical modelling tool kits, but the simpler descriptive subsystem models and the descriptive statistical models that academic geographers abandoned long ago. Yet these 'old tools' are virtually the only instantly applicable parts of quantitative geography that have any relevance to current needs. Indeed they are now being used successfully by marketing and management consultants.

It is certainly very ironic that probably the greatest single development (measured in terms of dollar turnover) in quantitative geography in the past decade has been the development of what previously would have been termed a regional taxonomic method. The resulting classifications of small area census data in Britain, the U.S.A., and Canada have revolutionized the marketing of goods by mail, changed the whole basis of services such as credit scoring, and placed great emphasis on the application of simple-minded geographical analysis techniques. These 'geo-demographic' systems, whilst based on an ecological fallacy and lacking any strong theoretical justification, have been immensely successful. They actually work despite their poor pedigree, and a large number of different users seem reasonably happy. It is possible to predict with a certain degree of precision, for the first time, differences in the propensities of people to consume products, to buy services, to generate bad debt, etc. The level of understanding is probably 'poor' in many instances but it is the best that can currently be obtained. If geographers want to demonstrate that they can do 'better' based on 'good science' and 'soundly based theory', etc., then these developments would be both worthwhile and welcomed if they can be proven in practice. The challenge exists and a large part of the world outside of academia might well be amazed at what geographers could do—if they really tried!

In the short term, the cost of seeking applied relevancy might well be a reduction in the emphasis on explanation and the tacit acceptance of a different and inferior form of understanding. However, even data-driven models need street credibility. All the better if the model can be explained by reference to some conceptualization of a process. So in essence the change amounts to no more than a switch back to an inductive paradigm to complement the *status quo* and give models in human geography a greater degree of relevance and marketability.

Data Base Modelling

A good way of explaining what is required is to think in terms of a statistical modelling system designed to do no more than represent relationships found to exist in a data base relating to some particular function. In many different situations, and for many different purposes, interest is often focused on building models for the description and prediction of a response variable. (e.g. bad debt, uptake of a service) from whatever other data happen to be available in a data base. It is often recognized that geography is an important predictor variable that needs to be incorporated as contextual information. The data sets are usually held at the micro-level (viz. households, persons), and can contain several million cases with a tremendous wealth of personal and household characteristics.

The objective is limited to finding a 'good'-performing model that can be used to describe existing data subject profiles, and to predict either the appearance of future characteristics or the demand for, or changes in, the uptake of services of various kinds. For example, given this personal profile what income would be expected? Can a model be found that predicts house prices? What factors most influence the purchases of product 'x' or service 'y'? If you have a theory then it can probably be tested, but the importance is not theory formulation but business and predictive efficiency. The goodness-of-fit statistic now takes on a new meaning; money lost! It is interesting that this emphasis on money is not new and was always traditionally implicit in planning uses of models. Finally, the basic data base modelling philosophy applies to both government and private sector organizations.

Types of Model

Any serious attempt at developing computer models orientated around implicitly geographic data bases might well be focused around three basic sets of model types: (1) exploratory models; (2) recycled statistical and system analytical models; (3) new developments.

Model set A: data base exploration models One family of data base models has already been developed. They try to identify best- or good-performing predictive models by crunching through all the data available in a data base in a highly automated and efficient manner. As computer data bases continue to multiply and cover more and more aspects of modern society so the requirement for a general-purpose, automated, fool-proof, 'Data-Base Modelling System' capable of working at the household and unit postcode level increases. The problem is technically interesting in that the model has to be able to deal with whatever data exist, it is seldom possible to add variables and it might be dangerous to willy-nilly drop any, just in case they turn out to be useful! It has to be able to handle dependent variables which may be either continuous, dichotomous, or categorical; cope with a few hundred predictor variables with mixtures of measurement scales; and remain undaunted at the prospect of multi-level contextual geographical data. The possible existence of non-linear relationships and interaction effects cannot be dismissed. Additionally, all real-world data bases also contain a fair degree of noise in the form of spuriously coded information and downright lies! Finally, there is the ultimate problem of providing results that can be readily communicated to third parties, often possessed of a non-technical disposition.

A family of Data-Base Models (DBM/1, DBM/2, DBM/3) have been developed to meet this specification. They are generalizations of the crude binary segmentation technology offered by the statistical package known as OSIRIS IV. The advantages in a data base modelling context are profound. For instance, their ability to handle categorical variables with more than thirty-two codes allows demographic systems to be directly incorporated into the household models as a single predictor variable. Such models will always outperform standard statistical packages such as GLIM because of their greatly flexibility. No doubt this DBM technology can be developed further.

Model set B: recycled statistical and system analytical models There is an obvious application for many old and well-established statistical models. Ideally they need to be clothed in the computational technology of the 1980s rather than the 1960s. For example, well-known problems with zonal data regression models can be completely avoided! The simpler methods of parameter estimation should also not be used. Robust estimators based on minimal assumption sets can be readily obtained via Monte Carlo methods.

It is also possible to think in terms of recycling some mathematical models. Indeed some are already being viewed in an analogous light. For instance, the shopping model has undergone a remarkable revival. In the 1970s problems over data supply largely killed off spatial interaction shopping models as a practical planning tool. It continued to be of theoretical interest mainly because of its mathematical properties when recast into a dynamic form (see Wilson's chapter) and its re-invention in a form suitable for generalized linear modelling. Today, new data capture channels are resulting in interaction data, especially for shopping and financial services, being available and of considerable interest to developers, banks etc. There is now an opportunity to operationalize many of the previous totally theoretical variants, especially the dynamic and optimization versions. Whether levels of performance are actually good enough to be useful is an aspect that requires further attention. The indications are that, compared with the alternatives, many of the geography models are considerable improvements (see Wilson again).

A number of other system analytical models are also of interest. Particular mention can be made of micro-analytical simulation models. It should be possible to apply this technology to model the dynamics of client trading data at the individual level and use it to investigate various sales or marketing strategies. Related methods of synthetic data estimation using bi-proportional and information theoretical estimation of missing contingency tables from marginal totals also have obvious applications; for example, in estimating data from partial data and in predicting responses. The principles embodied in accounts-based models might also find useful applications. As could simultaneous-equation models. Black-box models also seem to have a major potential usefulness. Their emphasis on description and prediction is no longer a problem but a particularly attractive selling point.

Model set C: new approaches based on fifth-generation computer thinking It is perhaps inevitable that the focus on data, on application, on performance, on best-fitting empirical models will also stimulate new approaches to model design. The traditional recipes have changed little in the past twenty years. They emphasize the role of theory in model specification. If theory either does not exist or is not so important, then new methods are required that can build well-performing models from data. The DBM philosophy offers one route, although it is limited in application. As computer time becomes cheaper, so it seems that other prospects become feasible. The idea of developing a machine-based, data-driven, Automatic Modelling System (Openshaw, 1986) is a very appealing prospect. It offers the means of searching the universe of all possible models that can be constructed from a limited but large set of pieces for 'better'-performing models. The computer hardware needed to routinize this model design strategy is already in existence and long CPU times are not a problem. Indeed in commercial applications run times lasting several months can be justified if there are signs of steady and regular improvements in performance.

The basic AMS principle of relaxing all human constraints on the model design process by identifying the universe of all possibilities which can then be either searched or enumerated, can also be applied to other areas of geographical analysis. The concept of a Geographical Analysis Machine (GAM) is another possibility. Instead of having to explicitly formulate a precise single hypothesis regarding something of interest, why not instead evaluate the universe of all possible alternative hypotheses that can be parameterized and then identify either the best *N* or use the distribution of all possible results to test the viability of one or two particular outcomes. This strategy already forms the heart of a cluster-busting approach for the spatial analysis of disease patterns. The same principles of searching through a universe of all possible alternative results can also be used as a search engine for investigating data bases for new customers or for irregularities in profiles or responses.

Some Predictions for 2007 (or maybe 2027)

Finally, some speculations of the state of models in human geography twenty years on might be useful. The models available in *AD* 2007 seem likely to reflect the extent to which geographers decide to respond to the challenges of information technology. Assuming they do, then it is likely that models in geography will still exist; if geographers do not respond then it is possible that the future geographical modellers will be computer scientists and all links with geography will have been severed. Either way, it seems that few of the current models will survive the next twenty years in their present form. The development of fifth-generation (5GL) computer languages and techniques could easily result in a far-reaching re-interpretation of existing models in terms of new languages (probably AIDA) based on different principles. This should offer new insights and therefore stimulate the development of new forms of model. The continued development of computer hardware and data base technology in the context of a global information economy will also, perhaps inevitably, stimulate the adoption of a different modelling philosophy; perhaps a form of adduction or a hybrid inductive-deductivism. It seems that there could well be four broad types of models of particular interest to geographers.

Some modelling systems will involve spatial analysis techniques in the form of Automated Pattern Analysers (APA), perhaps based on a generalization of the prototype geographical analysis machines now being experimented with. These techniques will be charged with spatial pattern analysis and in particular the identification of meaningful empirical anomalies. A family of related methods will be concerned with data compression via classification and pattern recognition techiques. Advanced versions of Data Base Models for fully automated data analysis could be another speciality.

The second category will almost certainly be a development of automated modelling systems based on AI techniques. It is conceivable that constantly running model-breeding machines will be inventing new models for data sets using variants of genetic optimization to convert standard macro-model templates and gene pools into application specific models. The form of the models will not be of any particular interest compared with their performance. These automatic model-developing systems will probably become an integral part of many key data base and computer systems in the early years of the next century.

A third category of models might be termed auto-theory new knowledge inferencing tools. The idea would be to develop AMS type models into new knowledge and theories of various human systems of interest via machine-based deductive systems. It appears impossible at present but it is probably only a matter of time.

A final category might well be based on a re-expression of some existing system-based models. Of particular interest are likely to be dynamic models of both aggregate spatial systems and micro–macro relationships with a temporal component.

Conclusions

Traditionally, it seems then that many of the best geographical modellers have had virtually no interest or propensity to perform empirical applications, and often only limited computing skills. Their abilities to create new theoretical models have far exceeded their abilities to provide empirical work or to discover real-world applications for their technology. At a time when support for purely academic disciplines is being reviewed, it is most unfortunate that nearly all existing models in human geography were neither conceived of as practical tools nor exist as part of live computer systems. Whilst the total inventory of models in human geography is now large most are not operational, few have been tested on data, there is very little software, and there is no obvious market for many of them outside of the academic world.

To remedy this deficiency a return to a more inductive and data-orientated philosophy is recommended. Although the history of science suggests that this approach has not been as fruitful as hypothetico-deductive methods, the situation in geography is now sufficiently serious and different from other sciences to consider a change in philosophy worth investigating. Furthermore, it has been noted that developments in information technology and computer hardware both favour a more extensive, computationally based, exploratory, applied, and data-orientated approach to model building, testing, and application. There is actually a growing demand for certain types of geographical model; a situation without parallel in history! Quantitative geographers need to respond soon.

Indeed there are many data-related spatial analysis and computer modelling possibilities that are becoming available for the adventurous geographer. There are vast potential new markets if geographers are able to sell themselves, merchandise their products, and adopt a less restrictive *modus operandi*. But the lessons of the past must be heeded, the old mistakes avoided and new approaches, models, and technologies devised to meet the rather different challenges if models in geography are to be diverted from their contemporary evolutionary cul-de-sac and adapted for use in data-rich theory-poor environments. It is, however, as has been emphasized, not an all-or-nothing situation. It is not necessary to totally abandon the past and start again. Rather it is a matter of complementing current interests with a greater inductive orientation as a means of gaining access to data.

Today flexibility rules instead of dogma, and Wilson's pragmatic eclectism rather than tunnel vision. Many geographers have already learnt the relevant skills and the available tool kits are well understood and reasonably adequate. The loss of a largely spuriously close link with theory may be sad, but it also opens up a veritable Pandora's box of new and exciting possibilities initially for inductive-based computer models in human geography and subsequently for more and better theories, if we care to look and are brave enough to try!

7 Quantitative Theory Construction in Human Geography

Bill Macmillan

'Theory construction has to be central to our concerns.'

David Harvey (Chapter 15 below)

Introduction

As an amateur gardener of very little talent, I know something about wildernesses. After twenty-odd years of modelling, geography's garden is not quite in that state but it is certainly in need of attention. All sorts of things have been planted in the past two decades and some of them have borne fruit. Much else has flourished briefly then died back. But in the view of many critics, the husbandry of modellers has allowed the place to run wild: weeding has been neglected so that the best of the new planting cannot be seen to advantage, fine specimens of traditional species have become overgrown, and insufficient ground has been set aside for the cultivation of new varieties. The response of some critics has been to conduct a campaign of slash and burn. Others have simply turned their backs and got on with cultivating their own patch. As for the model builders themselves, some would claim that they have been wielding the pruning shears to good effect for years, enabling sturdy plants to grow. Others would argue that overworking analogies like this is one of the things that gives modellers a bad name.

At the risk of annoying this last group, this chapter can be described as an attempt to hack away some of the undergrowth that prevents us from seeing clearly what models are, and why at least one activity that is described as modelling is of enduring importance, namely, quantitative theorizing. In contrast to Chorley and Haggett's eclectic view of modelling in the original *Models* volume, where they identify more than twenty different types of model, I believe it makes sense to distinguish only three connected uses of that much-abused word.

I intend to look at some criticisms of modelling which come from outside the modelling community and some which are best thought of as internal and, to a degree, personal. The external criticisms selected for scrutiny are ones that identify shortcomings in particular classes of models and tend to present objections in practice as objections in principle. The internal criticisms centre on the lack of clarity of purpose in the modelling enterprise. What emerges is a set of challenges to human geographical modellers to mend their methodological ways and to alter the portfolio of systems that they study.

Criticisms of Modelling

According to one group of critics, the major challenge for human geographers is to construct theories to account for the great historical processes of demographic, social, economic and political change. Modellers, it is argued, tend to deal with matters that are of comparatively little consequence such as journey-to-work flows, diffusion processes, agricultural land-use patterns and the like. In his chapter, David Harvey sums up this view by suggesting that the output from twenty-odd years of modelling 'in toto . . . adds up to little more than the proverbial hill of beans'.

What is more, from this perspective, the restriction of modelling to minor problems is not simply a matter of choice – it is fundamental. The argument is that the types of problem susceptible to modelling are characterized by independent, replicable events, whereas those that are not susceptible are characterized by non-repetitious (non-reproducible) historical processes, and it is these that are of real interest.

Another objection is that even in those areas where model builders can operate, they often fail to construct theories and, therefore, to provide explanations. This problem appears to have originated in the use of 'the mathematical process languages that lay most easily to hand', with which, according to Harvey again, 'it proved . . . very difficult to evolve any real theoretical argumentation'. What model builders do, it seems, is to produce representations of systems: what they do not do is offer much in the way of an explanation of the structure and behaviour of those systems. The objective (especially in areas like urban modelling) appears to be emulation rather than explanation. And if Sayer (1984) is to be believed, there is a problem here that is inescapable: in his view, mathematics is an acausal language and is therefore intrinsically incapable of being used in the formulation of causal explanations.

One virtue of the modelling work of the emulation variety is that it deals with real-world systems. This cannot be said, according to other critics, of those explicit attempts at theorizing that human geographical modellers have indulged in. In areas like location theory, for example, attention has been restricted to unreal or hypothetical landscapes and again, this is a fundamental and not a temperamental, restriction. Part of the argument, as articulated by Kennedy (1979), is that what matters most about real geographical systems is the configurational – the individual conditions obtaining in particular places at particular times – rather than the immanent – the law-like processes that are operating against this backcloth. Thus, if our modelling efforts enable us to understand a simple, general process, we still do not understand much about the naughty world (about particular, complex landscapes) and no amount of additional model building will help us to do so.

There is a related objection of particular concern in a human geographical context: it is not just the landscapes in models that are hypothetical, it is the agents who people them. Location and land-use models rely on the notion of rational economic behaviour, which bears as much resemblance to real behaviour as does the homogeneous plain to real landscapes. Adapting Kennedy's phrase, the real world is populated by naughty people, so modelling can be no substitute for a proper behavioural geography.

Even if all these points could be answered, there is another kind of objection that is raised by some critics. It is that the deficiencies of models can be understood only by examining the social context in which they were forged. When this is done, it is claimed, their poverty and partiality tend to be revealed.

To summarize, then, the critics whose views have been outlined, are dissatisfied because they want theories to be produced, not just representations; they want the subjects of those theories to be real rather than hypothetical systems; they want the real systems in question to be major systems; and they want the social context of the whole enterprise to be understood. There are clearly both substantive and methodological problems here, and they are not easily separable.

Responses

All of the above criticisms have something in them but most make unwarranted inferences about modelling in general from the apparent shortcomings of models of particular types. There are real difficulties involved in successful modelling but none of them are of such a magnitude as to require the abandonment of the modelling enterprise. Nevertheless, the shortcomings identified by critics have to be addressed, and to do that both the substantive and methodological issues must be tackled. This is attempted in outline below.

The methodological issues are discussed first in order to clarify the terms for the substantive debate. This involves disentangling some of the semantical knots that the words 'model' and 'theory' have been tied into. The methodological challenge is seen as one of moving from modelling to quantitative theorizing.

The substantive problems are seen as requiring both a change of focus away from patterns of activity towards processes of accumulation (a change which is already occurring and has long been pursued by some writers) and a move from spatial economic to geographical theory development.

Bill Macmillan

From Models to Quantitative Theories

So, what is a model? This question has been posed so often that it seems implausible that another attempt to answer it will add much to our understanding. Yet that understanding remains inadequate. In these circumstances, it is as well to consider whether or not we have been asking the right question. If, instead, we start by asking the two questions 'What is a model for a theory?' and 'What is a model of a theory?', we may make more headway. To answer these questions, though, we must be clear first about the meaning of that equally misused term 'theory'.

What is a Theory?

According to Rudner, 'A theory is a systematically related set of statements, including some law-like generalisations, that is empirically testable' (1966: 10). The three elements of this definition – on form, content and testability – all require some explanation.

The systematic relationship amongst the statements of a theory is a deductive relationship. That is, the set of statements contains axioms and theorems, where the latter can be derived from the former through the application of rules of derivation. The rules of derivation may be applied to the form of the statements without regard to their content. Content and form are connected by rules of interpretation. The separability of form and content is worth dwelling on. The abstract or uninterpreted form of a theory is known as its calculus and, generally speaking, a calculus will admit to many different interpretations. This is not to say that theories are abstract (quite the contrary) or that new theories should be produced by reinterpreting existing calculi (as Harvey argued in 1969 in *Explanation in Geography*). It is merely to make the uncontroversial point that the principles of logic apply to (and judgements about the validity of arguments are made on) the form of a set of statements and not its content.

The next part of the definition asserts that theories contain law-like generalizations. A law-like generalization is a universal empirical proposition – empirical in that it makes claims about the world and universal in that those claims refer to all members of some class of entities. Again it is important to be clear about what this does and does not entail. The requirement of universality means that the claims made are supposed to apply at all times and in all places. It is easy to misconstrue this requirement as precluding the possibility of theorizing about entities whose range is restricted in time and space, or, equivalently, as indicating that a different conception of 'theory' is needed in order to treat such entities. Whilst universality does require the claim to apply at all times and in all places, that is clearly not the same thing as asserting that examples of the entities in question are to be found universally. Thus, for example, to claim something about all newly industrializing countries

(NICs) is to do more than to assert that whenever a country exhibiting the defining characteristics of an NIC is found, then that country will have the accompanying characteristics being claimed for all NICs. All explanation involves making an appeal to such law-like generalizations, albeit implicitly in most circumstances.

In addition to being universal empirical propositions, law-like generalizations are hypothetical. That is, they have the form 'if C then E', where C represents certain general conditions and E an event or state-of-affairs which would follow from those conditions, should they materialize. For example, consider the following proposition: given a destitute farming family whose only assets are their farm implements (they have no land or food and cannot sell any labour), and given the further condition that there is a market in which they can exchange those implements for food, then they will make the exchange. This claim is hypothetical because it only says what would happen if certain conditions were fulfilled. For it to be applicable to the real world, it would be necessary to add a proposition to the effect that the conditions are indeed fulfilled. For it to be tested, it would be necessary to identify circumstances in which such a proposition is true.

This brings us to the testability part of the definition of 'theory'. Testability is a difficult concept to deal with. It is sensible to begin by making a distinction between observational, dispositional and theoretical terms. The terms 'implements', 'land' and 'food' are observational – they refer to things that can be observed. But ownership of these things and the ability to sell labour are dispositional (the act of labouring and the exchange of money can be observed but the ability to sell labour cannot). Dispositional terms are, nevertheless, connected directly with observable phenomena: to say that something is consumable, elastic, mobile, etc., is to attribute to it the ability to behave in certain observable ways. Theoretical terms, on the other hand, refer to entities or attributes which are neither observable nor have the disposition to produce observable behaviour. Theories are characterized by their inclusion of theoretical terms so theory testing is inherently problematic.

For a theory to be tested, its theoretical terms have to be replaced, in some way, by observational terms. Thus, for example, to test a theory in particle physics it is common practice to translate theory statements about particle trajectories into observation statements about traces in cloud chambers. The inaccessibility to observation typified by sub-atomic particles is of no great interest in human geography but there is another kind of unobservability that is. Rational economic agents, homogeneous plains, perfectly divisible commodities and the like are all unobservable phenomena. They are so-called ideal types. The problems posed by the presence of ideal types and the difficulties of replacing them by real types will be considered later.

If these problems did not exist, and if satisfactory observation statements could be found to cope with every unobservable, but supposedly real, theoretical entity, theory testing would still be a difficult business.

Given the universality of the claims made in theories, it is normally the case that not all members of the class of entities to which the theory refers will be observable: some will have ceased to exist and some will not have come into existence. Confirmation of such theories by exhaustive testing is, therefore, impossible in principle. If the theory refers to a unique entity, such as the world economy, the same argument applies – information about future behaviour and much past behaviour is simply unavailable. Falsification, on the other hand, appears to be possible in principle. However, the theory dependence and fallibility of observation mean that even falsification cannot be achieved conclusively (see, for example, Chalmers, 1982).

These legitimate concerns have to be put into perspective. No theoretical work in human geography is sufficiently well developed for us to worry about the conclusiveness of empirical tests. If we could say a theory was well confirmed, even if there was some contrary evidence, we would have good reason to be pleased.

Much more could be said about all these issues but this thumbnail sketch will have to suffice. The only additional point that needs to be made – and it is a point of some consequence – is that a quantitative theory is simply a theory that includes quantitative propositions.

What is a Model?

What, then, is meant by a model for a theory and a model of a theory? A model for a theory can be thought of as a prototype for a theory. For example, the theory of heat diffusion in a plate and the theory of gravitation have both been used as models for a theory of migration (Hotelling, 1979 and Ravenstein, 1885). Such 'modelling' involves a re-interpretation of terms and relations and rests on the separability of the form and content of theories. Whether or not it is desirable in practice to use a 'model for a theory' as an aid to theory development is immaterial to the definition. Such an approach is no more than an elaborate form of reasoning by analogy and has all the dangers that that entails.

As a model for a theory is itself a theory, a model in this sense is a linguistic entity (i.e. a set of propositions). A model of a theory, on the other hand, is a non-linguistic entity: it is a system of the type to which the theory refers – an empirical system if the theory is empirical and a formal system if the theory is formal (i.e. concerned with formal entities such as numbers, or points and lines).[1] Any 'possible realization' of a theory is a model of that theory. Thus, for example, any human popula-

[1] A system can be defined as a sequence $<S; R_1, \ldots, R_n>$ where S is a set of entities and $R_1, \ldots, R_n$ are relations defined on Cartesian products of S. An empirical relational system is a system in which S consists of empirical entities and a formal relational system is one in which S is formal.

tion whose history conforms to that described in demographic transition theory is a model of that theory. Again, the existence or otherwise of real systems that are models of particular theories is immaterial to the definition. A theory that has no real models is one that refers only to ideal types.

Models as Representations

The third sense of the word 'model' is that of representation. One thing, a folded tablecloth, to use an example from Cosgrove's chapter, can be said to be a model of another, the chalk downs. The Weberian wooden triangle with pulleys and strings can be said to be a model of a simple spatial economy and so on. The former is a model of the latter in the sense that it can be used to represent the latter. Such representations work (if they do work) because true propositions about the prototype entity can be translated into true propositions about the model. Typically, only a proper subset of the propositions that are true of the prototype are true of the model, and vice-versa.

The relationship between this sense of model and the other two is depicted in figure 7.1. Given a calculus that yields Theory 1 under interpretation 1 and Theory 2 under interpretation 2, Theory 1 is a *model for* Theory 2 and vice-versa. Any system that conforms to Theory 1 is a realization or *model of* that theory and System 1 is one such model. The general conditions that are part of Theory 1 are satisfied in specific ways by System 1: for example, if Theory 1 refers to a class of systems characterized by a parameter *g*, System 1 might be a realization of Theory 1 under the specific condition that g=0.001, say. Similarly, System 2 is a realization or model of Theory 2, where again the general conditions that are part of Theory 2 are satisfied in a particular way. Moreover, as there is a set of propositions (in Theory 1) that are true of

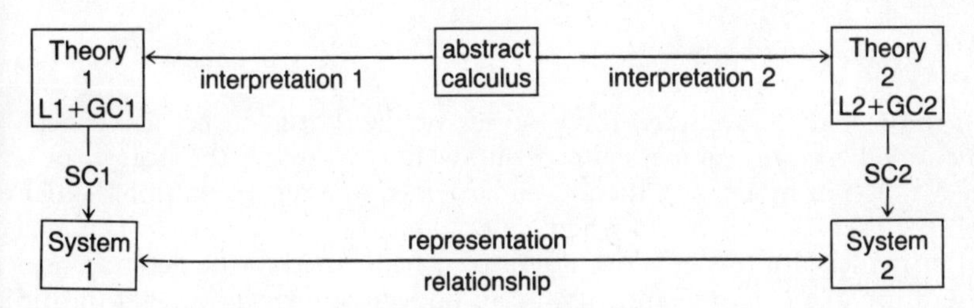

Figure 7.1 Schematic view of the relationship between three senses of the word 'model': Theory 1 is a prototype or *model for* Theory 2 and vice-versa; System 1 is a realization or *model of* Theory 1 (as Model 2 is of Theory 2); and System 1 is a *representation* of System 2 and vice-versa, provided that the specific conditions satisfied by System 1 translate into the specific conditions satisfied by System 2. L, GC, and SC stand for law-like generalizations, general conditions, and specific conditions, respectively.

System 1, and as these propositions can be re-interpretted (through the replacement of interpretation 1 by interpretation 2) so that they are true of System 2, System 1 can act as a model or *representation* of System 2 and vice-versa. For the representation relationship to work though, the specific conditions concerning Systems 1 and 2 must be translatable into each other, using the translation entailed in the connection between interpretations 1 and 2.

All of the above comments apply to theories in general but it is worth specializing the discussion briefly to talk about the particular attributes of quantitative theories. What makes an empirical theory quantitative is that models of it can be represented numerically. In terms of figure 7.1, this means that if Theory 1 is quantitative, each of its realizations is an empirical system (such as System 1) which can be represented by a numerical system (such as System 2). At the same time, this numerical system is a model of Theory 2 which is a numerical reinterpretation of Theory 1. These more specialized relationships are shown in figure 7.2.

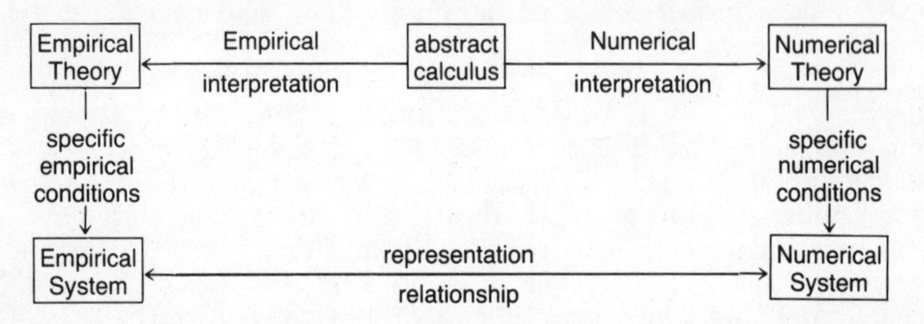

Figure 7.2 Schematic view of the relationship between an abstract calculus (e.g. an uninterpreted piece of algebra), empirical and numerical interpretations of that calculus, and models of those interpreted forms (an empirical system on the one hand and a numerical system on the other).

Models in Geography

In the last three sections, three senses of the word 'model' have been identified: the prototype sense of model for a theory, the notion of a realization or model of a theory, and the idea of a representation. Which of these senses is intended when geographers talk of 'models'? The idea of a 'model of a theory' does not appear to be used in the geographical literature at all; more's the pity, as an understanding of the relationship between theories and their subjects is crucial to an understanding both of the general difficulties of theory development and the special difficulties associated with the use of ideal types. On the other hand, both the prototype concept and the representation concept are invoked at different times, as the Harvey and Cosgrove references respectively demonstrate. The commonest sense in which the term 'model' is used, though, is none of these. Rather, it approximates to the concept referred to here by the label 'quantitative theory'. To be more precise, a model in this sense approximates either to a quantitative theory or to a quantitative theory together with a set of specific conditions. This usage of 'model' is unfortunate because it makes it hard even to distinguish between a quantitative theory and a realization of that theory, let alone to analyse that relationship.

The major reason why 'models' of this kind only approximate to being quantitative theories is that their form is not explicitly deductive. Instead, they tend to consist of a set of algebraic expressions, an interpretation of the terms contained in those expressions, and an informal commentary. For example, many contemporary works on location, land use and spatial interaction involve the statement of a mathematical programming problem, an interpretation of that problem's variables and parameters, and a commentary covering such things as the nature of the theory which the 'model' is supposed to be formalizing, operationalizing or generalizing, and the meanings that are supposed to be attached to the problem's objective function and constraints.

With this kind of 'model' it is sometimes difficult to see exactly what all the constituent propositions are and just what the relationships between them are supposed to be. There is a tendency for logical inconsistencies to be present and for the axiomatic basis of the system to be obscure. Nevertheless, in principle at least, such a system can be recast into an explicitly deductive form: it can be shown to be reducible to a quantitative theory. Why then is it called a model and not a theory? Part of the reason for this is the mistaken view that the word 'theory' applies only to informal, non-quantitative notions, whereas 'model' applies to formulations of higher status in terms of formality and precision. But another part of the reason may be a reluctance to acknowledge that 'models' are theoretical and, therefore, falsifiable. If we were to take a long hard look at the law-like generalizations on which 'models' rest – typically, in the above contexts, rather simplistic propositions about the behaviour of economic agents – we would be loath to offer them for testing because of a strong prior conviction that they would fail.

The most important reason for preferring not to use the term 'theory', though, seems to be a lack of theoretical purpose, or a subordination of that purpose to other objectives. That is, models tend not to be designed primarily as general, explanatory devices but as tools to describe individual systems, to predict their behaviour, and, sometimes, to suggest or test policies for modifying that behaviour.

Where the objective is description and prediction, interest tends to focus not on general law-like propositions but on particular conditions. In this context it seems sensible to hang on to the idea that the role of models is to emulate individual real systems so that predictions about those systems can be made. That is, it seems sensible to take the Haines-Young view that models are predictive devices (see his chapter), or to be more accurate, the view that predictive devices are models. However, this simply will not do. A prediction is a proposition – a proposition about the future. For it to be well grounded it must be derivable from a set of more basic propositions. In other words, it must be a theorem derivable from the axioms of some theory. A quantitative prediction is no more than a theorem derivable from a quantitative theory. But cannot physical systems like the Weberian wooden triangle, or a piece of laboratory apparatus, or a computer configured by some algorithm to behave in a certain way, be said to predict? No, they cannot: things behave, theories predict. A model (in the sense of a representation) cannot predict but it can be made to behave in ways which can enable predictions to be formulated. True propositions about the model's behaviour can be translated into propositions about the system of interest. The truth of the prediction will depend upon the quality of the representation, and that depends not just on the matching of specific conditions but on the existence of a common theoretical environment within which the two systems can be embedded (a common calculus which, under two separate interpretations, will yield statements that are true of the system of interest and its representation). For this reason, the folded tablecloth is a poor representation of the chalk downs if predictions are required of changes in the shape of the downs over time. It may not be clearly understood just what the form of the calculus is that provides the bridge between the system of interest and its representation (the reason for building the representation may be to help unravel this mystery) but there must be an underlying structural connection. Arguing by analogy is attractive because of the possible existence of such a connection, but hazardous because its properties, if it does exist, are unknown.

Where the objectives of the 'modelling' exercise are normative, evaluative elements are added to the predictive engine. But as I have just argued, what predicts is not the model-as-a-thing but the theory in which it is embedded. The better the theory is, the better the predictions. And the better the predictions, the better the policy indications should be (in terms of any given objectives). Thus, even in policy models, the quality of the theory ought to be of central concern and the possibility of testing the theory using the outcomes of implemented policies ought to be acknowledged (as Bennett argues in his contribution).

To summarize, I can do no better than repeat the quotation from Harvey's chapter cited under this chapter's title: 'theory construction has to be central to our concerns'. Model building is implicitly, and ought to be explicitly, about quantitative theory construction, even in a normative context. What is more, a deliberately axiomatic approach ought to be adopted. This would have the advantage of making it necessary to expose to scrutiny both the assumptions of the theory and the logic of the arguments constructed on those assumptions. In turn, this should make it easier to eliminate logical inconsistencies, refine assumptions and clarify the theory's domain (the set of real systems to which it applies – if any).

The Criticisms Revisited

It is now time to return to the criticisms outlined in the introduction. In brief, they were: (1) that modellers tend to concentrate on emulation rather than explanation; (2) that when they do attempt to theorize they restrict their attention to hypothetical systems; (3) that they do not (and perhaps cannot) tackle questions of real consequence; and (4) that theories cannot be understood without reference to their social context.

Emulation vs. Explanation

The first of these criticisms has already been dealt with, in effect, in the last few paragraphs. The view that modellers sometimes concentrate on emulation rather than explanation is well founded. However, it does not follow from this that the use of mathematical formulations ought to be discouraged. It is not the employment of mathematics that elevates emulation over explanation but the objectives and methods of its employer. There is nothing about mathematics *per se* which makes it incapable of yielding explanations. Indeed, if quantitative properties and relationships are important features of some phenomenon then an explanation of that phenomenon will require a quantitative theory and such a theory requires a mathematical formulation. In short, the argument that modelling and explanation are different kinds of activity ignores the fact that 'modelling' can and should take the form of 'quantitative theorizing'.

To argue that there is nothing about mathematics that makes it incapable of yielding explanations is to argue against Sayer's view that mathematics cannot provide causal explanations because it is an a-causal language (Sayer, 1984). In Sayer's view, mathematics 'lacks the categories of "producing", "generating" or "forcing" which we take to indicate causality' (op. cit. page 162). One can take issue with this view on two counts.

First, it involves an animistic use of language which serves to obscure rather than clarify the nature of causal relations. As Sayer is well aware, there are alternative definitions of 'cause' which do not involve such categories. According to him, 'the conventional theory of causation . . . focuses on regular sequences of events' (op. cit. page 162), or in other words, on 'regular associations (or "constant conjunctions") of causes and effects' (op. cit. page 100). But this notion, due to Hume, is surely not the 'conventional theory of causation'. Its vulnerability to the argument that there are constant conjunctions that are not causal (such as night following day) and that causal connections exist that are not constant conjunctions (such as striking causing matches to light normally but not invariably) led to its replacement by Mill's view of cause as sufficient condition. With the latter view of causality, there is certainly no difficulty in principle preventing causal explanations being formulated mathematically. I would argue that this view is preferable to that advocated by the realists, as represented by Sayer, but this is not the occasion to go into the argument in detail.

Second, even if one insists on the idea that cause involves such notions as producing and forcing, the fact that mathematics contains no such terms is quite irrelevant. Mathematics contains no empirical terms whatsoever. To formulate a piece of mathematics which can be interpreted as an explanation involving 'forcing' is no more difficult than to do the same with some other interpretation in mind. This is not to say that any old equation can be interpreted in causal terms, but that mathematical formulations that can carry such interpretations are possible. One can agree with Sayer that too much attention is paid to correlation and too little to causation, and that the former is sometimes confused with the latter, without agreeing that mathematics cannot provide causal explanations.

If these arguments are accepted, what Sayer's position appears to reduce to is essentially that presented earlier as the unfortunate tendency to use mathematics instrumentally – the tendency to emulate or simulate rather than to seek to explain.

One final point that Sayer makes (along with many other writers) is that one has to take great care over the quantifiability of social phenomena and this is certainly true. Some phenomena will not admit to quantification ('intransitive preferences' is the standard example). However, this does not mean that these phenomena are not amenable to formal analysis (the demonstration that intransitive preferences are not representable numerically is itself a formal exercise). Mathematics is best thought of as a language in which formal arguments can be expressed, whether or not they are wholly quantitative.

Hypothetical vs. Real Systems

Now let us consider the criticism that, when model builders do turn their hand to theory construction, they restrict their attention to hypothetical systems. The systems in question, in a human geographical context, tend to be characterized by homogeneous plains on which instant activity-landscapes are formed. Classical location and land-use theory and the whole of the New Urban Economics focus on such systems.

Work in both of these fields can be thought of as involving intellectual experiments whose value is to explain how certain phenomena could occur under simple conditions. The explanations involve an appeal to some law-like proposition which describes the mechanism that drives the landscape formation process (such as profit maximization). The most obvious deficiency of this work is that it neglects the complexity of the conditions that occur in real places at particular times. Now, the essence of Kennedy's naughty world argument is that the key to understanding real landscapes lies in the complexity of particular conditions (the configurational) rather than in law-like generalizations (the immanent) and as modelling entails simplification of conditions, it is incapable of yielding this understanding. The flaw in this argument is that *the belief that modelling entails simplification is itself a simplification.*

There are two major reasons for adopting simplifying assumptions. The first is to enable the effects of an individual factor acting in isolation to be demonstrated. For example, von Thunen wanted to demonstrate the effects of distance from market on land use, effects 'which we see but dimly in reality' (Hall, 1966), so he suppressed other variables. And again, Lösch was keen to ensure that 'no spatial differences . . . lie concealed in what we assume' (Lösch, 1954) in order to show that concentrations of activity arise for reasons other than the peculiarities of particular locations.

The second reason for simplifying is technical – it is to enable analytical methods to be exploited. Thus, the New Urban Economics is characterized by a continuous representation of space and the depiction of cities as monocentric. This allows continuous functions to be employed which can be analysed using classical calculus and the calculus of variations. In turn, this yields axioms on the circumstances and behaviour of individual agents from which general theorems on the spatial structure of various phenomena can be derived (ghetto formation, the effects on land use of externalities such as pollution and congestion, and so on). It is the simplicity of the conditions (monocentricity and spatial homogeneity) which enables general theorems about spatial patterns to be produced (in the form of analytical solutions). In terms of the discussion on the nature of theories, such a procedure relies on ideal, rather than real, types.

In both of these cases, simplifying assumptions are used as a matter of choice not necessity. The first case can be disposed of quickly. The decision to demonstrate the effect of one factor acting alone derives from the desire to make a particular theoretical point. Once the point has been made, the reason for pursuing that particular simplification no longer exists. The second case – the use of ideal types – is the one that concerns the critics. The charge they make, in effect, is that the employment of ideal types renders the resultant theories irrelevant to the real world. Those producing the theories would contest that view, but could hardly contest the assertion that the use of ideal types makes testing highly problematic.

What they might do instead is simply dismiss testing as inappropriate, on the grounds that the systems they have produced are deliberately hypothetical; their development constitutes an intellectual experiment rather than an attempt to formulate a set of propositions that are supposed to be true of real systems. But this is to concede the argument to the critics. An alternative tack is to regard the theory as applying approximately to a very limited set of real systems. Thus, von Thünen's theory has been applied to (and, by implication, tested on) a number of areas that approximate to having an isolated city set in a homogeneous plain (see, for example, Hay, 1984 and Griffin, 1973). Again, though, this is not much of a defence.

Fortunately, the continuous space approach represented by classical location theory and the New Urban Economics is not the only one that is available. The simple conditions that characterize these fields are not necessary in quantitative theories. Complexity can be introduced, although to do so normally entails sacrificing the possibility of producing theorems that describe spatial patterns in general terms. With complex conditions (whose description almost always requires a discrete representation of space), mathematical programming formulations tend to be used and particular numerical solutions are implicitly sought rather than general analytical solutions. It is important to note, though, that the statement of a numerical solution still takes the form of a theorem, albeit one that is restricted in its application to systems with particular parameter values. With this approach, then, the ability to say something general about a set of hypothetical systems is traded for the chance of saying something specific about individual real systems. It is less general than the classical continuous space approach but no less theoretical. It is an approach which eschews simplification in favour of accurate description.

Of course, all description and explanation involves simplification, but the point of the above argument is that *the use of mathematics does not entail any special simplification beyond that which is involved in the use of a natural language* (such as English). Extra simplification can be employed as a device but it is not essential. It follows that modellers – or quantitative theorists – are not restricted to studying strictly hypothetical systems. By resorting to numerical methods and computation it is possible to begin to capture the complexity and individuality of real systems.

Before leaving this point it should be stressed that there is really nothing new about it. The authors of the classics of location theory were all interested in analysing real systems but were shackled by a rather limited range of techniques for calculating location and land-use patterns. It was this as much as anything that led them to use what we now regard as unrealistic assumptions. Modern versions of the theories are capable of handling much more complex conditions because of the availability of much more sophisticated analytical and computational techniques. The contrast in the technical environments is perhaps best illustrated by comparing the Weberian analogue computer (the famous wooden triangle) with modern digital computers. The sorts of conditions that can now be handled in, say, agricultural land-use theory, include multiple markets, variations in fertility, the existence of transportation networks, unpredictable variations in yield and changes in these and other factors through time. In addition, these theories contain relatively sophisticated propositions about the behaviour of individual economic agents by way of law-like generalizations (covering such things as supply responses to price variations, uncertainty over yields and prices, and short-sighted adaptive behaviour).

So much for the complexity – or otherwise – of the conditions in quantitative theories. What of the law-like generalizations? Those who argue in favour of a proper behavioural geography are more concerned with the nature of the latter. Characterizing spatial economies as being populated by rational economic agents with perfect foresight is rightly regarded as being unsatisfactory. But, as has just been indicated, modern work has gone well beyond these neo-classical assumptions.

Nevertheless, it can still be argued that the behavioural basis of modelling work leaves a lot to be desired. The assumptions made are certainly not as sophisticated or realistic as they could be, when taken solely as descriptions of the circumstances and behaviour of individual agents. For geographers, though, what matters is not the behaviour of individual agents *per se* but what happens to the spatial economy as the result of the actions of sets of agents. It is easy to criticize the view of the agent in location and land-use theory. The difficult trick is to improve that view and, at the same time, show what the consequences of the new view are for the structure of the spatial economy at large. Thus, for example, to argue that decision makers face uncertainties and are short-sighted in their behaviour is not enough – it is necessary also to show what the spatial economy would look like if populated by such agents. The problem of moving from ideal types of agents to real ones requires a process of attrition. Arguing for better views of agent behaviour in isolation does no more than shift the theoretical problem from axiom quality to theorem derivability.

Major vs. Minor Systems

The two remaining criticisms are closely related. Both emphasize the historical. The one to be considered in this section was referred to in the introduction as the 'hill of beans' criticism. The other – which involves arguments about the historical contexts within which theories are produced – will be considered very briefly in the next section.

The 'hill of beans' view is that after twenty-odd years of modelling we have rather little to show for our efforts. Moreover, further modelling is unlikely to improve the situation significantly. It is accepted that we can now predict the behaviour of repetitious processes more accurately (diffusion, journey to work, agricultural land usage, etc.) but it is argued that modellers have said nothing, and possibly can say nothing, about great historical processes. The latter part of the argument is based on the view that modelling in general and systems theory in particular is inescapably a-historical.

This view appears to be held because of the odd way the systems concept has been used in geography. In other disciplines – and certainly in the systems theory literature (see for example Padulo and Arbib, 1974) – systems have histories. To a first approximation, a system is an empirical entity that is describable by a set of differential or difference equations. Anyone with a working knowledge of such equations will know that they permit the present state of a system to be explained only in terms of some past state and a law of motion. The fact that the term 'system' has been applied loosely in a-temporal contexts in geography is unfortunate, but it does not alter the fundamentally *historical* nature of the systems concept as it is generally understood. The notion that modelling is restricted to repetitious processes is also mistaken. Some of the most interesting modelling work of recent years has been concerned with chaotic systems whose future behaviour is always unlike their past behaviour.

Whilst the above argument addresses the objection in principle to historical, quantitative theorizing, it leaves untouched the objection in practice that little work of real consequence has been produced. Although one could take issue with Harvey about the size of the contribution modellers have made, it is hard to deny that there is some substance in his 'hill of beans' charge. In so far as human geographical modellers have been concerned with the real world at all, they have concentrated on contemporary, western, urban issues, usually in a planning context, with the greatest effort being devoted to questions of land use and transportation. It is possible, of course, to theorize quantitatively about systems of quite different kinds. Industrial, agricultural and settlement systems have all received attention but again very largely in a contemporary, western context. What is more, almost all of the work that has been done in these fields is set firmly within the paradigm of neo-classical economics.

Geography could certainly be enriched – and maybe the world made a little less poor – by altering the portfolio of subjects commonly studied by modellers to include some of the processes of development and underdevelopment. Two relatively recent publications by development economists have lit paths of inquiry which we would do well to go down. Kelley and Williamson's *What Drives Third World City Growth* (1984) cries out for geographical elaboration. The field of urban dynamics has been growing in sophistication and importance over the past few years, and although it has not yet come to grips with rapid third world urbanization, it has the potential to do so. The second publication that could provide a guiding light for us is Ravallion's *Markets and Famines* (1987). Again, some work has been done by geographical modellers on agriculture in risky environments, but the problem of famines has scarcely been touched.

I am not arguing that we should simply reproduce works such as these with spatial knobs on. Rather, I believe there is considerable scope in these areas for developing genuinely new and potentially useful spatial economic theories. But beyond this, the study of famine (and the same could be said of desertification, deforestation, and a number of other issues) appears to demand a peculiarly geographical form of analysis. Work on such issues could catalyse the development of a whole new class of human/environment interaction models, cutting across the traditional human/physical disciplinary divide and wrapping together the spatial analysis and human/environment views on the proper direction of geographical inquiry.

As for the problem of understanding the great historical transformations referred to by Harvey, it is certainly the case that modellers have not yet made a significant contribution. But there is no reason why – as quantitative theorists – they should not do so in the future. The phenomena themselves are not peculiarly qualitative, so are not inappropriate subjects for quantitative theorizing. They do involve major qualitative changes but, as recent mathematical developments have demonstrated, such changes can be analysed effectively using non-linear system theoretic methods. On the other hand, whilst there is no impediment to the use of mathematics, there may be no immediate virtue in it either. Before a successful quantitative theory can be constructed, it is necessary to have a set of ideas involving quantitative relationships. Familiarity with the language of mathematics may help such ideas to germinate but is not a substitute for familiarity with the empirical issues. The message for modellers is that there are issues here of considerable fascination which quantitative theorizing could help to unravel. For historical geographers the message is not to ignore the extra theoretical dimension that quantitative propositions and formal reasoning can provide.

It would be wrong to give the impression that this is completely uncharted territory, but it would be equally wrong to ignore the fact that the historical dimension has been woefully neglected by modellers. Marxist economic perspectives have been largely ignored (with one or two notable exceptions) but so too has neo-classical growth theory. And the great irony of all this is that by following the weight of empirical evidence on the importance of the historical perspective and the weight of technical evidence of the success of the historical approach of systems theory, modellers could resolve many of the problems that plague their a-historical work in fields such as central place theory, urban land-use theory, and industrial location theory. Much progress could be made simply by abandoning the presumption that equilibria exist, in favour of formulating descriptions of processes which may or may not tend to equilibria. Similarly, by turning away from patterns of activity towards processes of accumulation, great strides could be made.

The Context of Validation vs. the Context of Discovery

Finally, the problem of the historical context of theory development has to be mentioned, although it will not be fully addressed. That is a task that requires a chapter of its own at the very least. There is, however, one point that I have not seen made elsewhere in the geographical literature, that seems vital to the debate. It is that there are three distinct areas of inquiry which tend to be confused in arguments about geographical theories and theory development.

First, there is a set of questions that relate to the so-called context of validation, which have to do with the logic of inquiry – with the problem of acceptance or rejection of hypotheses and explanations. Second, there are questions that are concerned with the context of discovery, which have to do with the conditions, social and otherwise, under which theories are developed. The latter questions are empirical – they belong to the sociology of the discipline – whilst the former are methodological. Third, there are questions that can be thought of as being concerned with the context of application – with the social uses to which the theories are put. These questions have both an empirical and an ethical dimension.

The focus of the meta-theoretical debate in geography has shifted from the methodological, at the start of the modelling period, to the sociological and ethical. Thus, the nature of the questions posed and criticisms made have both changed. It is important not to confuse one kind of criticism with another. As I have argued in this chapter, some serious methodological problems remain to be solved by the modelling community. But this is not to deny that 'intellectual inquiry is part of – and ultimately responsible to – the conduct of practical life' (Gregory, 1978: 182). If we are to improve our theories we cannot ignore their logical and empirical qualities but we cannot surely confine our notion of 'improvement' to these attributes alone. By the same token, though, doubts about modelling raised in the contexts of discovery or application are not transferable to the context of validation: dissatisfaction with the ways models have been produced and used in the past cannot be taken to imply anything about the qualities of mathematics as a medium for theoretical argument.

Conclusions

At the start of this chapter a number of criticisms of models and modelling were outlined. Before addressing them, an attempt was made to clarify the word 'model' by giving it three related definitions, all of which rely on a prior understanding of the word 'theory'. According to these definitions, models can be prototypes for theories, realizations of theories, or representations. The concept of a 'quantitative theory' was also discussed, as was the use of the word 'model' to convey this concept. The point of this semantic treasure hunt was to arrive at the proposition that models should be conceived of, and developed as, quantitative theories (even in a normative context). Furthermore, it was argued that a more explicitly deductive approach should be adopted in the formulation of quantitative theories, in order to facilitate the scrutiny and refinement of their arguments and assumptions.

On looking at the criticisms, it was acknowledged that human geographical modellers have sometimes produced representations rather than theories, neglected real systems in favour of hypothetical systems, and restricted their attention to a limited range of subjects. But the ideas that explanation and modelling (in the guise of quantitative theorizing) are somehow inconsistent, that modelling is incapable of dealing with real systems because it relies on simplification, and that modelling is essentially a-historical, were all firmly rejected. Nevertheless, it was argued that far too little attention has been given to history, and that a change of focus is required from patterns of activity and spatial interaction to processes of accumulation. It was also argued that more of us should look beyond the contemporary western world for our subjects and that we should seek to develop distinctively geographical (rather than spatial economic) theory.

To summarize, the critics deserve attention because of the importance of some of their arguments but modellers should not be too defensive. Quantitative theorizing in human geography still has much to offer. Indeed, with a greater clarity of purpose and a changing portfolio of subjects, modelling ought to command a central place in the discipline for another twenty years at least.

8 Modelling in Human Geography: Evaluating the Quality and Feeling the Width

Sally Macgill

Why do geographers model? How well do they do it? These two questions provide the chosen foci for the brief commentary below, being of an appropriately fundamental kind, yet far from closed in terms of the modelling community's implicit collective response to date (compare, in particular, the chapter of Wilson, Openshaw and Macmillan in this volume).

The first question is initially merely a vehicle for taking stock of the range of possible justifications for modelling: for acknowledging why modellers are doing what they are doing and, *inter alia*, for prompting broader awareness of reasons that are too readily ignored in some unduly narrow views. Beyond this, the question is also a basis for encouraging wider acceptance of the innate breadth and multiplicity of legitimate modelling rationales: an essential foundation if different approaches and arguments (as exemplified by, for example, the Wilson and Openshaw chapters) are to be viewed more in terms of differences in chosen emphasis and driving motivation, than aggressive (and ultimately destructive) competition, or ambiguous contradiction. But this *a priori* tolerance of different modelling rationales is not an advocacy of blind or uncritical acceptance of their claims; there is also an imperative need for careful assessment of the quality of what is done through different approaches. Hence the inclusion of the second, and to the author currently more compelling, question. The commentary below takes stock of elements of a possible framework for quality assessment, and identifies a pressing need for more extensive and systematic evaluation of a related kind in future.

Why do geographers model?

1 In many instances the justification is overwhelmingly pragmatic – a model-based approach is often the only possible means for arriving at any kind of quantification or formal measurement of unobserved or unobservable phenomena – hence the wealth of model-based estimations, forecasts, simulations, interpolations, and generations of data and other indicators.

2 In other instances, modelling is undertaken within a more implicitly inductive philosophy, serving a crucial signposting function in structuring, exploring, organizing and otherwise making sense of available or obtainable data, for example through discriminating patterns and correlations.
3 Alternatively, models can be used as 'laboratories' for the surrogate observation of systems of interest which cannot be observed directly, and for experimenting in estimating the effects and consequences of possible changes to particular components, or generating future scenarios of the evolution and end states of systems of interest.
4 More prescriptively, models can be developed and designed with an explicit decision support role – most obviously, though by no means exclusively, in the case of 'optimisation' or 'alternative choice' models.
5 The pedagogical value of modelling is uppermost in other instances – a basis for improving understanding of causal mechanisms, of relationships between micro and macro properties, of the enabling and constraining effects of identifiable structural characteristics, of interplay between a system and its environment, or of identification of 'critical' levels of system components.
6 Others in the modelling community are compelled fundamentally by the intellectual stimulus that modelling provides for them. Here models serve as frameworks (or forestructures) within which theoretical statements can be formally represented and their empirical validity then put under scrutiny.
7 On the basis of any, or all, of the foregoing considerations, modelling can be invaluable in fixing frames of reference for comparative work – providing some insurance against the pitfalls of not comparing like with like.
8 And modelling can also provide a currency of accountability and a source of linguistic economy (and perhaps social solidarity too) among people who understand their language and have shared values and meanings. (For a more detailed treatment of the question 'Why do geographers model?' see Macgill, 1986)

Any of the eight rationales above might be deemed valid justification for modelling, and many modelling initiatives are manifestly indicative of several of them – after all, many are differences in emphasis, not in kind. There are also, of course, stark differences between different modelling initiatives in terms of which rationales come into play and which are excluded. But beyond recognition of the multiplicity of legitimate rationales, the second question calls for well-founded evaluation of the quality of what is being offered. Other commentators (see Cullen, 1985 and Wynne, 1984) have drawn attention to: the possible 'inaccuracy' of model predictions as compared to reality; unrealistic and therefore unattainable data demands of modelling; the possibility of thinking being narrowly trapped within the conceptual framework implied in a given model structure; the possibility of models being ignorantly taken as objects of blind reverence; the possibility of the technical language of modelling being inaccessible to some crucial audiences, who may then feel improperly excluded. Rather than engaging here with this particular style of discourse, we review instead the beginnings of a formal framework through which more routine systematic assessment might proceed (see Funtowicz and Ravetz, 1986, 1987). This will be viewed below as a basis for general but wide-ranging evaluation of the realism of model claims and achievenents, and as a vehicle for encouraging properly informed and sober interpretation amongst various audiences of what different modelling initiatives can and cannot deliver.

The framework is summarized in table 8.1. The column headings correspond to individual criteria through which the quality of a modelling initiative might be assessed (there is not a one-to-one correspondence between the headings and the modelling rationales indicated earlier, as the latter are not distinct). A range of possible ranking modes is given under each heading. Most of the headings are self-explanatory, though it is worth noting the different 'levels' of evaluation to which they are directed, encompassing applied (column 1), methodological (column 2), epistemological (columns 3, 4, 5) and sociological (columns 6, 7) aspects of the development of knowledge through modelling (or for that matter other) ventures.

The first column calls for an assessment of the relevance of model outputs to what one wants to know (to some 'real-world' problem). As is widely appreciated, model resolutions can be frustratingly deficient – models valid only for short-term projections may be called on to produce long-term scenarios; zonal systems may be too coarse (national for regional; regional for local). There can also be ambiguity and lack of consensus over what the appropriate measures or indicators for a given problem actually are. Or there can be an invalid transfer of models from one context to another. There is also the question of whether the product of some modelling initiative could have been 'better' derived some other way. Different entries in the first column are proposed in order to encourage honest disclosure, if as yet in somewhat broad terms, of the relevance of model outputs to what one wants to know.

The second column, labelled robustness, asks about the degree of variability in the numerical results of models given changes in data inputs, parameter values and mathematical specification. Such changes would be understood to range over sensible alternatives (in other words, within the domain of legitimate uncertainty over such aspects), though often ranging to probe the existence and impact of critical values, and with answers framed in formal probability terms. The normal way to explore such matters would be via some form of sensitivity analysis. The extensiveness, or otherwise, of such testing may itself constitute a further additional criterion for the assessment framework as a whole.

The quality of data inputs to any modelling initiative can obviously have a crucial bearing on the quality of the final product, and this constitutes a source of quality imperfection that operates at a different

Table 8.1 A framework for quality assessment

Relevance (1)	*Robustness* (2)	*Data quality* (3)	*Theoretical strength* (4)	*State of the field* (5)	*Extent of review* (6)	*Degree of acceptance* (7)	*Score* (8)
Direct	Firm	Bespoke	Laws	Mature	Wide	Total	4
Indirect	Resilient	Historic/ field data	Well tested theories	Advanced	Fair	High	3
Convenience	Elastic	Calculated data	Emerging theories	Intermediate	Limited	Medium	2
Symbolic	Weak	Educated guesses	Hypotheses/ analogies	Embryonic	Little	Low	1
Spurious	Wild	Uneducated guesses	Working definitions	No opinion	None	None	0
Unknown	Unknown	Unknown	Unknown	Unknown	Unknown	Unknown	–

Source: Adapted and extended from Funtowicz and Ravetz, 1987. See also Macgill and Sheldrick, 1987.

level than that implied in the way accuracy-robustness has been considered above and reflected in column 2. For example, poor data can vitiate the validity of even the most exhaustive sensitivity analysis. In principle, the quality of data inputs can be extremely variable, ranging from reliable primary data to secondary data of doubtful origin, and including proxy measures and sheer guesswork. The individual descriptors in the third column reflect this possible variety. Trade-offs between 'data quality' and some of the criteria represented in other column headings are also worth noting: for example, the demands for policy relevance (column 1), or the data demands imposed by a particular theoretical model structure (column 4) can in some cases only be met if there is a 'sacrifice' of data quality ideals.

Reference has already been made to the potential role of modelling in contributing to the further development of theory. A converse of this point is to ask what level of theoretical development is embodied in a given modelling initiative. Possibilities are suggested in the fourth column. Again the range given here is somewhat coarse, and more refined specifications can be envisaged: compare the given entries in the table with the more sophisticated appraisal in Macmillan's chapter of the position of theory in human geographical modelling endeavour.

The fifth column refers to a related, though distinct, aspect from the fourth, specifically, to what can be expected in the light of the state-of-the-art of a given field of study. One cannot expect a model to embody well-tested theories if the field within which it is being developed and tested is itself as yet in an embryonic state. Conversely, questions may well be asked of a model embodying merely speculative theory if the field as a whole is at an 'advanced' stage of development. In general, then, some broad indication of the state of development of the field is in order, and the descriptors given in column 5 span an appropriate range of possibilities.

Finally, sociological aspects of the development of knowledge – the extent of review and outcome – are important (columns 6 and 7). Since legitimate assessment of each of the four preceding counts demands a degree of review within some peer community (though defining a peer community might be problematic in some cases), the extent of this, and the outcome, can be usefully acknowledged.

There are undoubtedly other criteria against which the quality of modelling initiatives might be usefully judged, as well as useful disaggregations of those which already appear in table 8.1. But the table as given will be accepted here, as a useful and concise basis for brief reflective comment about the claims and achievements of modelling.

Specifically, the table can be used as a basis of a formal evaluation framework by virtue of the normative ordering among the entries which constitute each scale (the higher being 'better' than the lower). In particular, scores can be used as a concise way of representing individual elements of each row (see final column). For any given modelling initiative, then, a string of scores can be determined. For example, the string

(4,2,2,1,2,4,1) would denote a modelling initiative which is directly relevant to a given problem (a score of 4 from the first column), whose robustness is decidedly questionable (a score of 2 from the second), whose data is 'calculated' (a score of 2), whose theory is best described as 'hypothetical' (a score of 1), whose parent field is as yet 'intermediate' in its state of development (a score of 2), which has had wide review (a score of 4), and low acceptance (a score of 1). Rather a mixed effort, and well short of the 'ideal' string (maximum distinction on all possible counts) of (4,4,4,4,4,4,4).

The crucial point in all this, however, is not to expect the utopian ideal string of top scores from all modelling initiatives. On the contrary, it is to determine the true string fitting to any given modelling initiative and to appreciate its real significance so as to fix more firmly one's understanding of what is being offered (the process of deriving the appropriate scores can be a research activity in itself, often generating new lines of reflection for all concerned).

Specifically, the string of scores will represent, in effect, a qualitative assessment of the reliability of a modelling initiative. As such it can be compared with, used to highlight the limitations of, and ultimately complement, conventional goodness-of-fit statistics which represent, in effect, a quantitative assessment of the reliability of a modelling initiative. If both qualitative and quantitative assessment indicators are high, then we can indeed have confidence in the modelling initiative vis-à-vis the reality supposedly represented or emulated. If either indicator is weak, then our confidence should accordingly be diminished.

Beyond these remarks, the following observations are offered:

1 Deriving the string of scores fitting to any given modelling initiative is a means of reflecting on and communicating their assessed quality to others.
2 Comparisons of the strings of scores for different modelling initiatives could be very insightful, setting the quality characteristics of individual initiatives into relief against others. After all, different initiatives consciously, or unconsciously, strive for distinction on different counts.
3 Claims for modelling must be in line with (the legitimately assessable quality of) what can be delivered. Rhetoric that is out of line with formal evaluation should be rejected. Overenthusiastic claims for modelling and unduly damning criticisms each in their own way inadvertently circumvent this 'test'.
4 Poorer performance than is 'necessary' on any of the given counts should be rejected, whether manifest in terms of poor data, impoverished theory, or whatever.
5 There should be no shame about a less distinguished performance than is possible on any of the given counts. One should not expect delivery of the impossible. Low scores on certain counts may be a legitimate reflection of reality, not of inherently poor endeavour on the part of those involved.

6 Weaknesses on certain aspects may confound the attainment of distinguished performance on others: for example, irremediable data deficiencies, or endeavour in a field which is as yet embryonic in its development.
7 'Unusual' strings of scores should be viewed as possible pointers of something questionable (claims for strong theory in embryonic field, for example).

In general, the given framework also suggests a particular way of defining 'progress' in human geographical modelling: namely in terms of improving the string of scores achieved by a given modelling initiative over time. It is neither a simple nor assured process, for new endeavour may uncover areas of hitherto unacknowledged ignorance, or meet with new barriers. But the attempt to push forward from a given position must continue.

Different modellers within the compass of human geography are not only currently located in different positions within the framework of table 8.1, but also obviously hold different convictions as to how future upward movement is best achieved. For example, Wilson's argument emphasizes the need for an improved theoretical hold, through a structured pluralism, in order to grasp more fully the multiplicity of forces and influences bearing on a given problem area. (A logical corollary is perhaps that more narrowly focused theory of the past has made bogus claims to distinction.) Openshaw, on the other hand, argues for modellers to launch themselves at the increasing quantity and better quality of data currently available (in turn relegating the role of theory and identifying both with wasted opportunities in data-based developments over the past decade and an overinvestment of intellectual capital in theoretical issues. It is not the place here to seek to resolve these rather different views of paths to a productive future for different modelling communities, although it is worth noting the serious concern expressed by Gatrell during the conference discussion on the Openshaw paper about the relegation of theoretical endeavour. We merely re-emphasize instead the need for appropriately astute assessment of the emerging products of these, and of course other, continuing modelling initiatives, and in turn, the future need for improved frameworks within which such assessment can be made.

Twenty years on from the present, the coarseness of the framework displayed in table 8.1 may well be manifestly revealed. It can at present claim no more than the status of a rough checklist for evaluating model quality. But even as such, it can claim also to begin to meet a longstanding need for formal and systematic qualitative assessment of otherwise unchecked claims and criticisms of modelling initiatives.

PART III
Modelling and the Development of Applied Geography

9 Climate, Models and Geography

Ann Henderson-Sellers

Introduction

This is the decade in which climate is coming into its own. In 1957 there were fewer than fifty climate modellers; now there are many thousands. Whilst modern climatology expands and encompasses new disciplines, the relationship between geography and the study of climate is moribund. A review of the past twenty years reveals a sorry saga of bad timing, mismatched terminology and a lack of mutual understanding. Although numerical climate modelling was dominated originally by physicists and mathematicians new recruits from other disciplines are now being sought actively. A recognition of the potential of analogue models is a recent development, as is also the acknowledgement that temporal, and especially spatial, analysis of data and results are fundamental to the improvement of parameterization and, hence, prediction skills. Major international programmes are currently being planned in which geographers could and should participate. Involvement will require recognizing and embracing new techniques, including fractal geometry, tesseral arithmetic, aspects of machine 'intelligence' and computer 'vision'. The alternative is that traditional geographers will become entirely excluded from the 'climate community' and techniques of geographical analysis, including cartographic truths, will be rediscovered by these 'new climatologists'.

Prejudices, Bad Timing and Changing Definitions

As I intend to try to view my subject through other people's spectacles, I shall begin with some definitions. In the opinion of many raw undergraduates I have quizzed, geography is, by a resounding majority vote, 'my favourite subject at school', while the question 'what do you think of when I say model?', generally produces a schoolboy smirk. Climate is a bit trickier; it either generates a defensive, 'our school missed that out' or 'we had a weather station beside the playing field'. Colleagues, on the other hand, have more mature views. Geography is 'spatial, and probably historical, analysis of man/environment interactions', modelling is 'micro Basic on the 9th floor' and climate is 'covered in the second year climatology course'.

My own definitions are simpler. Geography, at least all I can remember of it, is epitomized by 'hot, dry summers and warm, wet winters', underlining both my early separation from the subject and, even then, my prejudice. Climate is when the BBC 'phone up to ascertain my views on the 'unexpected' drought or the 'sudden' cold snap, and models are persuasive creatures offering considerable temptation and illusory rewards.

My mental images seem to be closer to those of the freshman than of the mature geographer. Peter Gould, a real, professional geographer, describes, in his superb book *The Geographer at Work*, his 'passionate love affair' with Geographia, a 'delicious and seductive wench' (Gould, 1985a: xv). I'm afraid my viewpoint is rather different (figure 9.1). I see 'Mr Middle-Aged Geographer' as tempted, first, by the slinky but dominant 'Climatic Determinism' who offers sinful seduction but demeaning consequences. The voluptuous 'Climate Model' is very attractive but too expensive in that she has to be paid for in both cash and understanding. The safest course is to return home to the 'old woman', 'Climatology'.

Speaking of climatology in 1978 in an address reported in the *Bulletin of the American Meteorological Society* (10, 1171–4), Professor Kenneth Hare said,

> you hardly heard the word professionally in the 1940s. It was a layman's word. Climatologists were the halt and the lame . . . in the British service you actually had to be medically disabled in order to get into the climatological division. . . . It was clearly not the age of climate. Now it is. It's the respectable thing to do. . . . This is obviously the decade in which climate is coming into its own.

Climate certainly is 'coming into its own'. This may be seen by a glance at the current journals in the National Meteorological Library. The *Journal of Applied Meteorology* became the *Journal of Climate and Applied Meteorology* in 1983 and the *Journal of Climate* in 1988; *Archiv für Meteorologie Geophysik und Bioklimatologie*, Series B became *Theoretical and Applied Climatology* in 1986; the journal *Climatic Change* was first published in 1979 followed by the *Journal of Climatology* in 1981, *Climate Dynamics* in 1986 and *Paleoclimatology* in 1987.

Smagorinsky (1983) noted that at the international conference on numerical weather prediction held in Stockholm in June 1957, which might be considered the first international gathering of climate modellers, the total world's expertise could be accommodated in about 40 seats. Today it requires many thousands.

While modern climatology grows, the relationship between geography and climate seems to have been declining since the time of C. E. P. Brooks. Like many relationships this one has been prey to bad timing and lack of mutual understanding. The villain of the piece is the digital computer. Just when other budding partnerships were developing in the

Figure 9.1 Mr Middle-Aged Geographer contemplates the possibilities of climatic determinism, the 'minor' truth of climax vegetation and the expense of a climate model. On the whole he's safer with good old climatology.

1950s and 1960s the revolutionized and quantified geography was ignored by atmospheric scientists because they had to focus their full attention on the difficult task of solving the equations of motion fast enough for forecasts to beat the real world events.

In the late 1960s and early 1970s there was another period of missed opportunity when meteorologists opted to overwrite operationally retrieved satellite data immediately after use. This decision was forced on them by computer storage costs but has resulted in the very late development of climate archives from satellite sources. When, fairly recently, these were recognized as crucially important, the demand arose *not* from the traditional climatologists but from the climate modelling community (e.g. Schiffer and Rossow, 1985). Thus by the 1980s 'traditional geographers' are almost entirely excluded from the climate community and spatial and temporal techniques of analysis are being re-invented by those who are the 'new climatologists' (e.g. Preisendorfer and Barnett, 1983; Rossow and Garder, 1984).

While geography may well have been ripe for (r)evolution in the 1960s (Chorley and Haggett, 1967; Gould, 1985a) climatology was not. Observational data for the vast majority of the earth (the oceans and cryosphere) had to await consistent, global coverage by satellites, which did not really occur until the mid-1970s, and the concept of climate as an 'ensemble mean of weather' could not become established until meteorological models had been shown to work. This mismatch is clearly illustrated in particular by a survey of references to climate topics in the meteorological literature. Figure 9.2 shows that the 'drought' in climate/climatology topics being considered by professional meteorologists is coincident with the quantitative and modelling revolutions in geography. Just at the time new-style geography took shape, atmospheric scientists became highly exclusive and introverted; needing to conquer short-term dynamic and radiative problems before re-opening the question of climate. Figure 9.2 also shows that my personal preconception about the terms 'climate' and 'climatology' was incorrect. The word climatology was never used extensively, but persists even in the rarefied atmosphere of the *Quarterly Journal of the Royal Meteorological Society*.

The second mismatch is in terminology. In Geography Departments, climatology seems to be to do exclusively, or at best primarily, with the atmosphere. As I sit through presentation after presentation in which the lower halves, or sometimes even the lower thirds only, of slides are described while dramatic cloudscapes, dust storms and precipitation are ignored, I often wonder if the psalmist who wrote, 'I will lift up mine eyes unto the hills' was a geographer. There does seem to be a rather strong sense of climate being predetermined and external to all the other systems which are studied. On the other hand, the Global Atmospheric Research Programme defined climate as being enormously wide (GARP, 1975) encompassing the atmosphere, hydrosphere, cryosphere, land surface and biomass (figure 9.3). Thus 'geography' becomes a subset of 'climatology' rather than the other way around.

The third, and perhaps most exasperating, example of bad timing is in the recognition of the importance of feedbacks. This concept was one of the essential ingredients of the introduction of systems theory into geography. The climatic science was, however, much less mature than the geomorphology of the 1960s. The definition of a 'climate sensitivity parameter' did not come until the early 1970s (e.g. Schneider and Dickinson, 1974) and the recognition of the potentially powerful synergism of two or more climatic feedback effects is very recent (Dickinson, 1985; Wigley and Schlesinger, 1985; Hansen et al., 1985). Two important areas of climatic feedback are the water vapour greenhouse feedback

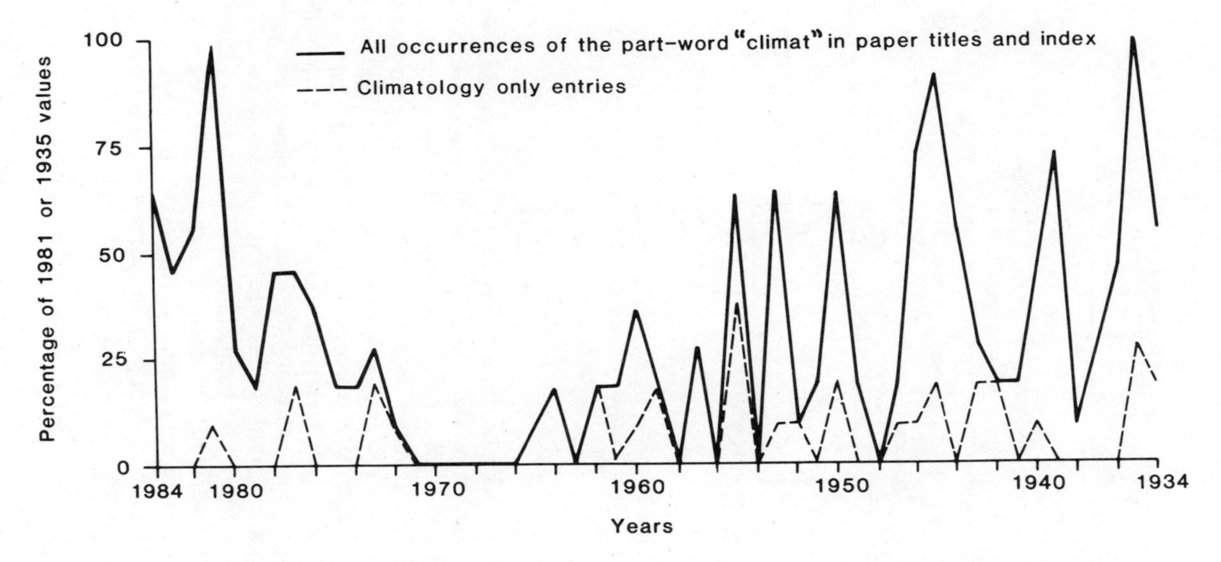

Figure 9.2 Entries in paper titles and the index of the *Quarterly Journal of the Royal Meteorological Society* including the part word 'climat' for the 50 years 1934–84. Also shown are entries including the word 'climatology'. All are given as a percentage of the 1981 and 1935 'climat' totals which are the same.

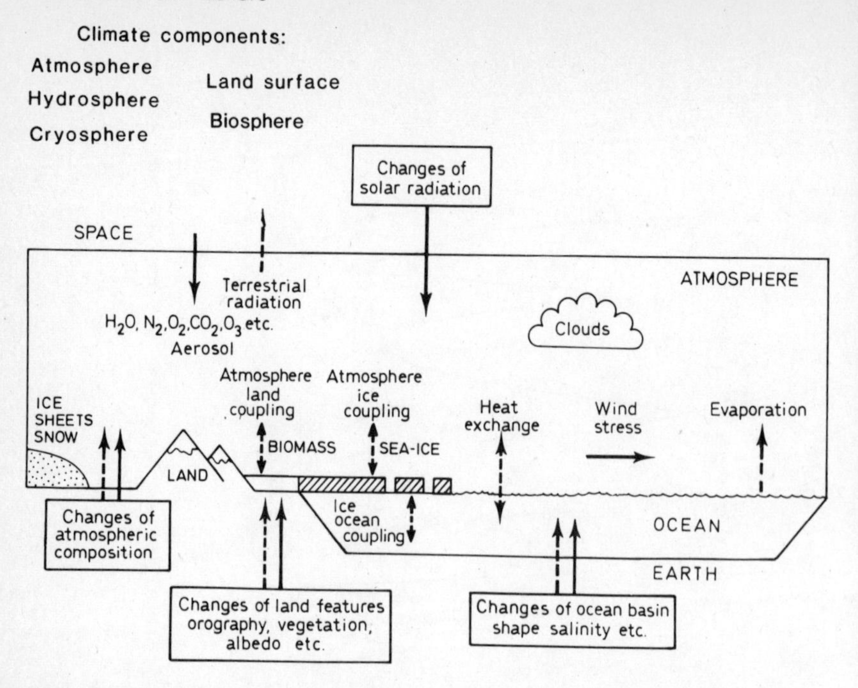

Figure 9.3 Schematic illustration of the components of the climate system. Full arrows are examples of external processes and dashed arrows are examples of internal processes (adapted from GARP, 1975).

(e.g. Ramanathan, 1981) and the cloud feedback (e.g. Ramanathan et al., 1983). If two processes produce feedback effects of opposite sign it is important that one process is not considered in the absence of the other. An example is the effect that clouds have on the radiative heating of the atmosphere; long-wave radiation causes a comparatively rapid cooling at the cloud top, whereas the absorption of solar radiation results in heating. To consider the effect of clouds on only one of the two radiation fields may be worse than neglecting the effect of clouds entirely. It may surprise geographers that diagrams such as figure 9.4 are only just becoming common in the climate literature.

In this chapter I am going to describe some geographically specific climate studies which, in my opinion, could and should be being undertaken by geographers. First, though, it's important to consider the history and current status of climate modelling.

Models, Modellers and Validation

Of the three types of model (analogue, physical and mathematical), the physical model is the least useful in climate studies because it is still

(a) FEEDBACK LOOPS

POSITIVE FEEDBACK

POSITIVE FEEDBACK

X → add Δ X

NEGATIVE FEEDBACK

X → add Δ X

Subtract Δ′X ← NEGATIVE FEEDBACK

(b) FACTORS ASSOCIATED WITH CLOUD FEEDBACK

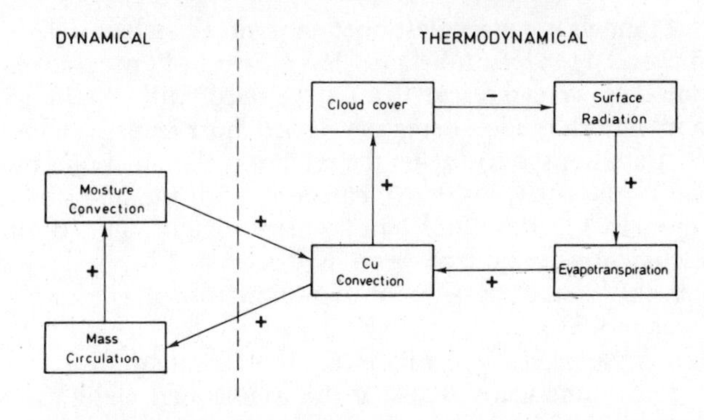

(c) TYPES OF CLOUD FEEDBACK

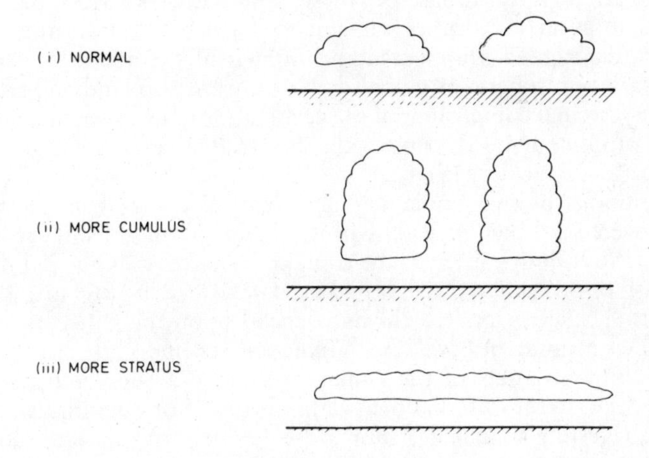

Figure 9.4 **(a)** Flow diagram illustrating positive and negative feedbacks. **(b)** Specific examples of dynamic and thermodynamic feedbacks and their directions in the case of a change in the amount of cumulus convection. **(c)** The exact nature of an increased amount of cloud is unclear. The cloud could either be more extensive vertically or more extensive horizontally (after Henderson-Sellers and Robinson, 1986).

impossible to induce a scaled-down atmosphere to hang onto a spinning model globe. This is not to decry the considerable usefulness of rotating annulus models (Hide, 1953; Frenzen, 1955; Fultz, 1961) and more recently, laboratory models of ocean circulation. The mathematical, more usually termed numerical, climate model is by far the dominant model type, although the exploitation of analogue models of climate could, and should in my view, be pursued more vigorously than is presently the case.

A 'Potted' History of Numerical Climate Modelling

As climate models are often described in terms of an hierarchy (figure 9.5 and Shine and Henderson-Sellers, 1983), it is often assumed that the simpler models were the first to be developed, with the more complex general circulation models being developed most recently. This is not the case. The first atmospheric general circulation climate models were being developed in the early 1960s concurrently with the first radiative–convective models. On the other hand, energy balance climate models, as they are currently recognized, were not described in the literature until 1969 and the first discussion of two-dimensional statistical dynamic models was in 1970.

The first atmospheric general circulation climate models were derived directly from numerical models of the atmosphere designed for short-term weather forecasting (Smagorinsky, 1983). These had been developed during the 1950s, and around 1960 ideas were being formulated for longer period integrations of these numerical weather prediction schemes. It is in fact rather difficult to identify the transition point. Scientists concerned with extending numerical prediction schemes to encompass hemispheric or global domains, were also studying the radiative and thermal equilibrium of the earth–atmosphere system, and these studies prompted the design of the first radiative–convective models (Manabe and Strickler, 1964).

Descriptions of two fundamentally identical energy balance climate models were published in 1969 within a couple of months of each other (Sellers, 1969; Budyko, 1969: the coincidence is between the publication in English of Budyko's model; it had been described in 1968 in a Russian publication). These models did not depend upon the concepts already established in numerical weather prediction schemes, but attempted to simulate the essentials of the climate system in a simpler way. These models drew upon observational data derived from descriptive climatology suggesting that major climatic zones are roughly latitudinal. A direct consequence of this intrinsically simpler parameterization was that these models could be applied to changes over a longer timescale than the atmospheric general circulation climate models of the time. It was work on energy balance models, in which the possibility of alternative stable climatic states for the earth were identified, which prompted much of the

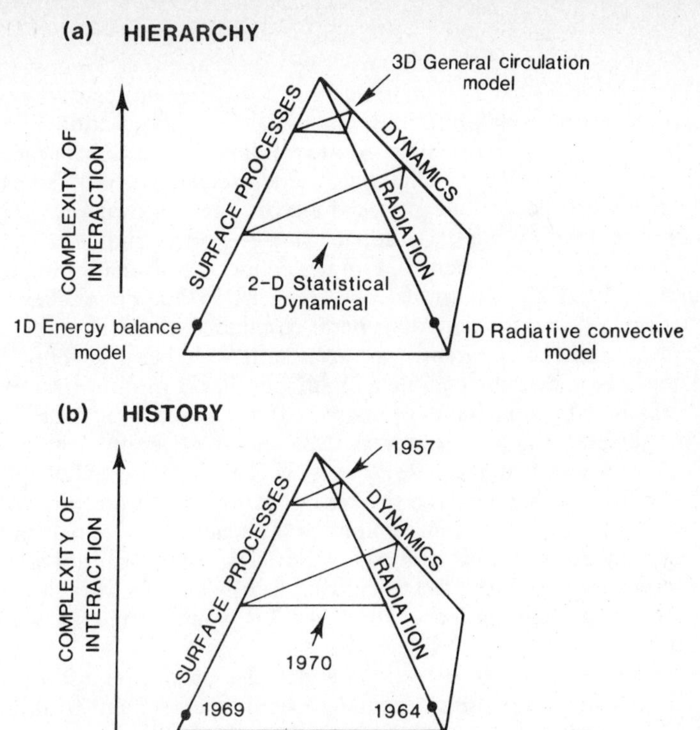

Figure 9.5 The climate modelling pyramid. The position of a model on the pyramid indicates the complexity with which the three primary processes interact. The base of the pyramid can be considered hollow since here there is essentially no interaction between the primary processes. Progression up the pyramid leads to greater interaction between each primary process. The vertical axis is not intended to be quantitative. **(a)** The positions of the four basic model types. **(b)** Particular climate models are seen to be based upon different methods of incorporating the primary processes and the level of complexity of the interactions.

interest in simulation of climatic change over geological timescales (e.g. North et al., 1981). Concurrently radiative–convective models, usually globally averaged, were being applied to questions of atmospheric disturbance including the impact of volcanic eruptions and the possible effects of increasing atmospheric CO_2 and other trace gases (e.g. Hansen et al., 1981).

The desire to improve numerical weather forecasting also prompted the fourth type of climate model: the statistical–dynamic model (e.g. Potter et al., 1981). A primary goal for dynamical climatologists was seen to be a need to account for the observed state of averaged atmospheric motion, temperature and moisture on timescales shorter than seasonal, but longer than those characteristic of depression systems. One group of climate modellers preferred to design relatively simple low-resolution statistical dynamical models to be used to illuminate the nature of the interaction between forced stationary long waves and travelling weather systems (e.g. Green, 1970). Theoretical study of large-scale atmospheric eddies and their transfer properties, combined with observational work, led to the parameterizations employed in two-dimensional climate models.

By 1980 this diverse range of climate models seemed to be in danger of being overshadowed by one dominant type: the atmospheric general circulation climate model. Although single-minded individuals persevered with the development of simpler models, considerable funding and almost all the computational power used by climate modellers was being consumed by atmospheric general circulation climate models. Six years later a series of apparently correct results were being generated for the wrong reason by these complex, highly non-linear models, prompting many modelling groups to move backward, hierarchically, in order to try to isolate the essential processes responsible for their results, which were observed from more comprehensive models. The desire to make climate models more realistic has led to the incorporation of novel topics within the framework of climate modelling and, hence, to the realization, however unwelcome, that within a subdiscipline one cannot assume constancy in the variables prescribed from other domains. It is clear that the strategy of utilizing a hierarchy of models is more than sound, it is essential.

Since the climate system depends upon scales of motion and interactions ranging in scale from molecular to the planetary, and from timescales of nanoseconds to geological eras, parameterizations, whereby processes that cannot be treated explicitly are instead related to variables that are considered directly in the model, are a necessary part of the modelling process (Saltzman, 1978). In particular a decision is made generally very early in model construction about the range of spatial and temporal scales upon which the model will be focused. Figure 9.6 illustrates the difficulty faced by all modellers. The constraints of computer time and costs and data availability restrict the prognostic (or predictive) mode. Outside this range there are 'frozen' boundary conditions and 'random variability'. Thus the two examples shown in figure 9.6 illustrate the range of prognostic computations for (1) a medium-range weather forecasting model and (2) an energy balance model being used to examine the effect of Milankovitch variations on climate. In both cases longer timescales than those of concern to the modeller are considered as invariant, and shorter timescales are neglected as being random statistical fluctuations, the details of which are of too short a period to be of interest.

The Case for Analogue Climate Modelling

Carefully devised analogue models could offer a useful alternative to numerical climate modelling. Here there is room for stout-hearted new-

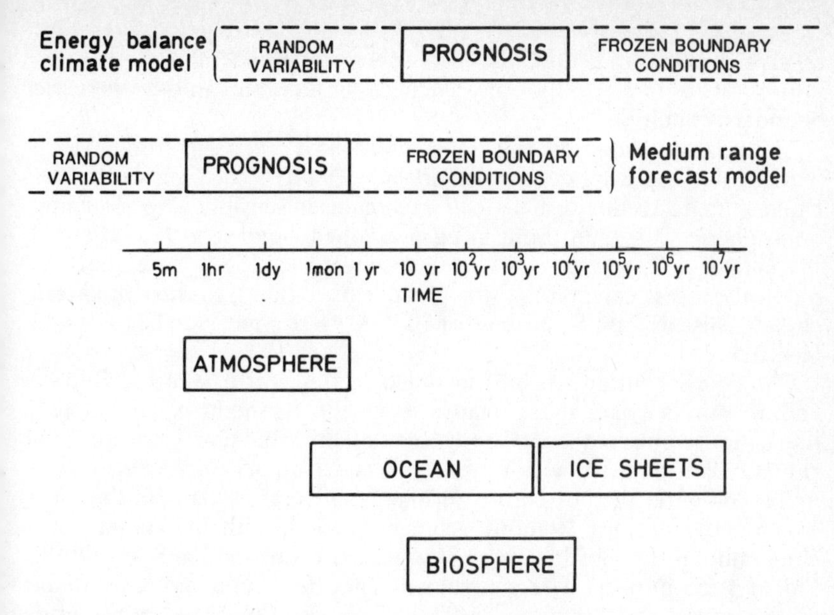

Figure 9.6. The importance of different temporal scales changes as a function of the type of model. The domain in which the model simulates the behaviour of the system is labelled 'prognosis'. It is expected that processes which fluctuate very rapidly compared with the prognostic timescales will contribute only small random variability to the model predictions, while processes which fluctuate very slowly compared with the prognostic timescale can be assumed to be constant. Two types of model are shown: an energy balance model and a medium-term weather forecast model (after Henderson-Sellers and McGuffie, 1987).

comers because the major effort in climate modelling over the past twenty years has been focused almost exclusively upon numerical modelling. I shall try to illustrate the potential of analogue models with a specific example which relates to an important climatological issue; the increasing levels of carbon dioxide and other trace gases in the atmosphere (NAS, 1982; WMO, 1986) and to a currently intractable problem of prediction for numerical models; the cloud/climate feedback effect (Stephens and Webster, 1979; Ohring and Clapp, 1980; Cess et al., 1982; Ramanathan et al., 1983; Webster and Stephens, 1984; Wilson and Mitchell, 1986).

Analysis of the instrumental record is restricted to the past 100–150 years (e.g. Wigley et al., 1980; Kellogg and Schware, 1981; Lough et al., 1983; Palutikof et al., 1984). A disadvantage of this instrumental method is that this period offers relatively small hemispherical temperature changes compared with those expected to result ultimately (i.e. in an equilibrium change) from future increases in atmospheric greenhouse gas concentrations. Thus instrumental scenarios should be taken as being indicative of changes and conditions to be anticipated during the early phase of CO_2-induced warming. On the other hand, the gradual, but steady, rise in atmospheric CO_2 levels will cause the climate to respond relatively slowly and, thus, the historical analogue method may, fortuitously, incorporate important changes in oceanic and cryospheric boundary conditions.

The relatively few investigations which have been undertaken with numerical climate models which include cloud prediction and hence cloud climate feedback find a greater climate sensitivity to doubling atmospheric CO_2 than those using prescribed cloud (e.g. Manabe and Wetherald, 1975). For example, Hansen et al. (1984) note that the physical process contributing the greatest uncertainty to their predicted climate sensitivity on a timescale of 10–100 years appears to be the cloud feedback.

Cloud observations are not included in the World Weather Record and, thus, other data sources must be sought. As might be anticipated, the quantity (but not necessarily the quality) and areal coverage of climatic data increase with time (e.g. Henderson-Sellers, 1986). Figure 9.7(a) compares the northern hemisphere temperature data, in the form of anomalies from a 1946–60 reference period, with the annual total cloud amount for northwestern Europe, the continental U.S.A. and the Indian subcontinent. The general tendency for cloud amounts to be greater in the warmer period (1934–53) is also illustrated in the inset plots in figure 9.7(b). The largest increase in cloud amount is seen in the U.S.A. The mean annual differences (warm minus cold) in cloud amount for the three regions studied are shown in figure 9.7(c). Cloud amount increases by between 0.3 and 1.0 tenths of sky cover over practically the entire U.S.A. and most of the Indian subcontinent. Over maritime northwestern Europe the increases are ~ 0.3 tenths of skycover.

These historical analogue results are in contrast to the few numerical model predictions of cloud changes in warming world experiments. Models have predicted decreases in total cloud amount of around 0.3 tenths of skycover in doubled CO_2 experiments (e.g. Hansen et al., 1984; Wetherald and Manabe, 1986). A possible, rather tantalizing, conclusion is that current general circulation climate model cloud prediction schemes tend to enhance temperature increases through cloud–climate feedback, whereas the historical data may suggest a negative feedback. The responses to these tentative suggestions in anonymous reviews by numerical climate modellers have been both hostile and revealing. They have stated, for example, 'no-one has any confidence in model-generated cloud changes', 'cloud parameterizations are perhaps the weakest aspect of current GCMs' and 'as is well recognized current GCMs have no regional prediction skill'. These statements are, indeed, accepted and acceptable within *numerical modelling* circles, but the rest of the scientific community, and certainly the man-in-the-street, might be forgiven for mistaking studies devoted to 'analysis of the Sahel drought', 'predictions of water resources in CO_2-enhanced climates' and 'reconstructions of the climate 18000 years BP', all of which utilize general circulation climate models, for being regionally specific. It is certainly hard in any of the very large suite of papers describing the results of numerical climate models to find the statement that, whilst being regionally specific, these studies have 'no regional prediction skill'.

There is, unfortunately, a hesitancy among many climate modellers about utilizing surface-retrieved cloud information. This reluctance is understandable in one sense, since surface observations of cloud suffer from many limitations. On the other hand, the level of agreement between trained observers is also better than the current level of agreement amongst a range of satellite-based cloud retrieval algorithms (cf. Rossow et al., 1985) and most modelling groups are quite prepared to display global maps of predicted cloud and compare these with surface observations of 'climatological' cloudiness. (Washington and Meehl (1984) use London (1957), Hansen et al. (1983) the collection by Berry et al. (1973), Potter and Gates (1984) use Berlyand and Strokina (1980) and Wilson and Mitchell (1986) choose observational data from Meleshko and Wetherald (1981) which are, in fact, a combination of an earlier version of Berlyand and Strokina data and the southern hemisphere data of Van Loon et al. (1972)).

Radiatively correct 'effective clouds' retrieved from satellite data can naturally be expected to produce radiatively correct fluxes, but validation of numerical climate models must not depend solely on such intercomparisons. An alternative and complementary strategy is to devise a method of computing 'surface-observed' clouds from numerical models. (This is actually no more difficult than attempting to compute 'satellite-observed' clouds from multi-layer spatially homogeneous model-predicted clouds). These cloud characteristics could be compared with current surface observations of clouds, including cloud type, amount, structure and possibly even basic optical properties. A preliminary example of the potential use of surface-observed cloudiness in the interpretation of radiative fluxes and climate model predictions was made by Ohring and Gruber (1983), when they introduced the concept of locationally specific climatologies determined from satellite data. More recently, inherently geographical studies based on satellite retrievals have become quite commonplace in the meteorological literature (e.g. Minnis and Harrison, 1984; Smith, 1986; Meisner and Arkin, 1987). There is every reason to believe that comparison of regional and seasonal cloudiness characteristics could aid the improvement of cloud prediction schemes in numerical models. In such a comparison and validation exercise spatial and temporal analysis and display techniques are of paramount importance. Here, in my opinion, is a genuinely geographical problem.

Regionally Specific Studies of the Land Surface: Monitoring and Modelling

The past decade has seen a general growth in the requirement for land-surface information within the realm of climate studies. The recog-

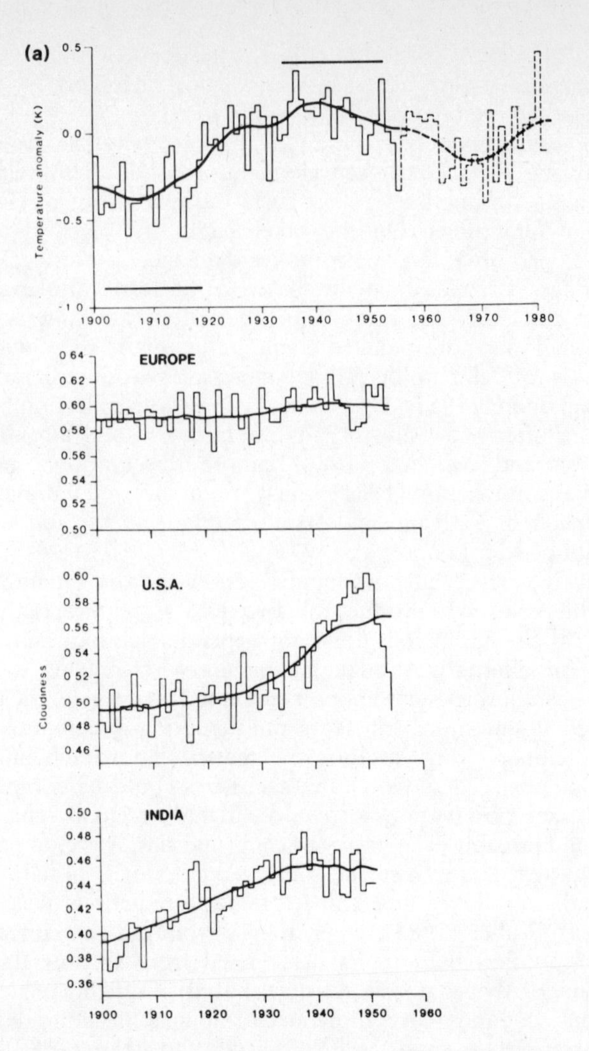

Figure 9.7 **(a)** Northern hemisphere mean surface air temperature variations (K) since 1900 shown as anomalies from 1946–60 reference period. The curve is of twenty-year filtered values. The warm and cold periods used here are marked (redrawn **Lough et al.**, 1983). Northwest Europe mean annual cloud amount (fifty-eight-station average) continental U.S.A. mean cloud amount (seventy-seven stations) and Indian subcontinent mean cloud amount (sixty stations) all for the period 1900–54. The curves are of twenty-year filtered values. **(b)** Location of the 195 stations used with (inset) examples of the seasonal regimes in cloudiness encompassed: Salt Lake City (40° 46′N, 111° 58′W), Aberdeen (57° 08′N, 2° 08′W), Port Said (31° 16′N, 32° 19′E) and Dwarka (22° 22′N, 69° 05′E). The twenty-year means and the extreme values for each month are shown. In each the warm period (1934–53) cloud amounts tend to be greater than those of the cold period (1901–20). **(c)** Annual differences in tenths of sky cover of total cloud amount predicted from the historical analogue study.

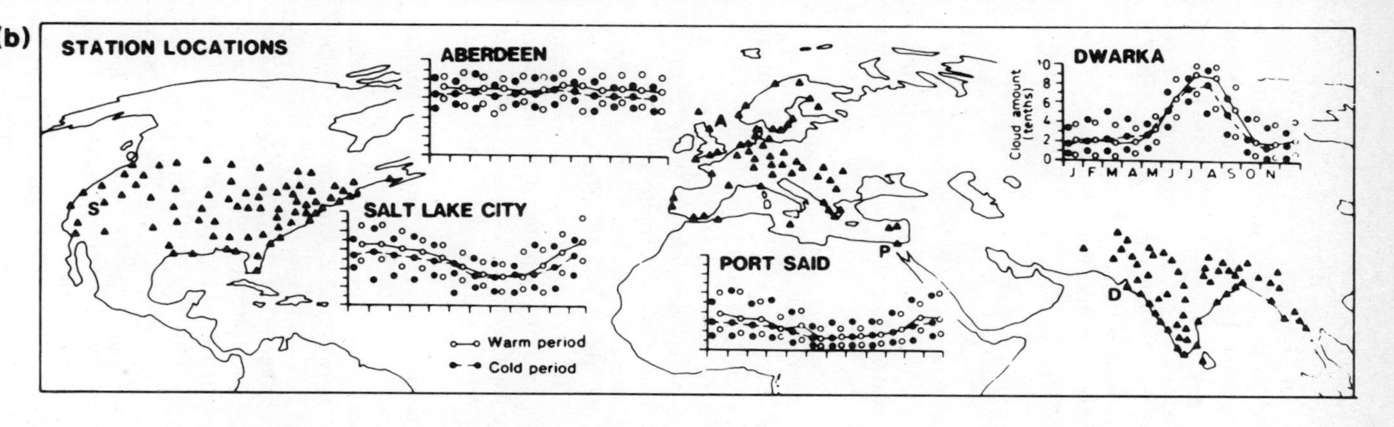
(b)
STATION LOCATIONS
ABERDEEN
SALT LAKE CITY
PORT SAID
DWARKA
Cloud amount (tenths)
J F M A M J J A S O N
Warm period
Cold period
S
A
P
D

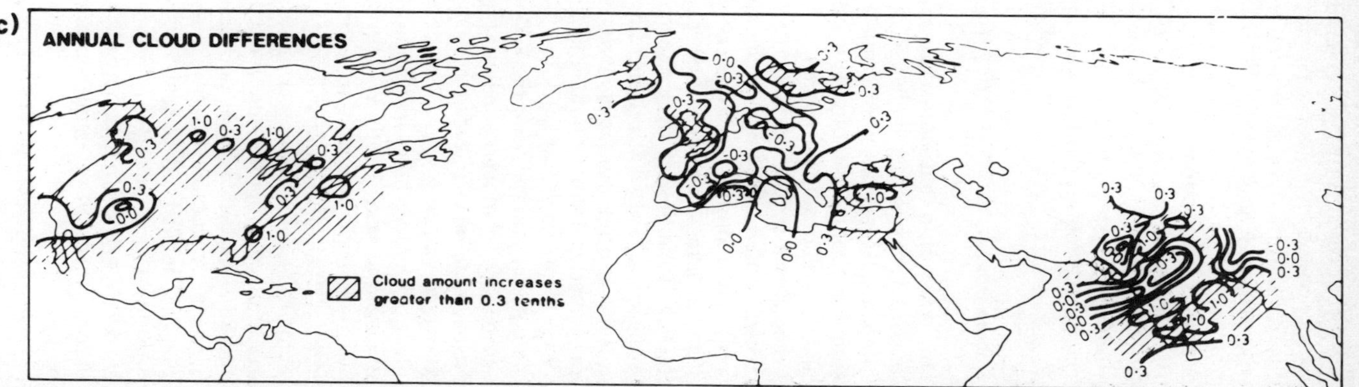
(c)
ANNUAL CLOUD DIFFERENCES
Cloud amount increases greater than 0.3 tenths

nition that the land-surface is a crucial, temporally and spatially varying component of the climate system has provided the impetus for research into climate modelling, monitoring and impact studies focused at the land-surface (e.g. Carson, 1982; Dickinson, 1984; World Climate Programme, 1985). The variety and relative rapidity of current anthropogenic modifications to the earth's surface, for example tropical deforestation and desertification of semi-arid regions (e.g. Sagan et al., 1979; Dregne, 1983), and the inherent vulnerability of agricultural production systems to climatic variations (Oram, 1985), present problems which require both modelling and monitoring approaches.

The 'Global Vegetation Index' (GVI)

The recognition that land-surface information could be determined from satellite data but that high-resolution retrievals such as those from the Landsat satellites did not permit adequate spatial coverage or temporal sampling prompted the idea of utilizing the meteorological satellites as land-surveillance instruments.

The global vegetation index (GVI) is constructed from the normalized spectral difference between channel 2 and channel 1 of the advanced very-high-resolution radiometers (AVHRR) carried onboard the NOAA polar orbiting satellites. The resulting normalized vegetation index (NDVI) is maximized over a seven-day period by simple substitution to produce the GVI. This product has been extolled by many and proposed as a monitoring tool, as an early warning characteristic and as an input for numerical climate models (e.g. Murphy, 1986; Tucker and Sellers, 1986; Henricksen and Durkin, 1986). It seems worthwhile, therefore, to evaluate the suitability of the GVI as an input of land cover characteristics in global climate modelling. Two aspects of the GVI were considered: the necessary maximization time period and the regional compatibility of recovered values.

A number of studies have indicated that a three- or four-week compositing of the basic weekly GVI product reduces the level of contamination by clouds, snow and scan-angle effects (*Users' Guide to Global Vegetation Index*, 1983; Justice et al., 1985). However, the time period selected seems to be somewhat arbitrary and it is not entirely clear what percentage of pixels remain contaminated at the end of three or four weeks. In addition, extended time period compositing may lead to confusion between vegetation activity and removal of contamination such as clouds. The GVI has been shown to exhibit considerable seasonal variations for particular vegetation formations (Tucker et al., 1983; Justice et al., 1985) and consequently vegetation-induced changes in the GVI may become obscured when compositing is carried out over an extended time period.

To investigate these problems further, the effect was calculated of increasing the compositing period on the number of GVI pixel values

Table 9.1 Impact of time period and length of compositing on the number of pixels replaced in successive composites

Period	Study region					
	North America (47°–67°N, 95°–115°W)		South America (3°–23°S, 40°–60°W)		Africa (0°–20°N, 10°–30°E)	
	No.	Percentage	No.	Percentage	No.	Percentage
May 17–23, 1982	–	–	–	–	–	–
May 24–30, 1982	167	41.75	358	89.50	280	70.0
May 31–June 06, 1982	161	40.25	47	11.75	96	24.0
June 07–13, 1982	367	91.75	57	14.25	113	28.25
July 19–25, 1982	–	–	–	–	–	–
July 26–Aug 01, 1982	140	35.00	281	70.25	283	70.75
Aug. 02–08, 1982	2	0.50	203	50.75	264	66.0
Aug. 09–15, 1982	0	0.0	185	46.25	140	35.0
Aug. 16–22, 1982	3	0.75	98	24.50	132	33.0
Aug. 23–29, 1982	155	38.75	19	4.75	214	53.5
Oct. 18–24, 1982	–	–	–	–	–	–
Oct. 25–31, 1982	383	95.75	121	30.25	103	25.75
Nov. 01–07, 1982	387	96.75	214	53.5	222	55.5
Nov. 08–14, 1982	56	14.00	44	11.0	108	27.0
Nov. 15–21, 1982	13	3.25	15	3.75	216	54.0
Nov. 22–28, 1982	0	0.0	158	39.5	53	13.25
Dec. 29–05, 1982	0	0.0	186	46.5	34	8.5

'maximized', for three study regions and over three different time periods (seventeen weeks in all) (table 9.1). In view of the general acceptance of a three- or four-week compositing period it is surprising to note the relatively high percentage of pixels still being maximized even after as much as five weeks' compositing.

Interpretation of this continuing alteration in pixel values of the GVI is difficult. As photosynthetic activity in a particular region increases (decreases), so the percentage of pixels exhibiting a change in clear-sky GVI value should increase (decrease). Thus, the continued increase in the number of pixels being maximized after the standard compositing period could be taken as an indication of increasing biophysical activity (although this can be confirmed only from data on ground truth). All methods of classification are likely to be less successful in marginal areas where vegetation is stressed or dying and/or bare soil composes part of the pixel area. Where the vegetative cover is green but discontinuous, the relationship between the GVI and the land cover is weakened, leading to problems in interpretation (Harris, 1986). These difficulties were underlined by Sellers (1985: 1365 and 1366): 'the presence of even a small proportion of bare ground [as opposed to an even distribution of vegetation] may seriously complicate the interpretation of multispectral data' and 'the presence of even a small fraction of dead leaves in the canopy would appear to reduce [the] vegetation index . . . drastically'.

Variations in the GVI resulting from contamination effects (e.g. Wardley, 1984) may be of a comparable if not greater magnitude than those arising from variations in photosynthetic activity both within and between species. Extended period compositing to remove cloud contamination must inevitably confuse a number of these effects. In the absence of data on ground truth and adequate correction techniques, composite GVI maps *for the globe*, such as those shown in Justice et al. (1985), can be little more than qualitative indicators of biophysical activity.

These results suggest that the selection of a suitable temporal compositing period is not so straightforward as is commonly assumed, being dependent upon both spatial location and the time of year. Unless considerable care is taken, noise removal will begin to remove the signal too.

An indication of the problems of generalizing globally on the basis of regional evaluations is revealed by figure 9.8, which illustrates the range in typical GVI values for different vegetation types. It should be remembered that these values have been selected as being representative of as broadly homogeneous regions as possible. Immediately apparent is not only the seasonal range in mean GVI values for particular vegetation types, but also the considerable range of values for particular vegetation types at a particular time period (figure 9.8). This must lead to problems of accurate classification, although these might be ameliorated by inclusion of further periods of data. It is also apparent that the interpretation of particular values of GVI is regionally specific, since different GVI values occur for the same vegetation types at different locations. If the characteristic spectral signatures of specific vegetation types are not transferable then extrapolation from the regional to the continental scale and to the global domain seems likely to be at worst impossible and, at best, highly ambiguous. Thus while the GVI may well capture seasonality, it is very difficult to incorporate this feature into a global model because the characterization is not global.

All of these considerations are in essence inherently geographical. The classification techniques and the comparisons required use geographic methods and data, and yet there seems to be poor understanding of spatial concepts among many of those generating and using these data. For example, the GVI data were originally supplied in the form of two hemispheric (1024 × 1024 pixels) polar stereographic arrays; a format closely related to the original orbital data (cf. Fye, 1978). These mapped data have a resolution of 15 km at the equator, decreasing to 25 km at 60°N. Recently the archive has begun to be re-mapped onto a Mercator projection (Ohring, 1985, personal communication). Finally, since it seems very likely that incorporation of fully interactive land-surface sub-models into general circulation climate models will increase their sensitivity (Dickinson, 1984; Wilson et al., 1987), the GVI in its present form would increase model variability due to its own inherent variability.

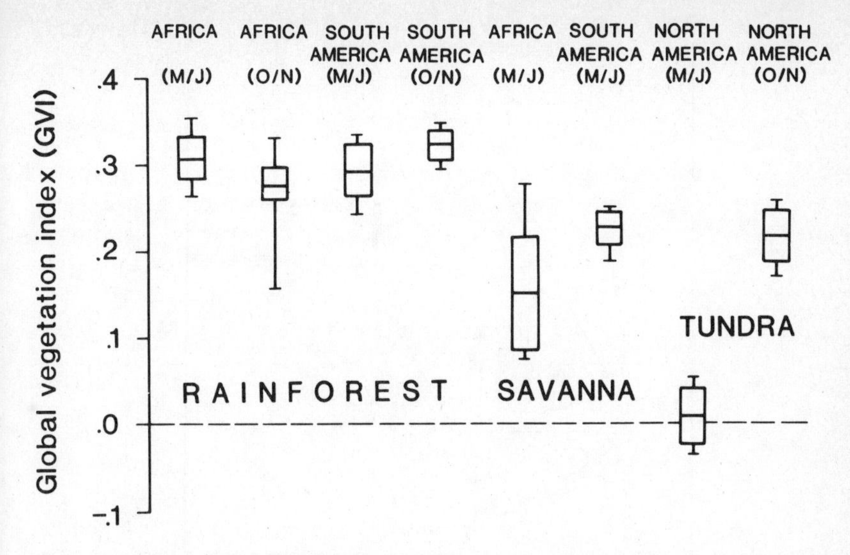

Figure 9.8 Box-whisker diagrams showing the mean (centre of box), interquartile range (box top and base) and extremes of derived seasonal GVI values. All values are four-week composites, 1982. The variability is too large to make *global* ecotype classification a viable possibility.

Land Surface Parameterization in Climate Models

The history of land-surface parameterization in climate models is rather short and not especially glorious. In very early numerical schemes the land surface was simply a reflector of solar radiation and an emitter of infrared radiation. The first hydrological parameterization scheme, which was employed by Manabe (1969), has been termed the 'bucket model'. In this highly simplified scheme the 'soil' had a 'field capacity' of 15 cm. The bucket filled with water when precipitation exceeded evaporation and, if the bucket became full, the overflow was termed runoff. Subsequently an up-market version of the bucket model was developed, sometimes termed the Budyko model, which drew upon climatological relationships between daily averaged evapotranspiration and daily averaged potential evapotranspiration (e.g. Penman, 1948; Thornthwaite and Mather, 1955; Budyko, 1974). In these models (figure 9.9(a)) evapotranspiration occurred at its potential rate when the soil was at or close to saturation. If soil moisture dropped below some critical value, the actual evapotranspiration was given as a proportion of potential evapotranspiration; the proportionality factor, β, being set equal to the ratio of the current soil moisture to the critical soil moisture.

Unfortunately the empirical results upon which the β parameterization is based are diurnally averaged data. This parameterization scheme is, therefore, useful in general circulation climate models, which use diura

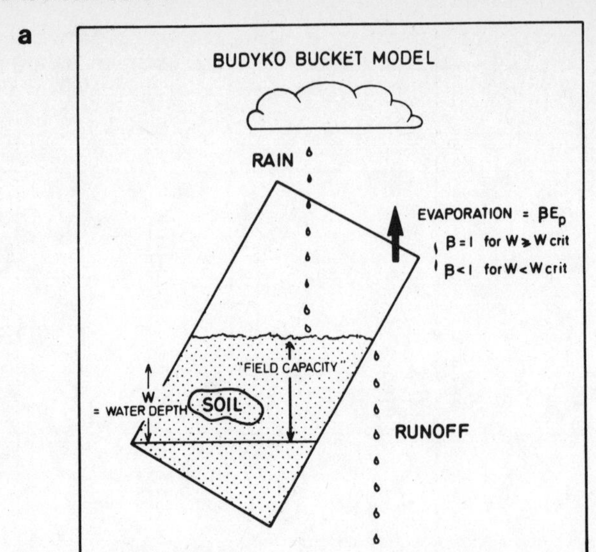

b

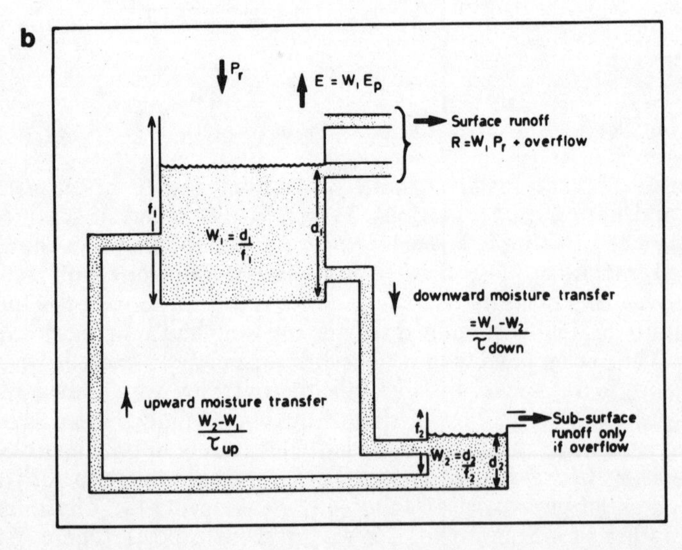

Figure 9.9 Land surface parameterization schemes currently used in global climate models: **(a)** bucket, **(b)** two soil layer, **(c)** (biosphere/atmosphere transfer scheme) BATS.

nally averaged solar heating, but is inappropriate for any climate model which incorporates the diurnal variation in solar radiation at the surface. Some improved variants of the β parameterization, or Budyko scheme, have included a second, lower, soil layer (e.g. Hansen et al., 1983 and figure 9.9(b)). Hunt (1985) reviewed single- and multi-layer soil parac

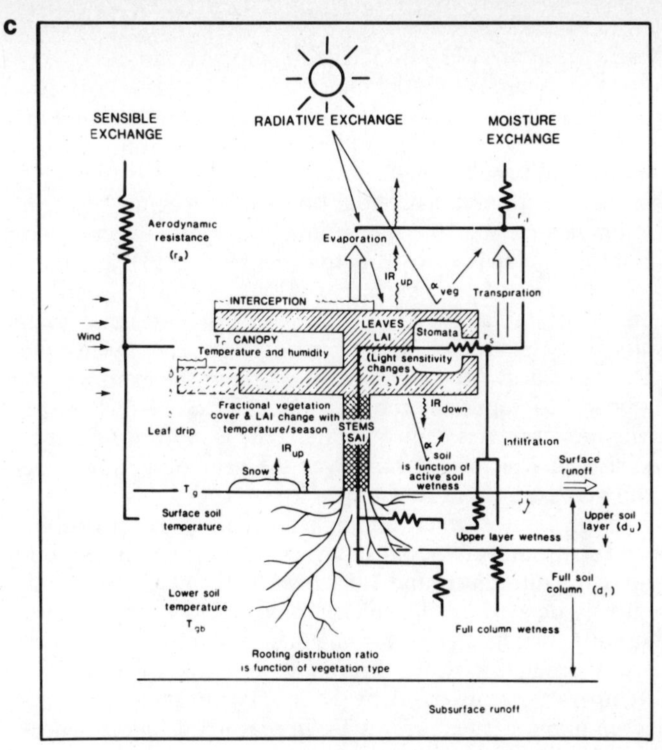

meterization schemes for climate modelling, considering in particular their applicability to drought-prone areas, and found a two-layer model preferable. Once the diurnal cycle is, however, included it is essential that realistic descriptions of the sensible and latent heat fluxes be made at each time step and that explicit, separate computation is made of the foliage energy and water budgets. To this end, Dickinson (1984) developed a solution to the soil water parameterization based upon comparison with a much more detailed, 100-layer, soil model. This scheme includes in its representation the diffusion limitation of evaporation which often occurs around midday. Dickinson (1984) also deals separately with interception of precipitation (liquid and solid) by vegetation and subsequent re-evaporative losses and leaf drip, with moisture uptake by plant roots, which can be differently distributed amongst the active and full soil columns, and stomatal resistance to transpiration. The processes incorporated in the scheme are shown in figure 9.9(c). The sensitivity of the complete land-surface parameterization scheme has been described by M. F. Wilson et al. (1987). In common with the 'simple biospheric model' of Sellers et al. (1986) the biosphere/atmosphere transfer scheme (BATS) permits simulation of ecotypes ranging from bare ground to tropical forest.

The sensitivity of these land-surface parameterizations currently incorporated into general circulation climate models can be examined by integrating a land-surface model independently of other features. External features must be prescribed. Figure 9.10 shows this forcing for the example described here: solar radiation with a sinusoidal variation during the day and zero at night except for a diurnally averaged model which has mean diurnal (twenty-four hour) solar forcing (figure 9.10(a)), precipitation at a rate of 10^{-6} m s^{-1} during every tenth time step for the duration of the time step, thirty minutes (figure 9.10(b)), and the daily air temperature (figure 9.10(c)) such that the mean is 300K with a diurnal range of 6K. All simulations are for fifty days, the last ten days and final day results being examined and, generally, the last twenty-four hour cycle being described. All the simulations are, in the examples discussed here, supposed to represent moist tropical rainforest. The prescribed rainfall regime simulates approximately the conditions in the central Amazon basin; 3154 mm of rain over the year occurring in relatively intense showers equivalent to 1.8 mm h^{-1}.

The resulting soil surface temperatures differ considerably (figure 9.11). The Dickinson soil model and the two-layer soil model both show maximum temperatures around 1400 h while the peak in the soil surface temperature is delayed in the bucket model and even further delayed when diurnally averaged solar forcing is imposed on it. The soil surface temperatures predicted by the BATS model agree fairly well with the canopy temperatures observed by Shuttleworth et al. (1985). All the night-time minimum temperatures are in reasonable agreement with one another but the peak day-time temperatures vary considerably and the diurnal soil surface temperature ranges in the five models are 3.3K, 14.4K, 12.7K, 5.7K, 2.0K.

Land-surface parameterization schemes such as these are being developed, of necessity, by climate modellers. Any such scheme for use in a general circulation climate model must consider grid elements, usually $\sim 3° \times \sim 7°$ in size as single entities. Ecologists and biogeographers, who might be expected to have a better grasp of the problems and, hence, the parameterizations to be employed, are naturally somewhat hesitant about the concept of such a 'big (500 km × 500 km) leaf model'. They could be encouraged, for example, by figure 9.12, which shows all the BATS surface energy fluxes. Note the repartition of sensible into latent heat during and immediately following the showers. Evaporation of intercepted precipitation causes latent fluxes greater than the absorbed solar energy. The increase is compensated for by increased negative fluxes of sensible heat. Observations made by Shuttleworth et al. (1985) in Amazonia show that total evaporation from a fully wetted canopy often exceeds 'potential evaporation', the additional energy being obtained from air drawn into the wetted area from adjacent dry canopy regions.

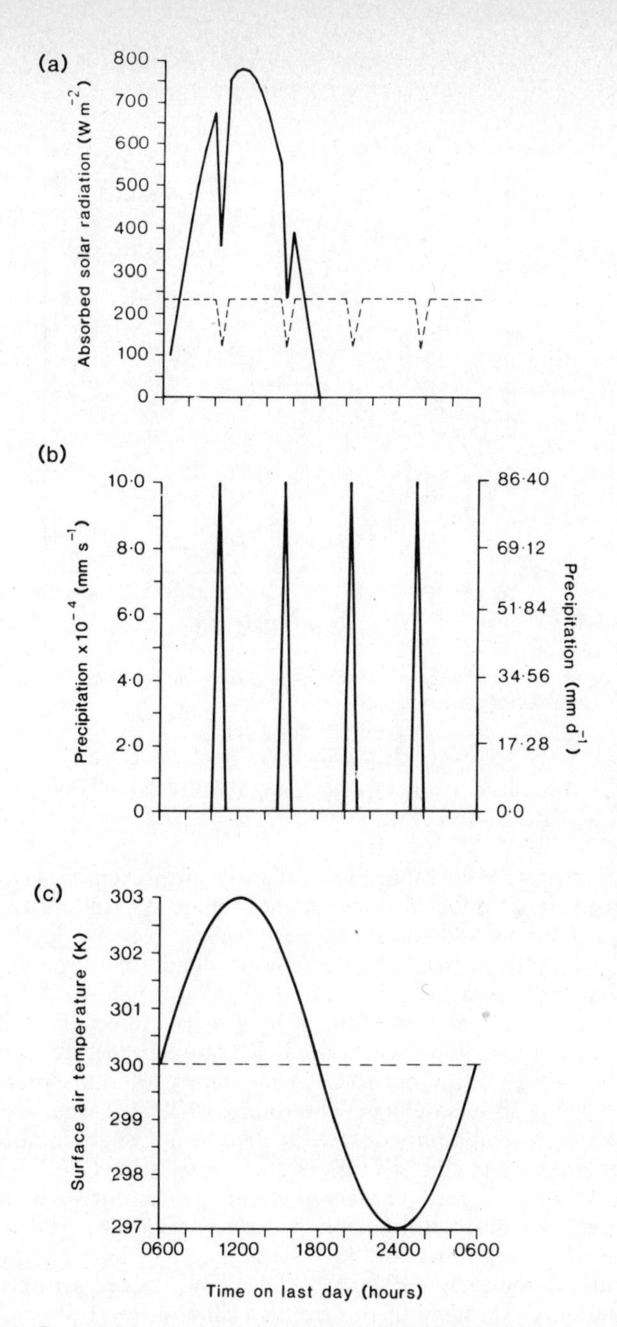

Figure 9.10 Prescribed parameters imposed to mimic the conditions forcing the land surface in a tropical forest for simulation sensitivity tests. **(a)** Absorbed solar radiation, **(b)** precipitation and **(c)** air temperature (dashed lines for diurnally averaged bucket scheme).

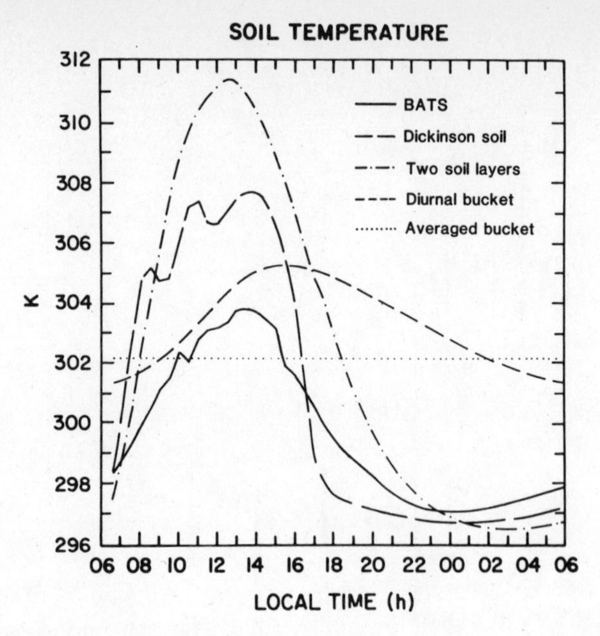

Figure 9.11 Soil surface temperatures simulated by a variety of current land-surface schemes for tropical forest conditions

Global Climate Information and Geography: Do they Have a Future Together?

There is currently a wide range of investigative projects in progress which are designed to aid understanding of the climate system and to provide data for validation and improved parameterization of global climate models. Two major international programmes not only focus upon, but are also named, 'climatological'.

The International Satellite Cloud *Climatology* Project (ISCCP) began formally on 1 July 1983 (Schiffer and Rossow, 1985). The goal is to produce a five-year, homogeneous data archive of the world's cloud cover as viewed from satellites. The timing of ISCCP was designed to coincide with the availability of satellite data from five geostationary and two polar-orbiting satellites. This satellite array is likely to be the best data source from the point of view of cloud retrieval for some time. The International Satellite Land Surface *Climatology* Project (ISLSCP) was begun in 1985. Its aims are to coordinate a series of experiments culminating, in the early 1990s, in a closely interlocked set of measurement programmes ranging in size from satellite retrieval of land-surface features at a scale of tens of metres up to parameter measurement on scales equivalent to the grid scale of general circulation climate models (figure 9.13). Such studies are intrinsically geographical (cf. Lawton,

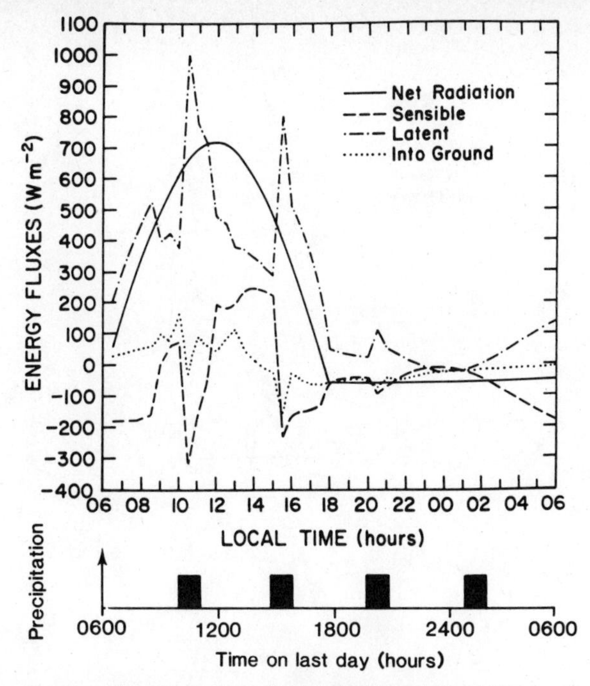

Figure 9.12 BATS surface fluxes from the tropical forest simulation

1983) and yet few, if any, geographers are contributing to the planning and analysis.

These and other programmes have regionally specific components (e.g. the monsoon experiment (MONEX) and the Tropical Ocean Global Atmosphere (TOGA) programme prompted by the recognition through climatological observations of the El Niño/Southern Oscillation (ENSO) phenomenon) and yet there are few, if any, geographers involved in observations, analysis or prediction. In a recent, special issue of the *International Journal of Remote Sensing* (volume 7, number 11, 1986) devoted to monitoring the grasslands of semi-arid Africa, only one of the thirty authors was affiliated to a Geography Department.

Where are the Geographers?

One effect of the quantitative revolution was that geographers became very enthusiastic about making measurements against which to test hypotheses. In the pre-satellite era the absence of macroclimatological measuring tools forced many physical geographers into subjects like small-basin hydrology, fluvial geomorphology and microclimatology. Furthermore, this revolution was, for many geographers, a statistical revolution. It may be possible to grasp the essence of statistics without

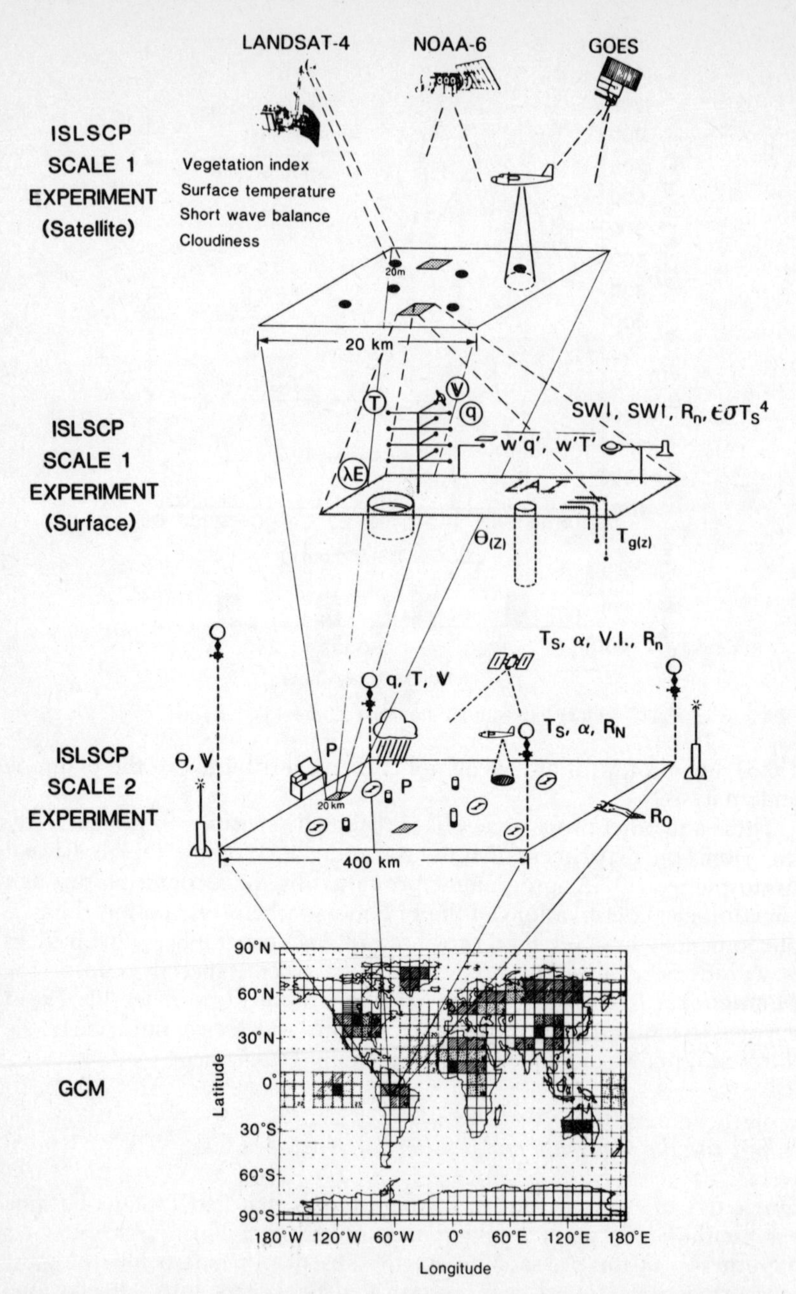

Figure 9.13 The variety of scales of observation envisaged as forming integral parts of the International Satellite Land Surface Climatology Project. Measurement at all of these scales of resolution is required in order to determine land-surface characteristics for general circulation model studies.

mathematical training, but modelling is an intrinsically theoretical activity; the tools are mathematical and, more importantly perhaps, computational. Hare (1977: 263) commented on the poverty of geographical training: in recent years we have 'swept geography departments into the social-science divisions of faculties of arts and sciences where, from playing second fiddle to geologists or literary critics, we learned to play second fiddle to economists and sociologists'.

The reluctance of geographers to become involved in modern climatology seems to be due, in part, to the belief that the physical world is mechanistic and that mathematical theories designed to capture its character cannot ever be used to represent the 'inherently unpredictable' (usually human) aspects of geography (cf. Johnston, 1986a; Goudie, 1986). Often atmospheric science is used as the supreme example of this assertion, the argument being that in a physically based science such as meteorology it is 'simply' a question of programming large computers to solve tens of thousands of equations in order to predict the state of the atmosphere tomorrow from the observed state today.

Inasmuch as weather forecast models are successful, this is a reasonable assertion, but it denies the random or probabilistic element. Climate is so highly non-linear with many processes interacting on different temporal and spatial scales that current research into chaos and transitivity resemble many aspects of social/human studies. 'Chaos' is a deterministically produced process that appears random, but is often not completely so (i.e. periodic variations can sometimes be detected) (e.g. Holden, 1986). 'Intransitivity' is the term used to describe the behaviour of a system which possesses two, or more, sets of long-term statistical properties, the choice of set being determined solely by the initial state. Simplified systems can be shown to exhibit different chaotic or intransitive solutions depending only on small changes in the equations describing the system. These studies are directly relevant to the discussion of climate predictability and sensitivity.

Fifty years ago the meteorological literature contained a considerable proportion of climatological and specifically geographical works. For example, in the *Quarterly Journal of the Royal Meteorological Society* can be found 'A cartographic study of drought' (Baldwin-Wiseman, 1934), 'The diurnal range of temperature and its geographical distribution' (Ashmore, 1939), 'The seasonal and geographical distribution of absolute drought in England' (Lewis, 1939) and 'The variability in Indian rainfall and the growth of population' (Whipple, 1942). Still more diverse subjects include 'Some human reactions to the great aurora of January 25–26, 1938' and, more topical still, 'Unemployment among British workers' in a 1935 letter.

By 1947, however, Professor Gordon Manley, in his Presidential Address to the Royal Meteorological Society, felt moved to consider 'The geographer's contribution to meteorology' by identifying positive contributions which he felt geographers could make to the advancement of the science of climatology. He says 'integration of observations, either in time or space, is a very proper function of the geographer' (Manley, 1947: 5), that 'maps are used every day by the meteorologists is an adequate reminder that he cannot fail to derive advantage from some cultivation of the geographical outlook,' (Manley, 1947: 6) and 'it does seem to me that a study of the fluctuations of climate, lying just within the instrumental period . . . should go far towards the understanding of the reasons for post-glacial changes' (Manley, 1947: 8). The opening sentence of Manley's last paragraph is still, perhaps surprisingly, valid, 'Yet it remains true that in the majority of our universities, the Department of Geography is the only place in which a synoptic chart is displayed.'

A Personal Summary

I have tried to identify aspects of the study of climate which are, in my opinion, geographical. In particular I have pointed to the urgent requirement for regionally specific analyses as a basis for improvements in data sets for input into global climate models, for validation of the results of these models and for the development of appropriate parameterizations. I believe that these are the responsibility of geographers, since they alone are likely to be trained in the range of disciplines required.

It is not true to state, in the context of climate study, that geographers 'missed the boat in the 1960s' but rather that in those revolutionary times there was no climate ship upon which to embark (cf. figure 9.2). Now there is, and the time is ripe for the involvement of geographers. I am concerned, however, that the new generation of geographers will perhaps not be as well equipped to make the contributions so urgently required as were geographers fifty years ago. Although it is not, in my opinion, necessary to understand every, or even many, of the mechanical aspects of, say, a general circulation model in order to comment constructively upon most of the features which I have described, I am inclined to agree with Minshull (1975: 148) that there has been in many areas of geography 'an orgy of model borrowing and misapplication', which has neither educated nor developed the users.

The geographers, and there are many, who refuse to become involved in mathematical manipulation of data and rules because they believe that either a satisfactory mathematical language is not available or, the other side of the same coin, that their particular problem is not mathematically tractable, are fooling no-one but themselves. There is now a wide range of manipulative tools available for adventurous geographers which are as similar to 'conventional mathematics' as clouds are to cleavage. For example, expert systems are exploiting one aspect of 'artificial intelligence' called fuzzy logic. Geographic information systems and geographical space search systems are both computer-based 'models' of human thought (Smith and Pazner, 1984; Smith and Peuquet, 1985).

More recently, fractal geometry has been applied to a wide range of real-world phenomena with some considerable success (e.g. Peitgen and

Richter, 1986; Lovejoy et al., 1986). Of particular interest in the context of climate and geography are the assertions that cloud and rain areas have a constant fractal dimension (Lovejoy, 1982; Lovejoy and Schertzer, 1986) and that realistic landscapes can be generated using specific fractal geometries (Mandelbrot, 1977; Goodchild, 1982). If the latter is true then there is a possibility that the results of geomorphological processes can be modelled using fractal geometries. Similarly if clouds do exhibit scaleless behaviour then fractal geometry can, perhaps, be applied to the difficult problem of cloud prediction.

Tesseral arithmetic is the child (albeit, perhaps unwanted) of geography. Tesseral addressing depends upon tessellation of the plane and subsequent manipulation is a much more natural form of arithmetic for spatial operations than the normal Cartesian process (Diaz and Bell, 1986). In particular, the spatial operations – translation, rotation and scaling – can be performed on the locational data structure. Just as ordinary arithmetic models the geometry of the number line, tesseral arithmetic models the geometry of the number plane. As ordinary arithmetic implies a translation back and forth along the number line, tesseral arithmetic translates on the plane, but in any direction. Similarly, as ordinary multiplication has the effect of scaling, and may change direction right or left of zero on the number line, so does tesseral multiplication scale, and may change the direction of the line between zero and the number multiplied, this time by any angle. Coupled with this is the advantage that tesseral algorithms are computationally more efficient than their Cartesian equivalents and they completely avoid trigonometry in the calculations. More importantly this data-addressing scheme has the additional advantage of handling both vector and raster data forms.

Tesseral arithmetic, using quad tesselation, has been shown to be highly successful for regional data base manipulation but the question of addressing global data sets still remains. A novel scheme has recently been proposed in which decimal latitudes and longitudes are interleaved to give locations on the sphere which has applications to the technology of cartographic plotting, because knowledge of spatial location can considerably reduce the 'pen up' time and so increase the speed of map productions. These, and other, exciting and fundamental developments, including machine vision and expert systems, which relate directly to the spatial (and temporal) analysis techniques, are truly characteristic of geography and warrant urgent consideration by geographers. The question is whether geography and geographers are now, or want to become, aware of these developments (figure 9.14).

I have tried to identify areas of climate research where there is currently a clear requirement for geographical (that is, spatial and temporal) analysis. These problems, which relate particularly to data archiving, data synthesis and intercomparison of data and predictions, *must* be addressed and solved in the near future by climate modellers. If geographers do not have enough interest and/or courage to join in this rapidly growing field others will (e.g. Jarvis and McNaughton, 1986) and papers

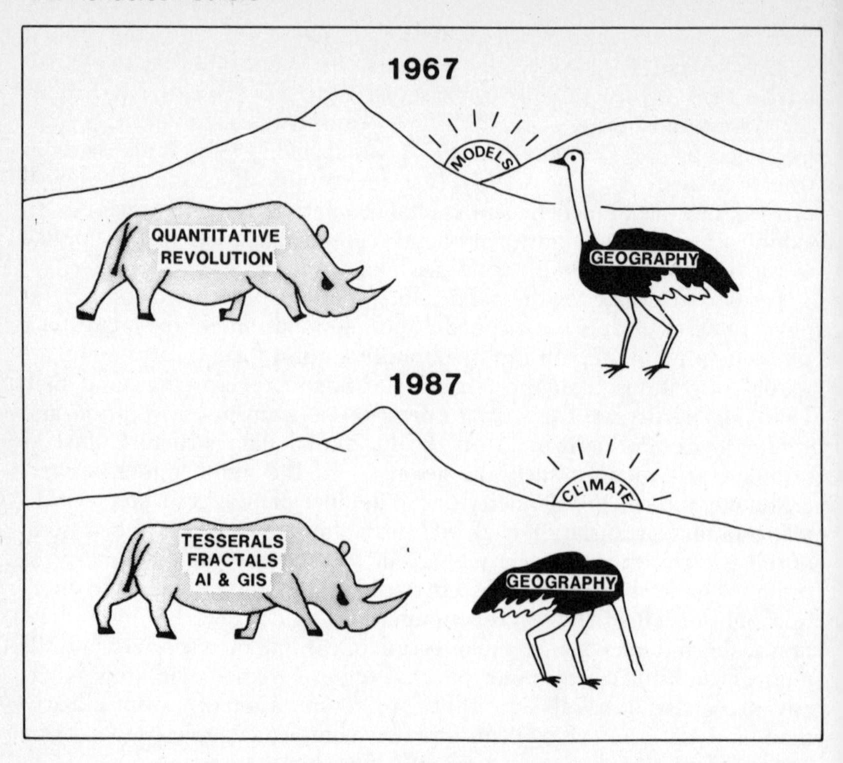

Figure 9.14 In 1967 the 'attack' by the quantitative revolution coincided with the 'dawn of modelling' but in 1987 when the Oxford conference took place, the question to be asked was: will the new revolution be recognized or has geography's attitude changed?

with titles such as 'Selection of a map grid for data analysis and archival' (Rossow and Garder, 1984), which describes and extols the virtues of using an equal area grid rather than the generally accepted equal angle (equal latitude, longitude) map grid of the numerical climate modellers, will become commonplace as the climate community rediscovers basic geographical and cartographical truths.

Acknowledgements

Sandra Mather is thanked for drafting all the figures except 'Mr Middle-Aged Geographer' (figure 9.1) which Dr K. McGuffie drew. I am grateful to Graham Thomas and Dr R. E. Dickinson with whom, respectively, I have been working on interpreting the GVI and developing BATS; and to Professor J. G. Cogley, Dr A. M. Harvey and, especially, Dr D. K. Chester, all of whom commented contructively on an earlier version of this manuscript. This chapter appears in the form in which it was presented at the Oxford conference so developments since then are not covered.

10 Urban Modelling and Planning: Reflections, Retrodictions and Prescriptions

Michael Batty

> There are no such things as applied sciences, only applications of science.
>
> Louis Pasteur

Introduction

This chapter reflects upon the development of urban models for planning purposes which were first applied in the United States in the late 1950s. The notion that such activity constitutes a 'science' is questioned, for it is argued that the field entirely owes its conception and existence to the demands of public policy. The rise and decline of modelling in practice is traced, the types of models identified and some important achievements are noted. As the demands of policy changed and the rationale for models disappeared, the field continued, institutionalized in academia and shielded from the harsher worlds of practice. The reasons for this changing context are briefly sketched and the discussion then turns to a possible re-emergence of the field as this time those remaining in it respond to the new landscape of practice. This emergent context is quite different from that of a generation ago, and it is still unclear as to the applicability of models in this changed environment. Finally some prescriptions for the field are noted and the value of modelling as a mode of thinking in the computer age is emphasized.

Urban models first emerged in the late 1950s in the United States. They were preceded by the development of transportation models which developed in response to the growing need to cater for the automobile. The driving force for these developments was practice. There was no science as such on which to build, and the field was constructed from such bits and pieces of theory which the model-builders and planners could easily lay their hands on. The intellectual landscape to which these techniques related was a peculiar mix of economics and engineering, and models were 'invented' in a practical, policy-based context. If there is ever an area which contradicts Pasteur's view that science exists independently of its applications, it is urban modelling, for the field emerged in response to applications, and it remains an open question as to whether or not it would still have developed without the driving force of urban policy.

This intimate connection with practice means that the field will never be to able to escape the volatility of its social context. Urban planning and policy-making have changed dramatically over the past twenty-five years and the development and popularity of modelling is closely reflected in such changes. As the post-war boom turned to recession, and as the emphasis in western industrialized societies turned from strategic planning to short-term tactical management, as urban society began its transition from industrial to post-industrial, and as the economy began to radically restructure itself into one based on information, urban models moved from the centre stage of strategic urban planning to the edge, to disappear finally in the form they were originally cast a generation or more ago. The intellectual and practice-based landscape is still littered with their remains and much of the landscape has been moulded by their development, but the cutting edge which these techniques once represented in planning practice and theory is long gone. The purpose of this essay is to reflect on these developments and to explain why. But this is not simply a retrospective affair. The discussion will also be prescriptive and critical, and will attempt to assess the continuing role for such activity, particularly in the context of the emergent computer age. There is a new driving force to modelling which once again represents a way of thinking about and 'doing' planning. This we will attempt to assess.

It is not easy to provide balanced reflections on a field which one has been close to for so long. Indeed, good critiques are usually produced by those who stand at arm's length from the field, those whose views are fresher and keener, and who are less committed in the first instance. There is much to Max Planck's oft-quoted dictum that a scientific theory only finally dies when those who have cut their teeth on it disappear from the scene too. There have been dramatic changes in many areas of social policy-making and intellectual life over this period in which we are reflecting on the development of urban modelling. To introduce this drama, it is worth providing a personal example to illustrate the sorts of changes which have occurred, so that these might be put in perspective. Some three months before the Oxford conference, I began to teach a course which I had not studied in any detail since my student days; this was urban design, the interface between planning and architecture. Of course, I was well aware of the reactions against large-scale public architecture, especially in the form of housing which marked a transition from the boom years of the 1960s to the massive disengagement from public sector involvement in such areas in the 1980s. But what I was not prepared for was the almost total switch in architectural style, fashion, method – theory in fact, which has occurred over the past twenty years.

To those like myself who had been reared, indoctrinated even, on the tenets of modern architecture embodied in Louis Sullivan's hallowed phrase 'form follows function', who could not accept any form of ornamentation on buildings because such detail conflicted with the aesthetics of the day, the transition which has occurred in practice is a shock. In post-modern architecture, anything goes and form very definitely does *not* follow function. Architecture of the 1980s is a total reaction to the concrete towers of the 1960s and the emerging 'style' is pluralist to the point where style no longer has any meaning. Skyscapers are now being built in New York which are festooned with Gothic and Georgian ornament, and there has been a neo-classical revival at all levels from the most domestic to the most public of architecture. If you stand away from a field and only follow it in 'lay' terms for a generation, when you return it is bound to contain surprises, but the sort of thing now going on within architecture would have been unthinkable in the heyday of modernism. Cosgrove uses similar examples in his chapter in explaining the logic of the humanist approach in contrast to the quantitative as the appropriate basis for theoretical geography.

Architecture, as Denys Lasdun has remarked 'is a social art' (quoted in Knevitt, 1985) and its driving force is very definitely public acceptability. The social milieu in which it exists is so strong that no individual ingenuity and flair can counter the structural trends at work. A remarkable demonstration of the power of this driving force is involved in the proposal, now rejected, for the construction of a Mies van der Rohe building in London's Mansion House Square. The building originally designed by one of the world's great modernists before his death in the late 1960s, was then approved, but it took nearly twenty years to assemble the site. By the time this had happened, the ground rules had so changed that even one of the world's great architects could no longer fight the turning tide. At a more modest level, one could construct examples involving the application of urban models and planning techniques in the 1980s which would simply be regarded by planning practice as curios from the past. They would no longer be understood.

I tell this story because it illustrates the sort of dramatic change in the social context of ideas which we are reflecting upon here – where the intellectual terrain exists in response to an applied context, where the science exists because of applications, not despite them. It is easy to forget that the field in question – urban modelling – was developed in response to the burning policy questions of the day in the North America and Britain of the 1950s and 1960s. The field, like so many areas in the social sciences, has since been institutionalized and professionalized, and this can blind one to thinking that it exists separately from its context. To a degree, elements of the field have been adopted by academic geography and in the past ten or fifteen years a synergy has developed between geography and modelling. Modelling is now very much a theme which builds on geography's quantitative revolution, which in turn has been central to that discipline's development. But after I have developed my own assessment of the field, I will also argue that modelling can never be a central construct of geographical theory although it does represent a way of thinking and 'doing' geography (in Johnston's (1986b) phrase) which is important and has lasting value. The particular models of the 1960s and the articulations of the urban system which they embody are no longer widely accepted. The social context has changed, and it is unforgiving in its rejection of all that has gone before. Indeed, Pasteur's quote which introduces this chapter is very much cast in doubt when one examines this field, and perhaps it sows this doubt throughout geography itself, indeed possibly throughout the social sciences.

The intellectual landscape which we will survey is a very different one from that which dominated the 1960s. Many remnants of those days remain and remain useful, but their *coherence* has gone. As I was involved, like many speaking at the conference, in moulding a bit of the landscape, I would like to be able to explain how it has changed, of course because of the need to develop applicable knowledge for urban planning, but also because I cannot accept that my responses then or now to this area are in any sense 'wrong' or inappropriate. And more importantly, such reflection is necessary so that we can adapt such experiences and the formal intellectual apparatus which has been constructed, to the continually changing context.

First we will sketch the context and identify how this acts as a driving force to the field. A variety of approaches to urban modelling have emerged over the past twenty-five years, and from the vantage point of the 1980s it is easy to identify different schools of thought based on social physics, micro-economics, structural dynamics, and so on. Here we will note the achievements too which have been particularly impressive with respect to synthesis and the convergence of ideas. One of the most interesting features of the field has been its institutionalization, and because the conference is a reflection on models in geography, a little time will be spent in tracing the interaction between this field and mainstream human and theoretical geography. We will then examine how the context to urban modelling has changed, noting the emergence of practical problems and ideological challenges. But most of all, these changes have been spurred by the transition from industrial to post-industrial, by the rise of information technology and for modelling, we will argue, these trends represent a mixed blessing. The cult of information, as Roszak (1986) calls it, involves an obsession with data, and ideological change has led to the development of the private sector as the main means of transition to the information society. Computers have always been essential to modelling but what is currently emerging by no means marks a resurrection of this field in the form it was predicated a generation ago.

Articulating Applicable Models

In the 1950s when all this began, social science was in awe of the physical sciences. There seemed nothing more appropriate than 'big science'. To mimic big science was the implicit goal of many a social scientist and there were countless public statements of this quest which implied that a science of society was quite literally around the corner. What gave rise to this optimism was the optimism of society at that time. Dramatic advances in theoretical science which had demonstrated their power in nuclear physics were just sinking in, in society-at-large; the war had ended in a moral victory for the Allies; in trade cycle terms, the fourth Kondratieff marking the recession of the 1930s was over and fifth long wave spawned by the invention of the computer had begun. General systems theory, cybernetics, operations research, all heralded the prospect of new frameworks for synthesizing the ad hoc knowledge which dominated the social sciences. Economics in particular, following the remarkable example of Keynes in the 1930s, looked more and more like a science in its structure and application. Moreover, much of the intellectual apparatus being assembled looked as though it might work in everyday life.

In land-use and transportation planning, the real quest was to contain growth and to accommodate the automobile. New infrastructure was urgently required and the link between land-use and transportation was becoming more clearly articulated. The context was right in that demands by urban policy-makers for new transportation infrastructure set in train the development of methods for dealing with the emergent complexity of spatial interaction. And if the context was right, the computer made it possible. For the first time, the sort of extensive complexity characterizing cities (as well as other such systems in both the physical and social sciences) could be managed, manipulated and explored using computer models. In short, the computer presented a working environment or medium akin to the physical scientist's laboratory. Such analogies were not lost on those involved; indeed they were positively exploited (Dyckman, 1963). Transportation planning studies were started for several U.S. cities in the mid-1950s (Voorhees, 1955) and by the late 1950s the first land-use models were under construction. A wave of such models appeared in the early 1960s and the speed at which they came to dominate the planning scene is illustrated by the special issue of the *Journal of the American Institute of Planners* devoted to land-use modelling published in May 1965 (Harris, 1965).

The types of models then developed now appear to represent a rather narrow conception of the urban system: essentially distinct land uses, articulated in measurable economic and demographic activities formed the subject matter of these models which were designed to locate such activities in spatial units represented by zones at the level of census tracts. Interaction between such activities in the form of transportation and inter-industry linkages were central to the workings of the models.

Spatial interaction and trip-making were embodied in gravitational analogues while model structures were conceived along simple econometric lines, with some emphasis on algorithmic computer-orientated solution methods. The emphasis was more technique-based than substantive, but the models were informal and rather pragmatic in that they avoided mainstream statistical theory.

There was a key distinction between comprehensive models – models dealing with two or more sectors/activities of the urban system and their consequent interactions – and partial or single-activity models. Lowry's (1964) Pittsburgh model represented the former, while Lakshmanan and Hansen's (1965) retail model for Baltimore, the latter. These models were largely one-off, static descriptions of the urban system, calibrated or massaged to fit some base date at which data were available, and then deemed suitable for making forecasts. There was only limited emphasis on optimization modelling despite (or perhaps because of) the fact that the planning system in which such models were embedded was designed to find best or optimal solutions. Schlager's (1965) land-use plan design model was conceived in the operations research tradition as a linear programming model, but there was really only one optimization model which signalled anything like the sophistication of urban and regional theory which would come to characterize such efforts over the next two decades. This was the housing market model designed by Herbert and Stevens (1960) for the Penn–Jersey Transportation Study which reflected a unique synthesis of urban economic theory with transportation through the mechanisms of the land market modelled at the individual level in terms of utility maximization.

This lack of focus on optimization is intriguing. It was almost as if the strategic planning systems themselves were regarded as being superior to any formal method of optimization, and it was perhaps implicit recognition of the fact that planning, in the last analysis, was a satisficing rather than optimizing process. This reflected the consensus type of society in which this style of planning was embedded, and it also reflected the positivist flavour of the behavioural–social sciences in North America (Simon, 1977). Yet the models which were developed in no way represented a science in the making. Everything was in response to the policy context and the whole show was driven by publicly articulated needs. Indeed, as far back as the French Physiocrats in the 18th century and certainly back to Ravenstein (1885) and Reilly (1929), methods of social physics which were so important to the emergence of this field were driven by the demands of the market and/or urban–economic policy. Even the maverick attempt in the late 1960s by Forrester (1969) to apply his technique of systems dynamics to urban problems was grounded in a public-policy context.

It could be argued, perhaps, that all science is driven this way, and there is clearly a school of thought within the history of science which suggests that the structure of science is dependent on its social context. But in the traditional sciences the knowledge base does stand apart from its applications, as Pasteur implies, if only because the base seems more stable, less controversial and hence easier to institutionalize. The point really is that the emergent field of urban modelling in the 1960s was fundamentally dependent on applications and practice for its very existence. Its development was pragmatically conceived, hence highly responsive to its environment. At the time, another revolution was proceeding within geography – the quantitative revolution – but this was proceeding in a parallel world, initially with very little contact with transportation and land-use modelling. Quantitative geography was building a spatial science, slowly adapting location theory on the one hand, and formalizing spatial statistics on the other, and this was taking place worlds away in academia. In urban modelling the *modus operandi* was very different in that models were built and researched mainly in practice and were not characterized by any of the elegance associated with spatial statistics and location theory. Indeed, Wilson (1984), in writing about his own involvement with the quantitative revolution in geography, considers his work to have been well outside the mainstream, in planning rather than geography in this period.

In the late 1960s, after the first wave of ideas had been absorbed in the U.S., these techniques began to diffuse globally, first to the U.K. and then to other western industrialized societies, and eventually to the developing world (Mohan, 1979). In the U.K., a planning system had emerged which embodied a hierarchy of planning instruments, from regional policy to structure plans down to local and district plans. The emphasis on strategic thinking, on growth and on positive intervention in the land market provided a well-fitting context for the development of models. Looking back even the language being spoken at the time appeared model-based, sometimes by persons such as those preparing the Planning Advisory Group Report which led to the new strategic planning system, persons who have never heard of gravity models and location theory.

In the late 1960s, as urban modelling swept through the U.K., the models built were a little more sophisticated than their earlier U.S. counterparts. Some early difficulties had been resolved and there was not the wealth of applications or the variety that had characterized the U.S. experience. Consultants were much less involved in their development because the U.K. environment was much less supportive of such possibilities than that in North America, and thus the U.K. scene was dominated by research centres such as the Centre for Environmental Studies (CES) in London, the Urban Systems Research Unit in Reading, and the Centre for Land Use and Built Form Studies in Cambridge. By the mid-1970s the CES work had diffused to Leeds and a Planning Research Applications Group had been started by the CES. The object of many of these groups was to both research the emerging 'science' of urban modelling and to 'apply' such science, often in the form of advice to strategic planning authorities. Yet despite this arms'-length distancing from practice, the first models were spurred by the practical context: the Lakshmanan and Hansen shopping model was first applied by

McLoughlin et al. (1966) to evaluate the impact of a proposed out-of-town shopping centre in northwest England, while Cripps and Foot (1968) developed a more comprehensive 'Lowry-style' model for statutory planning in the county of Bedfordshire.

The 1970s saw the beginnings of the institutionalization of this area. A critical mass of researchers had been assembled and their interests lay in improving and researching the same models further. But the real irony was that modelling had been developed to the point of take-off at the very end of the post-war boom. As boom turned to recession, optimism turned to pessimism, idealism to disillusion. The story is well known, but to anticipate the reaction that set in, in the early 1970s, the policy context changed so dramatically that urban models could no longer inform policy-makers about the most important questions they had begun to ask. This growing practical disillusion, which was accompanied by a new cutting edge in the form of a resurrection of political economy in the social sciences, combined to change the context entirely. But alongside this there was technical and organizational disillusion in that some models were oversold, their data demands were too great, they were technically inconsistent, they produced poor or incoherent forecasts, and so on. There was a catalogue of practical disasters which were largely due to learning on-the-job, so to speak (Brewer, 1973). None of this helped improve either the image or the applicability of these techniques. Yet there was a good deal of important work with models started in the 1970s and some impressive achievements have resulted. To set these in context before the retreat from modelling is charted, we must take one step back and examine what exactly these urban models attempted to do so that we can appreciate their limitations.

Model-based Conceptions of the City System

Urban models are based on a conception of the urban system which is articulated in terms of urban activities. As demographic and economic mechanisms essentially underpin such model workings, the models are constructed to embody rudimentary economic processes reflecting the spatial demand and supply of labour and produced goods. The link in these models to the observable superficiality of the system – the physical configuration of buildings and land uses – is tentative, although a one–one correspondence, which is assumed to be unproblematic, exists between land uses and activities. In this sense then, although the models are essentially economic in focus, their treatment of the local economy is simplistic and they are often referred to by urban economists as 'physical' models (Anas, 1986).

Location in such models is essentially regarded as a way of accounting for spatial interaction either in terms of trip-making expressed as travel demands or in terms of inter-industry flows as reflected in the workings of the macro-economy. Essentially the models of the 1960s and early

1970s represent a fusion, albeit sophisticated, of gravitational concepts underpinning spatial interaction with macro-economic theory as reflected in input–output and economic base analysis. Demographic processes have occasionally been integrated into these frameworks but the labour market is normally represented in a somewhat naive fashion in such models. In structure, such models are largely based on linear mathematics at the level of coupling activities and are usually static in conception, embodying no possibilities for qualitative change. For example, models in the Lowry (1964) tradition which dominate the field, are essentially mechanical artifacts reflecting social physics based on Newton and classical economics based on Keynes.

Right from the inception of the field there has been a vibrant debate about the degree of comprehensiveness of such model structures. In practice there was some early pressure for partial models, particularly in retailing, but the main forces have been towards comprehensiveness and integration. The nature of the planning system into which such models were embedded, the quest to integrate land use with transportation, the desire by model-builders to follow the general systems dictum that 'the whole is greater than the sum of the parts', as well as the 'biggest is best' syndrome, all forced the pace towards large-scale modelling (Lee, 1973). Although at the time I argued that the experience was not dominated by large-scale thinking (Batty, 1975), in hindsight that style of thinking, emanating from the 1960s, was quite different from the partial, even parochial, certainly more self-interested thinking which characterizes public policy in the mid- to late-1980s.

After the first wave of models in the U.S. and almost as soon as modelling began in the U.K., many reviews appeared which were concerned with how the field should develop. Remarkably, again in hindsight, such reviews seldom addressed new models for new policy questions, for the policy context was regarded as stable: the agenda was essentially a research agenda dominated by discussion of the need for greater comprehensiveness, the need for dynamics, the need for better mathematics and statistics, and the continuing plea for better integration with economic theory. Since then, some progress has been made on all these fronts, some of it rather impressive in technical terms and some of it having lasting importance with respect to the art of modelling. But among model-builders, the emphasis has been on the models themselves, never on the context of application. It is an open question, however, whether model-builders could have influenced the degree to which these tools were applied had they concentrated their fire on policy rather than technical questions.

The easiest nuts to crack in urban modelling have been the most obvious, namely those involving statistical estimation. In some respects, advances in these areas have simply been a matter of learning the appropriate mathematics and statistics and applying it. But from this has also emerged the most dramatic accomplishment, and that relates to a theoretical synthesis of model estimation, specification and application around the underlying idea of optimization. Extremal methods initially to generate consistent models were introduced into spatial interaction and transportation modelling in the mid-1960s by Wilson (1967) amongst others. His entropy-maximizing methods were quickly seen as possible structures for statistical estimation through reconciling entropy with likelihood. Maximizing entropy or likelihood enables best estimation of such models, such estimation being conceived as an optimization process. The methodological link from entropy to likelihood to statistical optimization was quickly worked out, and then came the breakthrough. Calibration was none other than a kind of substantive optimization, and once the mathematical machinery was in place, could be used to optimize the model in other ways, in ways for example that planners and policy-makers might wish.

There is another link through to economic optimization which we will come to in a moment but the work of Murchland, Wilson, Evans, Coelho and Williams represents a peculiarly British contribution of great theoretical importance to this area (see the book by Wilson, Coelho, Macgill and Williams, 1981). In a sense, though, it came too late to save the field. The link between modelling, planning and optimization had finally been worked out. In fact it was up to others, particularly Brotchie, Dickey and Sharpe (1980), to apply such models with impressive results but in a different time and place. There remain even more extensions to be worked through. The link between model structure and optimization is a rich seam which enables models not only to be optimized but to be inverted and solved in diverse ways (Batty, 1986), and there remains a field day in such research for the technically minded. It is an irony of history that such good models finally exist which could well have produced excellent advice in their day had they been available. But that day has passed.

Research into making static models dynamic proceeded on two fronts. Adding time onto static structures, producing pseudo-dynamic models, has met with limited success for such developments represent no more than simply indexing the cross-section with respect to time. There has also been some cross-fertilization with the types of space–time models emerging from quantitative geography (Bennett, 1981), but the real drama in model development has been much more research-orientated. Embedding static model structures into quite well-worked-out dynamic frameworks which involved qualitative or structural changes in the urban system has been a major force. Simple non-linear equation systems which are coupled with one another can give rise to quite sharp discontinuities or bifurcations in the development trajectory of the system if model parameters change continuously across certain thresholds. The idea of a catastrophe and catastrophe theory provides a useful image of the types of effect such model systems can capture. Wilson (1981a) shows how such models can be used to generate discontinuities in locational development, accounting for the growth of new shopping centres, for example. Allen (1983) develops similar frameworks to show how new cities can form and even get started in the first place.

All this work is very much in the tradition of building an urban science through modelling, and although it has some important conceptual implications for urban policy-making, these types of models are not applicable in the same way as those sketched earlier. And may be it is no accident either that such approaches have become popular in the age of discontinuity, as technological change is forcing such discontinuities onto what were comparatively stable industrial cultures. Indeed, there are some who have drawn their inspiration for describing social change from these sources (see Prigogine and Stengers, 1984).

There are other achievements which must be charted, and the most important one remaining involves further extension of the optimization paradigm. The original Penn–Jersey model due to Herbert and Stevens (1960) involved optimization by maximizing utility in a spatial context. The entropy–likelihood–optimization paradigm was easily extended to embrace utility in the early 1970s and then there emerged a remarkable convergence of these styles of macro-urban modelling with micro. In the late 1960s a theory of utility maximization involving a random element of choice was introduced by McFadden, and then began the development of a new style of travel demand modelling referred to as discrete choice analysis. The optimization paradigm was formally extended to embrace random utility theory (Williams, 1977) and the link between micro- and macro-modelling, so long a source of difficulty and contention began to fall into place. Most of these developments occurred in the U.S. where there have been many practical applications in transportation planning, and which remains a fairly vibrant model-based area of public policy-making (Ben Akiva and Lerman, 1985). In this area too there have been contributions by geographers, particularly in the area of discrete or categorical modelling which also emerge from developments in mainstream applied statistics (Wrigley, 1985).

Some of these developments have resulted in quite sophisticated applications, but usually from the standpoint of applied research rather than policy-making. Anas (1982) has provided a synthetic model linking traditional spatial interaction, discrete choice theory and urban economic processes which use state-of-the-art estimation and optimization methods. Putman (1983) has developed a variety of integrated urban models building on the tradition of comprehensive modelling and also developing these in an applied context. There has been much work at the coal face on zoning, calibration, model structure and data description. But there have been very few attempts to extend the battery of methods and techniques now available to other sectors of the urban system or to other substantive questions. Policy questions have been rarely researched at other than the speculative level and, in hindsight, it is hard not to brand this field as the province of technicians. Much of the achievement has been thoughtful and careful but there has not been a distinct group of researchers ever tackling new substantive issues and new policy questions from a modelling perspective. This will be all too apparent when we discuss the retreat from modelling but before then, we will turn to the institutionalization of this field.

The Institutionalization of Modelling

The argument of this chapter is that modelling can never represent a science because its intellectual rationale is entirely determined by a volatile social context. It is driven by urban policy and, as policy is an inherently unstable affair, the structure of knowledge demanded will also reflect this volatility. Of course various bits and pieces of the field will have lasting importance, but the map of its intellectual terrain is continually changing, just as the image of architecture is dictated ultimately by public acceptability. Yet modelling has acquired a life of its own for social reasons. It has become institutionalized, in the U.K. within geography to some extent, and it has become partly professionalized. In this section we will attempt to explain this phenomenon.

The disciplines from which model-builders drew their inspiration in the early 1960s represented an eclectic set of subjects broadly spanning the range from economics through to engineering. These subjects were in themselves developed in an ad hoc manner comprising an amalgam of inconsistent theories and techniques, and the resulting models reflected this to an even greater degree. But in the 1950s and 1960s there was some real hope that emergent frameworks involving ideas in systems theory and cybernetics, new forms of applied mathematics in operations research, and the new medium of empirical testing – the computer – would all come together to produce a 'science of society'. Although the structure of these fields then, as now, looked as though they could never by systematized, the hope of such restructuring was uppermost in academic minds.

In the groves of academia itself, new subject areas such as planning and other social sciences spawned through a variety of social and intellectual spin-offs from more established areas were particularly attracted to these emerging ideas. These areas were extremely fragile in terms of their knowledge base, and systems theory and the like seemed to provide the path to intellectual salvation. There was also a growing core of researchers trained in little else, or at least having been exposed to little else, who required some stability to reinforce their interests. The field attracted bright and active minds, its advocates learnt fast and there were some impressive advances at the technical level such as those referred to above. The post-war boom which had created a massive expansion of the university system in all western industrialized societies was still working itself out in academia through increasing demands for journals, textbooks, research grants and such like which all fuelled the drive to institutionalization.

Yet again in hindsight, the institutionalization was half-baked, as indeed it was for many other areas of the social sciences. In general the field was not attractive enough to persuade enough research students to fly under its banner, although, this said, many other areas in the social sciences faced and continue to face the same prospect. But more importantly, the volatile practical context to such work also detracted from the vibrant growth which occurred on its inception in the 1960s. The field has continued in its original fashion in the more pragmatic environment of North America but in some continental European countries, for example France, it has never developed. Only in Britain does it appear to have attempted an institutionalization in areas such as geography and planning.

Nevertheless, examining the scale of work in this field, what has been produced is impressive. The Oxford conference itself indicates the seriousness with which the broader context of modelling and quantification is treated. In the narrow context of urban and regional modelling, at least ten textbooks have been produced between 1972 and 1982, three of these being of the more advanced type. Around ten technical monographs akin to textbooks have been written and there are many books which include important sections on the field. There are at least twenty books reflecting conference proceedings and there are thousands of articles. Indeed in the analysis of citations by Wrigley and Matthews (1986), two of the top five most cited books and papers in geography are associated with this field. But during the last five years, momentum in the field has decreased. The last textbook was written in 1982 (Foot, 1982) and although there are conference proceedings still being produced, my impressions are that the rate of production of modelling work, in the tradition sketched here at least, is slackening. There are, however, new directions which we will pick up later.

Developments in quantitative geography ran parallel to urban modelling, as indicated earlier. Quantitative geography was dominated by a concern for spatial statistics which initially was focused on developing appropriate statistical assumptions incorporating space. This was a very different tradition, not exposed to public policy issues at all but concentrating on issues such as spatial autocorrelation, point patterns, diffusion theory and latterly space–time modelling. In fact, quantitative geography has been largely statistical in emphasis and even substantive theories of location in the classical tradition have been more the prerogative of urban and regional economists than geographers. Nevertheless, as urban modelling became institutionalized within the confines of geography, cross-fertilization of ideas did begin, particularly in more recent developments involving discrete choice theory, urban dynamics and estimation theory for models.

The overlap between geography and this style of modelling is of interest. A measure of how these parallel worlds interact and intersect can be produced by examining the participants at a series of key conferences on urban modelling over the last twenty years. In fact, urban modelling is characterized by a string of conferences dating back to the early 1960s, starting with Philadelphia in 1964, Dartmouth 1967, Liver-

Table 10.1 An analysis of modellers by professional training and work affiliation

Conference	*Arch.*	*Geog.*	*Plan.*	*Econ.*	*Eng.*	*Math.*	*Sci.*	*Arts/ Other*
Dartmouth	8	7	5	15	40	7	11	7
1967	(0)	(0)	(16)	(5)	(12)	(0)	(3)	(63)
Cambridge	25	8	6	11	19	11	11	8
1974	(14)	(8)	(25)	(8)	(14)	(3)	(6)	(22)
Oxford	14	13	10	13	13	3	20	10
1980	(3)	(7)	(40)	(17)	(13)	(0)	(7)	(13)
Waterloo	9	11	6	17	35	11	11	0
1983	(2)	(11)	(26)	(9)	(28)	(0)	(9)	(15)
Total	13	9	6	14	32	8	12	5
(four events)	(4)	(6)	(24)	(9)	(17)	(1)	(6)	(34)

The first line associated with each conference gives the percentage of modellers in terms of disciplinary degree. The second line (figures in parentheses) gives the percentage by work affiliation. In the case of 'Other', this refers to practice.

pool 1970, Cambridge in 1972 and 1974, Oxford in 1980 and Waterloo in 1983. There have been more specialist conferences on urban dynamics, on spatial interaction and other topic areas but those listed represent events which attempted to provide state-of-the-art evaluations of the field. In fact, we will not examine the Philadelphia conference in 1964, nor the Liverpool 1970 and Cambridge 1972 conferences because these represented rather smaller, more exclusive events, and were less typical than the remaining four state-of-the-art meetings.

In table 10.1 we show the percentage of participants (persons giving papers and/or discussants) with respect to their main disciplinary training and (in parentheses) with respect to their then present work affiliation. There are some striking points: first the relatively high proportions of engineers by training and present affiliation; second the high proportion of persons now affiliated to academic planning institutions; third the decreasing proportion of others (bureaucrats, practitioners, consultants) represented at each of the successive meetings; and finally the low porportion of geographers by training and present affiliation contributing to the subject area. In fact this is also borne out by an analysis of publications of the ten texts. Only one is written by a person trained as a geographer, whereas three were written by persons now working in geography departments, none of whom was trained as a geographer. This may sound as if the field wishes to distance itself from geography, but this is not so. The analysis simply makes the point that the area is professionally eclectic. Although it may be institutionalizing itself in part in geography, it remains diverse, rooted to practical applications and indeed dependent upon practical needs for its very survival.

It is worth posing two hypothetical questions to conclude this section which represent retrodictions – predictions of what might have happened to this field if the social context and needs of urban policy-makers had been different. Firstly, would the field have emerged anyway as geographers and economists continued in their quest to build a spatial science? The answer is probably no. The eclecticism of the field is its great strength, but equally its vulnerability in that the field has been awash with ideas but the area has not attracted a very large or stable core following. Interestingly, urban economics has always stood aloof from urban modelling and consequently it is much less applicable but stronger. If there was a driving force intellectually in urban modelling, it was transportation which initially spawned or provided the seed beds for spatial interaction and discrete choice modelling.

The second question is more speculative and relates to the extent to which this field has provided any basic material for the science of geography. In fact, the strength of the field is not in its theorizing about cities – in this it is eclectic and draws on anything at hand – but in providing new ways of thinking about cities, and new ways of 'doing' planning. The computer age has provided a powerful medium for enabling us to articulate new insights into all kinds of material and social phenomena, and the method of creating such insights is in general a form of modelling. Exploring ideas on computers does represent a new kind of investigation, different from formal mathematical and scientific traditions. In one sense as in adopting any theoretical position, it takes an act of faith to accept the approach, but geography is in a unique position among many social sciences as having a sufficiently strong basis of modelling to inform the area without it dominating the intellectual superstructure.

Criticisms: the Retreat from Rationality

Before we digressed to talk about the institutionalization of modelling, we had painted a picture of the emergence of urban modelling as a response to growth created in the boom years of the 1950s and 1960s. But along with models emerged a strong planning system at least in Britain which represented a vehicle on which models were carried and applied. This highly structured approach to planning was itself a process of optimization rather than management, of problem-solving predicated on the rational decision model which has emerged in many fields of inquiry since the late 1930s. The rational decision model portrayed the planning process as one in which goals were set, problems defined, solutions generated by searching across a sample of alternatives characterizing the solution space, with the best plans chosen, then implemented after solutions had been rigorously evaluated against the prior set of goals. The process was cyclic, solutions were generated iteratively and revised continually, and the optimization paradigm was frequently invoked (Harris, 1967).

Urban models represented the *modus operandi* of a strategic planning system based on this rational process. What emerged was a model-based planning but not strictly speaking planning models. Planning was regarding as being almost too complex to be formally structured as a process of optimization and, as such, models came to be applied predictively within planning rather than planning being specified as the control function within sets of urban models (Batty, 1985). This is why optimization models never really caught on: such formality lay beyond the bounds of a highly complex, uncertain system, and in any case, the ultimate power of advice and solution lay in the professional credibility of planners and policy-makers themselves.

This style of planning characterized the application of land-use–transportation models in the U.S. in the 1960s (Boyce, Day and McDonald, 1970) and the early wave of structure plans in the 1970s in the U.K. Indeed, a Department of Environment (DoE) publication, *Using Models in Structure Planning*, produced in 1973, provided some guarded support for the use of models in planning. But it had taken twenty years for this context to be established, and by the time it all came on stream the world had begun to change. Boom turned to recession, the western economies began to overheat as the fifth Kondratieff entered the mature phase of its product cycle, deindustrialization emerged as a phenomenon, and the questions that planners came to be concerned with were no longer those which models could inform them about.

Growth turned to decline, but urban models were equally good at subtracting activity as adding it, and thus it was not the shift from positive to negative that provided the problem. It was the shift from *planning* growth to *managing* decline that was at issue. This marked not only a disillusion with modelling but with planning itself. Indeed, strategic planning, particularly in the U.K., and with it the vehicle which carried the models and the organization in which they were embedded, has disappeared. Questions in planning were no longer seen in terms of allocation or distribution across space but of competing for economic growth, of alleviating the worst effects of decline, of conserving what was under threat and of opening up the market to whatever economic activity showed any signs of life. Consequently the grand strategies of containment which dominated British planning for a century disappeared almost overnight.

Changes in the questions asked, abandoning strategic long-term planning for short-term expediencies, the growth of new paradigms seeking to answer 'bigger' questions than the modelling fraternity had ever considered were there for the asking, combined with practical disillusion through difficulties of learning how to do intelligent modelling in practice all conspired to quell the original enthusiasm for these new techniques. Models were also expensive and were mystical in their operation. In the reaction there was always an element of 'Ludditism'. The last large-scale urban models applied in British planning practice were the models developed by PRAG (Planning Research Appliations Group) in the Teesside area in the 1976–7. Some retail modelling continued but Development Control Policy Note 13 issued in 1977 by the DoE advised local authorities even to abandon these, for them only to be picked up by the retailing chains in the mid-1980s intent on increasing their market share. We will return to this point later, for any resurgence there has been of late for modelling relates to this type of private sector response.

By 1980, structure planning and regional policy were being dismantled to be replaced by ad hoc agreements and responses. New types of local area-based government agency had come to dominate the planning scene, cutting across the notion of comprehensive planning. The final demise of the metropolitan counties in 1986 represented the end of a long saga of attacks on the notion of comprehensive planning, and now it appears that this response is part of a wider change in attitudes and policies towards the welfare state itself. This is an interesting and complex argument which also marks the transition from industrial to post-industrial society but in all of this, which we cannot explore further here, models were long gone.

The good work accomplished in modelling during the 1970s, in particular the development of the optimization paradigm, has never come to be applied in practice. The questions had changed so radically by the time such work emerged that this work could be regarded as little else than advances in the science rather than application of modelling. A couple of examples will serve to impress the point. In the mid-1970s Sayer (1976) criticized one of the assumptions of the economic base model as being unidirectional in causation, of not embodying any feedback between basic and non-basic components of employment, the division into which was arbitrary. In fact, by the early 1980s, work had been done which resolved these criticisms in that model structures were available which incorporated every possible feedback loop (Batty, 1983). The division between basic and non-basic was inevitably arbitrary to some extent, but models did emerge in which this arbitrariness could be dealt with. However, Sayer's criticism now looks somewhat mild in comparison with the wholesale disappearance of the traditional basic sector from many areas. In the transition to a post-industrial, deindustrial urban society, the basic–non-basic split itself has changed, and the transition has been so swift that the components of what is now 'basic' to the economy are not at all clear.

The second example also relates to this transition to a different style of economy. Traditionally, questions of competition are handled in urban models by mechanisms involving substitution. For example in modal split, a new transport mode directly competes for its patronage with other modes. However, new forms of communications involving information transfer are now emerging which are increasing interaction, and rather than competing with existing modes as originally anticipated, these are increasing the capacity of every mode. Such additions and complementations are extremely difficult to model because they constitute qualitative changes of a novel and surprising kind. Indeed, there are many such examples which betray a subtlety to urban systems that most current theories, and certainly urban models, are not able to handle.

The kind of milieu of planning which has emerged is essentially pragmatic, reflecting a collapse of the consensus of the past fifty years and an abandonment of comprehensive thinking. Ideological change reflecting an increasing self-interest and technological change have changed the questions being asked so radically that most theories culled in the years prior to the current transition now seem irrelevant. Indeed, the speed at which new theories have been developed during the past twenty years marks this degree of change. The driving force for modelling the urban system in traditional spatial terms has gone, but in its place there is a new context. The new landscape is one of a weak public sector and increasingly dominant self-interest. The situation is one where, if models are demanded at all, it is to enhance self-interest, not to embody any of the goals for which they were originally designed. The questions being asked about spatial differentials are just as interesting as they ever were but the interest in using models to address them is quite different from those of a generation ago. The new rationale for modelling, if it can be seen as such, is one of narrow, sectoral interest, involving the bits and pieces of models that would appear to enable as much profit to be extracted as quickly as possible in spatial terms. In fact, the landscape is somewhat indifferent to modelling, as we will show. Nevertheless this is an applied context which is still fuelling a little research.

The Cult of Information

The social context in which we now find ourselves is dominated by self-interest. The notion of a collective or public interest and of planners working for it or towards it has more or less disappeared, and thus local agencies and governments have become caught up in this increase in the competitive edge. Thus ideological change conspires with technological change to make the theories and models, which had taken half a century to develop to the form in which they might be applicable, barely relevant. Not only have the questions changed but so have the answers. The configuration of activities in the post-industrial city is a very different affair from that it has replaced, and even the causal processes appear different. As noted above, the very classification of activities is now in doubt for, with the rise of the information sector, the growth of multinational corporatism, and the decline of manufacturing base, location theory must be rewritten if it is to be useful in planning the future city.

Space is still important and thus spatial models are still in principle useful vehicles for exploring planning policies. But telecommunications, if not replacing traditional transport infrastructure and modes, is elaborating spatial structure to a new level of complexity. Information networks, data bases and new methods of automation are changing the structure of parts of the city beyond recognition. The model-based conceptions referred to in an earlier section are no longer very appropriate, for not only have new activities based on information emerged but also new infrastructures based on telecommunications, and new methods of communication and causality have appeared. Information not energy has become the new source of power, and the computer represents its driving force.

There is a quiet irony in all of this. An information age centred on computer-based systems would seem to imply that an age of computer modelling was about to emerge. In fact quite the opposite has occurred. Although the computer spawned the urban model in the late 1950s, the power of computation to change the context in turn changed the very rationale that led to modelling. In the very early days in the 1960s some models failed because of limits on computer memory available. But such constraints quickly disappeared and, in the review of modelling already discussed in this chapter, the role of the computer other than that of a facilitator of models has rarely been mentioned. Nevertheless the massive diffusion of computers throughout society has the potential to make models more transparent and to enable communication of their operation and results through new modes of computation such as expert system and computer graphics. These are all prospects which are immediately realizable but, apart from occasional examples, there is little sign of the computer revolution having affected modelling in these or any other ways.

In fact it is a concern for information rather than modelling which marks the information age. In a recent survey of British local planning authorities, Bardon, Elliott and Stothers (1984) report that of 335 planning authorities found to be using computers (66 per cent of all authorities), nearly one half (49 per cent) were using computers to process development control data; 32 per cent were using computers to process census data while 14 per cent were involved with digital mapping in some form. In contrast only 3 per cent were using retail models and the most popular (and standard) forecasting technique used was population projection, with 12 per cent using such methods. These are striking figures which indicate that models are barely used at all in British planning at present. In contrast, there is substantial interest in information systems. However, one must ask the question as to how such information is processed once it has been stored, and it is here of course that models are most appropriate.

This cult of information is increasingly problematic (Roszak, 1986). Information is power, and many private agencies and firms have emerged who work in the market for information. Information systems have become big business and consultants specializing in mapping and analysis of information have grown rapidly in the past decade. It is easy to overestimate the size of this market and the number of firms involved. In Britain, perhaps there are a dozen like CACI, Pinpoint, MVA Systematica and so on who specialize in information analysis with a spatial basis, and there may be many more who are involved in occasional research work for other parts of the private and sometimes public sector. Some of these companies are involved in using models, but the models which have been picked up are only ever partial in scope, usually involving retailing or other forms of market share. Such models appear to be those of the first generation and are rarely those which represent the state-of-the-art. These companies are able to exploit the intellectual terrain and choose models which enable them to predict market share on a spatial basis. But the evidence on the scale of their use is ambiguous and the advantages to advancing the state-of-the-art unclear.

This is particularly true in the area of retail modelling where some retail companies have large research staffs using models. Yet there is some uncertainty as to whether the models can produce good predictions. They are based essentially on the Lakshmanan and Hansen shopping model which was first specified by Huff (1963) some twenty-five years ago. At least the researches over the intervening years have showed how such models can be improved in practice, but little of the insights into zoning, specification, estimation and optimization, which are all available possibilities now, have been incorporated. Clarke and Wilson (1987) see some hope for a resurgence of modelling in these developments but I remain sceptical. The practice which has emerged is even more unforgiving than that of earlier generations.

Data, Data Everywhere and Not a Thought to Think

The diffusion of computer technology and the rise of data banks which contain facts and information about individuals across whole arrays of issues is one of the more sinister features of the emerging information society. There is a sense in which computers attract data in that the availability of information systems leads to their rather unselective use. In the past, before the advent of universal computer technology, information was filtered by those whose job it was to store and process it, but the cult of information presumes that all information is good. Data banks are being filled daily with all sorts of information which a generation ago would have been discarded as worthless. This proliferation of data, and the many banks which now exist, are available only to those who have the power to buy it, or control it; and although there are clearly better data sets available than there were a generation ago to aid our understanding of cities, for example, access to these data is restricted. There is something rather strange about a situation where data can be collected about ourselves 'freely' but then only made available to us at a cost. The commodification of information is an issue that is becoming central to the information society, and threatens to become one of the most problematic features of the economic future (Goddard and Openshaw, 1987).

It might appear that one of the original problems of applying modelling in practice has disappeared in the intervening years. Models had always run into data problems, and lack of data, or at least the right kind of data, clearly did influence the development of the field. But lack of data has never really been a central issue, and in any case it is an open question as to whether any of the appropriate data is available now. Some spatial interaction data on retail behaviour is clearly now collected as a matter of routine by the large retailing companies, and for those building shopping models this is clearly very useful. Some panel data collected by market research companies would also be useful in some model-building but, in general, the increase in data available has also to be seen in the context of government reducing its organized collection and aggregation of data at many levels. It will clearly be necessary to explore these new sources of data, as Rhind suggests in his chapter, but they are unlikely to provide a new momentum for urban modelling.

Thus data availability has never been the key issue in modelling. As Shubik (1979) says: 'Ours is a data-rich and information-poor society'. The real critique of models relates to more substantive issues, to the questions which models are able to respond to, not to the data which are available. Indeed, there may be many questions for which data are not available, but this has rarely been the case in model development. Urban models were originally constructed around conceptions of the urban system which were readily measurable and quantifiable. The data were never perfect and there were gaps, but there was always enough to get going. At present, however, there are many issues pertaining to urban spatial structure which do require better data if they are to be understood, but the data being collected by the private sector in its penchant for continual data-gathering are unlikely to be very useful in this quest. Once again there is the obvious point that useful information can be collected only if some idea of what it is likely to be used for has been established. This sounds trite, but the cult of information seems to provide a context in which it is often forgotten. The issue is as much one of better theory as it always has been than of better data, as Harvey and Scott and Macmillan also argue in their chapters.

In fact, one of the criticisms of modelling is that models produce too much data and that their outputs are difficult to assess. This is clearly the case. Anyone who has ever operated a spatial interaction model will know that some way of summarizing – usually through statistical measures, averages and so on – is required if the predictions of such a model are to be evaluated. One of the problems is actually producing reasonable statistics which indicate how well such models do perform. Over the years there has been quite some debate about model performance and invariably good statistical performance (high R^2 values, for example) has not been borne out when the data predictions are examined in more comprehensive ways, by mapping for example. What the computer revolution has enabled is much better ways of summarizing data using text and graphics, and there is some work which indicates that the results of modelling can now be better communicated to model-builders as well as model-users using new forms of display technology and the like. We will return to these points in our conclusions. One thing, however, is clear, and that is that modelling is antipathetic to data for data's sake, but that the growth in computer technology has indeed become coincident with the cult of information.

The present obsession with data is thus quite inconsistent with good modelling. As Wildavsky (1973) once rather humorously remarked: 'if the data are thicker than your thumb . . . they are not likely to be comprehensible to anyone'. Shubik's (1969) observation that 'Most simulations have a value inversely related to the fourth power of the quantity of computer output' further impresses the need to impose great structure on data in order to ensure highly intelligible modelling in the computer age.

Conclusions: the Future is Not Like it Was

A generation ago, chapters like this were concluded with a long agenda about how we should improve urban modelling. I will not attempt anything so presumptuous here, but I will reinforce the value which modelling has as one form of communication and understanding which is now reasonably well established in the social sciences. My own work is now very much in the area of communicating model-based ideas, and data in intelligible forms using computer graphics, and the kind of styles one is now able to bring to modelling have opened up dramatic new prospects for communicating and handling complex issues. A generation ago, when you wished to communicate modelling ideas, you would sit down with a stack of printout and hopefully convince your audience that this was where it was all at. Today, computer graphics is a much more effective medium, and in spatial problems it is essential. Models have got to 'look' right to be acceptable, as well as 'compute' right. I will only note this in passing, but a new physicalism is emerging in the sciences based on the power of computers to enable visualization of ideas which have always remained complex, abstract and inaccessible to most people. Interestingly, similar trends appear at work in other branches of geography, for example in the climate models presented here by Henderson-Sellers.

There are of course technical and theoretical areas where models can continue to be improved a little. Enabling qualitative change to be handled, and devising new models but in the same tradition to simulate relationships between new categorical descriptions of the urban system, are worthy tasks. But these are likely to remain in the confines of academia. In practice there are situations emerging where models provide eminently sensible ways of communicating ideas. Modelling as a kind of bargaining process (Dutton and Kraemer, 1985) is a prospect which holds promise, counter-modelling in Greenberger, Crenson and Crissey's (1976) phrase still characterizes econometric modelling where models compete with respect to forecasts, and modelling as a way of summarizing and making data intelligible is an area which could dominate the applied scene. There are opportunities in district health authorities, in retailing, in utility management and so on, for modest applications (see Openshaw's chapter). But the drive from practice, from policy-making for new sorts of models is no longer there and shows little prospect of returning. The field can never be like it was again, and it is now being fashioned in a manner much more akin to the way operations research has developed. Operations research has never warranted an agenda for future research and applications, and one is not warranted for modelling here. Modelling has now been absorbed into the intellectual culture surrounding geography and planning. Its future prospects are unlikely to be determined by dramatic conversions to its ideology but more by its pragmatic value to researchers and policy-makers, sentiments which are echoed in many of the contributions to this book.

11 Applied Geographical Modelling: Some Personal Comments

John R. Beaumont

The chapters by both Michael Batty and Anne Henderson-Sellers (and other chapters in this volume) highlight the variety and the high quality of modelling research undertaken in geography over the last twenty years. Being rather critical, the progress and impacts have been more of a technical nature, rather than in the potential application areas. For the future health of this research it is important to look forward, and a number of important general issues are raised by both Batty and Henderson-Sellers. Indeed, it is important that the discussion is broadened (and unfortunately one of the potential difficulties of a conference examining progress within a particular discipline is that its self-reflection involves discussions among the converted).

Moreover, it is important that the context in which we undertake our research is understood. Batty, for instance, refers to the institutionalization of modelling arguing correctly that 'modelling can never represent a science because its intellectual rationale is entirely determined by a volatile social context'. From a personal point of view, having been away from geography and academia for five years, there is an appearance that the momentum of the modelling research of the 1960s, 1970s and early 1980s has been lost (at least, in the areas of urban modelling and planning of which I have some knowledge). This is confirmed by both discussions; Batty traces the decline of modelling in practice and Henderson-Sellers states, more generally, that 'the relationship between geography and the study of climate is moribund'. Batty and other authors are correct to express general concern about modellers' abilities to say anything meaningful about the problems facing society today. Openshaw (1986) states that 'it seems that the best people are studying completely the wrong problems', an assertion he repeats in this volume. Henderson-Sellers is inclined to agree with Minshull that there has been 'an orgy of model borrowing and misapplication', which has neither educated nor developed the users. The emphasis must be on applied problem-solving, rather than fitting a problem to a method!

Underlying the Batty and Henderson-Sellers chapters is an acknowledgement, which is very important in practice, that discussion and implementation of any modelling must involve the complete model-building process from purpose to evaluation. Elsewhere, I have argued,

> it is rather more straightforward to hide behind the 'workings' of a few numerical simulations, with a brief concluding paragraph alluding to potential future applications, than to have to face the vagaries of variable definitions and considerations of varying data availability and quality in the real world. (Beaumont, 1987: 1442)

An immense intellectual stimulation can be provided by tackling policy-related problems. Given the disasters in the southeast of England in November 1987, there is clearly some requirement for improved weather forecasting and communication.

Moreover, from my experiences, ultimately, the success and relevance of modelling applications are dependent primarily on communications between technical modellers who produce the studies, and management, who use the results. Unfortunately, failures to provide meaningful and applicable analyses are often more attributable to poor lines of communications between the analysts and management than to any technical shortcomings *per se*. The gulf of these two cultures was demonstrated at the recent ESRC-funded British workshop on '*Store Choice, Store Location and Market Analysis*' (see Wrigley, 1988) in which academics' modelling work was deemed to be in a language as widely comprehended as Serbo-Croat! A general worry arose because none of the authors at the conference really addressed the issue of linkages between commerce and industry, government and academia. Immediately prior to the Oxford conference the Department of Education and Science (1987) had published *Higher Education, Meeting the Challenge*, in which there were statements related to a desire for collaborative and actionable research.

In terms of models, clearly the users must understand the strengths and weaknesses of the analyses. Towards this end, the use of simulation modelling, for example the spatial interaction-based approach, as discussed by Wilson in this volume, offers many opportunities for increasing the use of the models developed by geographers. In the past couple of years many commentators have stressed the potential benefits to users of decision support systems and applications areas of artificial intelligence, primarily 'expert systems'. With expert systems, in which the computer makes decisions (and therefore the user in a sense is abdicating responsibilities), there is a need to be able to precisely define the problem of interest for operationalization. In this sense the opportunities for modellers working in 'physical geography' can be expected to be different from those working in 'human geography'. As physical geography models often incorporate representations of physical laws, there should be more opportunity to make use of artificial intelligence in an applied way. In comparison, evidence to date would indicate that the complexities of human behaviour and experience cannot be defined precisely, and therefore expert systems are unlikely to be generally successful in the immediate future. Decision support systems, which incorporate data and strong analytic engines, would be more appropriate for human geography, particularly building on modelling capabilities.

Whatever the type of system, for progress, it can be argued that the marketing of the expertise and experience of modellers becomes important. However,

> to equate marketing primarily with advertising or salesmanship is incorrect and unnecessary, because it is possible to select a style of marketing that develops traditional professional standards without hard selling. At the outset, however, for successful marketing, it is a prerequisite to ascertain the needs of the target markets and how they can be best satisfied. (Beaumont 1986: 421)

There are no doubts that modelling work from geography is applicable (see, for instance, Openshaw's and Rhind's chapters); the problem is to demonstrate the usefulness and relevance to users (and be willing to leave the relatively safe environment of academia for the harsher worlds of practice (see, for instance, Batty). There is an important need for education, because potential users of many of the models described in this volume do not have the necessary statistical/mathematical background to be able to comprehend them. This raises an important issue of presentation, particularly the removal of equations to a technical appendix, and the packaging of the software which makes full use of the advances in microcomputers and graphics capabilities. It is important to see the models as solutions to problems and therefore emphasize that aspect rather than the technical sophistication.

It is proposed that one could understand an evolution in terms of applications of geographical models in terms of the following four stages:

stage 1 pilot trials/'demonstrator' application;
stage 2 efficiency/convenience applications;
stage 3 effectiveness applications;
stage 4 formerly inconceivable applications.

In many respects, disappointingly, after twenty years we are really not in general beyond putting together demonstrator applications. However, as a recurrent stimulus to our future research, the possibilities of developing formerly inconceivable applications in the future must be exciting.

In conclusion, it is important to express a personal belief that modelling applications will become increasingly important, particularly as enormous amounts of data become readily available through advances in data capture technologies. The 'computer era' has finished; the 'information era' has arrived. While the speed of the modelling take-up is unclear, at this stage, one would expect the process to be slow if current practices are not changed (unless other disciplines become involved in our traditional areas!). Henderson-Sellers describes geographically specific climate studies which, in her opinion, could and should be being undertaken by geographers.

> Such studies are intrinsically geographical and yet few, if any, geographers are contributing to the planning and analysis. . . . If geographers do not have enough interest and/or courage to join in this rapidly growing field, others will.

The differences between 'pure' and 'applied' research are important. By the first, I mean no direct and specific immediately deliverable end is in sight except in enhanced comprehension of the details of a particular topic, say, the concept of structural stability; by the second, I mean that attention is focussed on particular problems that have been identified, and directly beneficial results should be produced. For geographical modelling to progress, it is important that the environment can be provided for both types of research to be undertaken positively. Fractal geometry, for example, is a technique that is interesting both urban modellers and climate modellers. While data capture technologies, enhancing data availability and quality, can open up model-based applications, there remains a need for better theory (see also Macmillan, and Harvey and Scott, in this volume).

At present, unfortunately, doubts exist because of the shortage of funding facing U.K. higher education (for a broader discussion of this issue, see Beaumont, 1988). Notwithstanding the funding constraints and the obvious duty for modellers to develop priorities particularly with potential end users, the way forward must require much more coordination in the modelling research area. Moreover, as academics, we must develop links with the organizations who can afford and guarantee large research budgets over a number of years. We must change our orientation and expectations – it is essential to think big, even if we start small! Batty (1987: 122), for instance argued after the Oxford conference that,

> it may well be that the only way to gain access to data for model based work in the future is by joining hands with . . . [the controllers of data], but in that case, it will not only be modelling which is changing but the very role of academia itself.

Thus, interestingly, the issues raised for the next twenty years are of a practical and developmental, rather than a research, orientation. The future is unknown, but it could be exciting, if geographers are willing to become involved (probably with new partners) in doing geography.

PART IV
Modelling and the Changing Technical Environment

12 Computing, Academic Geography and the World Outside

David Rhind

Past, Present and Future

Introduction

The past ten years – and particularly the past five – have seen a revolution in the use of computers in research and applications in geography. Many other disciplines have experienced the same changes. Thus Durbin (1987) argues that 'the dominating factor which determined the limits of what was considered as achievable [in statistical science] in the first half of this century was the primitive state of computing technology at the time'. He goes on to conclude that the newly available computing power should result in radically different ways of approaching statistics through exploratory data analysis, graphical presentation, model selection and validation – all as parts of a broader, more eclectic approach. In geography, however, the primary characteristics of this revolution have been as follows.

1 An enormous growth in the design, creation and use of data bases primarily for monitoring and inventorying purposes (see Tomlinson Associates, 1987) has occurred. Authoritative estimates suggest the market for systems and services of this type will be about $3000 m per annum in the U.S.A. by 1990.
2 A move away from calculating traditional parametric descriptive and inferential statistics towards use of simpler and more robust descriptors has been effected.
3 The numbers of academic geographers involved in this growth in use of geographical data bases have been minuscule compared with those of non-academic geographers and individuals from other disciplines.
4 The commercial sector, rather than governments or academia, has increasingly produced the new technical developments and fostered new applications.
5 Existing turn-key systems can, in skilled hands, carry out virtually all

Table 12.1 GIS functionality

1. *Data capture*. Both manual and automated data entry of points, line, grid cell encoding or entry by keyboard or voice of attribute data.
2. *Data validation*. Includes calibration, consistency and feasibility checking.
3. *Structuring* – e.g. skeletonization of line-scanned data, topological structuring of vector data.
4. *Editing*. Includes insertion, deletion and modification of both attribute codes and geometric elements, plus node snapping, sensor noise removal, etc.
5. *Structure conversion* – e.g. raster to vector and vice-versa.
6. *Abstraction* – e.g. generation of Thiessen polygons, of polygon centroids, grid interpolations from irregularly distributed X, Y, Z data.
7. *Geometric and projection conversion*. Includes both ad hoc 'rubber sheeting' and predefined algebraic transformations.
8. *Spatial definition* – panelling or clipping of the model, with extraction of new tiles (e.g. spanning multiples of map sheets).
9. *Generalization*. Includes re-sampling at coarser resolution – e.g. by point reduction in a line or by dissolution of small areas.
10. *Enhancement* – e.g. for detection of edges.
11. *Classification*. Recoding of the attributes of the entities to form classes – e.g. classification by spectral response, by attribute codes for choropleth mapping or stream ordering.
12. *Integration*. Relating data models by logical (Boolean) or arithmetic functions. Includes overlay operations.
13. *Selective retrieval*. Based upon attribute code combinations and/or spatial search.
14. *Statistical summaries* of the characteristics of entities or groups of them.
15. *Analytical modelling* – e.g. network analysis, location/allocation analysis.
16. *Program interfaces*. This facilitates the incorporation of user programs into the existing system.
17. *Display*. Generation of graphic images in map, oblique view and non-map graphical form.
18. *Data output*. Output of data in data transfer formats and data structures acceptable to other users and systems.
19. *Data base management tools* for managing access to and archiving of the data models – e.g. security protection, data base control of storage, retrieval and update.

Source: Green, Finch, Rhind and Anderson, 1985

the desirable manipulations (see table 12.1) though not yet in the manner, speed, comfort or safety required.

6. The formally published literature is a most misleading guide to the developments which have occurred. Much of the best work appears only in conference proceedings – or does not appear in print at all.
7. In addition to the statistical concerns of some researchers, some early developments began mainly as a process for automating cartographic production; the main thrust now, however, is the design, creation and exploitation of geographic data bases – from which the cartography is only a spin-off.
8. All of these developments have been underpinned by the rapid decrease in the cost of computing power, especially by the advent of the microcomputer.

This chapter examines these developments, anticipates future ones and discusses their implications for geography. The interpretations are, of course, predicated upon certain value judgements about the nature of our subject area; these are outlined below.

The Nature of my Geography

For me, the following statements – though trite – are both defensible and perceived truth:

1 Our subject matter lies in the area of man/environment relationships and in understanding how spatial differentiation has occurred.
2 Our skills are acquired through practice in this area and a broad understanding of the concepts of, and developments in, other, related disciplines.
3 It is desirable to exploit one's geographical skills and tools for societal good.
4 The generic initial questions which we, as geographers, normally wish to ask are:

		Task type
(a)	What is at location(s) X, Y (and Z)?	Inventorying/ Monitoring
(b)	Where is A (and B . . . not R not S . . .) true?	Inventorying/ Monitoring
(c)	What spatial pattern(s) exist(s) over our domain of interest, how did they come about and where are the anomalies?	Mixed
(d)	What if . . .?	Modelling

5 Following the asking of these questions, the analysis carried out may range anywhere from the qualitative to the formal and highly quantitative.

What do Geographers Need from Computing Tools?

Following on from these four generic questions, and based upon practical examples from the literature and numerous discussions, table 12.1 sets out the tools required for handling spatial data. This assemblage of tools is now often termed a Geographical Information System (or GIS). The key elements of it are that:

1 the analytical capabilities are integrated with the data collection, validation, editing, manipulation and reporting functions;
2 a single command language is used for the whole system;
3 the manner in which operations are carried out should be substantially

transparent to the user, though readily available to the expert;
4 a very wide range of data may be fed in without recourse to re-programming;
5 a 'tool box' or 'work bench' facility is available so that individual skilled users can create their own tools as macros from what already exists.

In general terms, such tools have also to be reliable, fast in operation, cheap, available on the variety of computer systems now extant, easy to use and fool-proof. No current system meets these requirements; the R and D needed to create such tools is set out in a later section. It is worth indicating how far we have come, however: contemporary systems providing most of this functionality are now available for PC AT systems for under £6000 and seem likely to become as widely used – and hence cheap – as dBase III within the next five years, i.e. of the order of £500.

The Computing Context

The past The raw cost of computing power has decreased by about an order of magnitude every five years over the past twenty years. Similar if not identical decreases in the cost of data storage have also occurred; this is probably of greater significance for geographical applications – at least until our modelling superstructures are matched to our retrieval capabilities (see below). It is instructive to cite examples: in 1967 the computing power in the University of Edinburgh consisted of one English Electric KDF9 which had only 32K words of main memory, was fed by punched cards and had virtually no operating system. In 1973 the most sophisticated computing system in British universities was the NUMAC one, serving the whole of the universities of Newcastle and Durham: this IBM 360/67 had 1 Mb of main memory. Today, machines with the same memory and raw processing power can be bought for less than £3000.

Perhaps the most significant development of all, however, was the advent of personal computing: the first of these microcomputers were launched on the market in 1974 and 1975 but the best-known (Apple) was formally launched in 1976 and the Apple II, Commodore Pet and Tandy TRS 80 were launched the next year. This development only became of real significance to applied geographers when conflated with another: the growth of both Wide Area and Local Area Networking. The former began with ARPANET circa 1974; so far as academics in the U.K. are concerned, the Joint Academic Network (or JANET) is now an indispensable tool for collaboration and communication. In my own department, for instance, we receive and transmit up to 100 messages every day to colleagues world-wide. The transfer of mail, data files and collaborative proposals, and the access to specialist facilities elsewhere (such as NOMIS) by these means, is now perfectly routine and highly desirable.

As a consequence primarily of the decrease in cost of computers, the availability of geographically referenced statistics and cartographic files began to increase: thus it is now cheaper to purchase Population Census data in computer form than on paper (parenthetically, this situation is unlikely to occur in cartographic-type data but that is also becoming more generally available). These developments all 'sucked in' increasing numbers of non-expert computer users and these users became increasingly intolerant of vendors who sought to 'lock them in' to proprietary operating systems, data base designs and hardware. As a consequence, pressure has grown to define international standards for connecting computers and peripherals (via Open Systems Interconnect), for standard operating systems (notably UNIX), for graphics commands (such as via GKS and PHIGS) and for data transfer (e.g. via ISO 8211 and in cartographic formats defined by OS (1987), AC (1988) and others).

The future A recent 'look ahead' by Dangermond and Morehouse (1987) has suggested the likely developments in hardware which will occur in the next five to ten years. They – like all other observers – predict that the trends of increasing processor speed and storage capacity per unit of expenditure will continue.

A VAX 11/750 was 'state-of-the-art' circa 1979 and models of these computers are (at the time of writing) installed in several British university geography departments. Though 'computing power' is an elusive concept, these are generally rated as about 0.5 million instructions per second (or MIPS). It seems certain that individual person workstations with far greater power will come into widespread use in the late 1980s: a new UNIX-based Sony micro, for instance, is claimed to be a 2 MIPS processor and sells for under $10,000. Dangermond and Morehouse (1987) predict twenty to thirty MIPS workstations by 1992, all at about the same cost as today's PCs (circa £1000 for a reasonably configured Amstrad); some observers expect 100 MIPS workstations by the mid-1990s or earlier. Superfast processors – with speeds of thousands of MIPS routinely available within the next five years – will become much more common. In particular, raw computing speeds will certainly increase dramatically as we move away from von Neumann sequential processors and gain access to parallel processors such as those based upon transputers. File servers holding data bases orders of magnitude larger than today's equivalents, and from which selections of data can be speedily and routinely made and transferred to a local workstation, will become the norm. When this occurs is partly determined by data transmission rates: those current today are typically the maximum of 64 Kbits/second on JANET but the LIVENET fibre-optic network across London already transmits data at 140 Mbits/second. Special-purpose hardware – such as the ICL CAFS or TRW Fast Data Finder – which provide much faster access to data will certainly become relevant to geographers. That these higher speeds are required is obvious from a study of table 12.2.

Table 12.2 Some typical volumes of geographical data

Gazetteer of all place names in Britain on the OS 1/50,000 scale maps	3.5
An OS 1/50,000 map sheet (N.B. 204 to give complete GB coverage)	c. 10–30
A typical OS 1/1,250 scale map (N.B. 210,000 to give complete GB coverage)	0.1–0.25
PACE topographic data of northern Europe (see Wiggins et al., 1987)	200
Soil map of the EEC digitized from 1/1 million scale	25
A Landsat Multi Spectral Scanning image of a 185 km 'square' area	30
A Landsat Thematic Mapper image of the same size area	350
Global one-time coverage of Landsat TM data (including overlaps)	4,000,000
All of the 1981 Population Census Small Area Statistics Data	1,000
Address, postcode and household classification for every household in GB	1,000

All volumes are approximate and are quoted in megabytes – i.e. millions of the equivalents of alphabetic characters. These data volumes are highly approximate since they are related to the data structure, the type and form of data compaction used and other factors.

In short, the hardware seems likely to diminish progressively as a serious problem, though our capacity to match this with expanded tasks, or the use of slow-running languages like PROLOG, may complicate the situation. What these changes *do* entail are major consequences for software design (reliability, extendability, documentation, and ease of use become more important than squeezing out more speed or using less memory) and in the human management and operational aspects of running computer systems. Most seriously of all, they complicate – in the U.K. at least – the question of copyright on data and of how to avoid nonsense created by even the relatively skilled user: as we show below, geographical data are usually voluminous, hedged with qualifications, imprecise in meaning and – when used in combination – of unknown validity.

Problems Arising from the Nature of Geographical Data

Notwithstanding the recent and projected developments in computing, problems remain with GIS. Certain of these stem from the intrinsic nature of geographical data. The latter include:

1 The modifiable areal unit problem (Openshaw, 1983).
2 The degree and form of spatial autocorrelation frequently present in spatially distributed data.
3 The effects of any spatial sampling scheme utilized (N.B. the disagreement between the Forestry Commission, Huntings and ITE on the bias induced by rival schemes).
4 The data volumes characteristic of 'real-world' data sets (see table 12.2). At present, many geographers maintain the fiction that data sets

for teaching purposes can be tiny subsets of those used for research: this is unreasonable, especially at late undergraduate and taught postgraduate level.

5 'Fuzziness' in terms of the extent of the areal units (e.g. where areas are abstracted to the extent of being represented by an area centroid or where boundaries have some sizeable locational error attached to them – a common situation).
6 Incompatibilities between spatial data sets which need to be linked (e.g. between definitions of unemployment in the Population Census and in Department of Employment standard series; between on-land and offshore data commonly referenced to National Grid and UTM coordinates respectively, each compiled from hand drawn maps).

Who is Active in this Area?

A few academic geographers (e.g. Tobler, 1957) have long been active users of computing resources. In addition, several U.K. university geography departments have also long been active in numerical modelling in one field or other with individuals such as Haggett, Kirkby and Wilson prominent. On the other hand, a very restricted number of such departments have evinced interest in GIS until the past two years: a survey at the beginning of the 1980s by the present author (Rhind, 1981) found little academic or even commercial involvement at that time, but considerable 'drive' from local and central government. The number of geography departments *now* active in the area is much larger but is still small compared with the new participants elsewhere (see Rhind, 1987). These include the following:

1 *Departments of national governments*. The Federal Republic of Germany has, for instance, spent £8 million on developing a national GIS. Indeed, the first such system – CGIS – was sponsored by the Canadian government (Tomlinson, Calkins and Marble, 1976).
2 *National mapping agencies* (Ordnance Survey (OS); U.S. Geological Survey (USGS); Institut für Angewandte Geodasie; Institute Géographique Nationale; NATMAP; Energy, Mines and Resources; Survey of India; Geodaetisk Institut, etc.). Though essentially part of national governments, these vary widely in their responsibilities, reporting mechanisms and type of operations. The great majority of them, however, have now adopted or are adopting a geographic data base approach, rather than the earlier cartographic orientation. As a consequence, large-scale availability of spatial data – as opposed to digital versions of pictures – is becoming a fact. In Great Britain, for instance, some 30,000 OS digital maps are now available as both a topographic framework and as raw data. The budget for this is considerable: £23 m has been spent on digital mapping and the OS annual turnover exceeds £50 m; it has over twenty research staff and

commissions significant R and D externally (OS, 1986). The National Mapping Division of USGS has over 100 research staff.

3 *Local government*. Wherever this tier of government has responsibility for land-based taxation (e.g. in the U.S.A.) and/or acts as the formal cadastral agent (e.g. in Australia and in much of mainland Europe), computer systems to handle such geographical data – often called Land Information Systems by land surveyors, or Geo-relational Information Systems by photogrammetrists – have been created. Major developments in this area are occurring in S.E. Asia – if only because of the need to establish land ownership rights to act as collateral for loans.
4 *The utilities*. These are the driving force in the U.K. and seem likely to spend in total well over £100 m in the next five years collecting OS digital map data and integrating it with their own networks and billing systems: British Gas is already committed to such an approach.
5 *Commercial organizations*. These notably include those firms selling small area marketing intelligence: in the U.K. these include Pinpoint, CACI and Mackintyre. Credit-rating agencies are increasingly entering the field and the growth of EFTPOS and increasing credit card use ensures that enterprises like Marks and Spencers can accumulate small area post-coded (hence geographical) data and exploit it for marketing intelligence purposes. In the U.S.A. firms such as the $3 billion turnover TRW run geographically referenced data bases of 130 million records on the creditworthiness of individuals and the characteristics of 8 million firms; they also monitor the locations of the U.S. Navy's ships in real time and have been involved in modelling nuclear assaults as part of the SDI programme! In general, they employ hundreds or even thousands of staff on such tasks.
6 *System vendors*. Considerable convergence is now being seen between the offerings from firms which began in discrete marketplaces – the statistical analysis systems (e.g. SAS), the image analysis facilities (e.g. I2S, GEMS), the CAD/CAM systems (e.g. Intergraph) and the few who began with the concept of geographical data bases (e.g. ESRI). There are now about fifteen reasonable GIS systems available worldwide (see Green, 1986) and many more are in prospect (e.g. from IBM and ICL). The investment in these systems has often amounted to millions of pounds and, in some cases, tens of millions: Intergraph, for instance, employs 250 staff on its mapping software alone and Siemens employs 150 software engineers extending their SICAD GIS. A particular growth area is in car navigation systems such as the Etak Navigator (White, 1987), Philips CARIN and the U.K. Autoguide. These are targeted at a worldwide market of over 500 million cars and require topologically structured map data.
7 *The military*. Both in terms of data set creation and in support of system creation, the military have been a major influence – perhaps the greatest. This is readily understood: each Tornado aircraft costs £17 m, whilst a simulator costs perhaps £3 m, and the data on which to

train pilots or steer Cruise missiles cost comparable sums. The incidental role of the U.S. military, in particular, has been critical in establishing the strength of their commercial GIS sector. Sometimes this has been beneficial to the world community (e.g. in the creation of World Data Base II by the CIA and its subsequent 'public domain' distribution by the U.S. National Technical Information Service).

8 *International agencies or societies.* These developments include the Bickmore world digital topographic data base proposal (ICA/IGU), the Global Environment Monitoring System (GEMS) and the Global Resources Information Database (GRID) of UNEP, the plans to establish global data bases in FAO, the investigations of the International Soil Science Society (IUSS) into a world soil data base, the activities of the International Union of Geologists (IUG) into a 'control' map of the world (CGMW), the IHMF–WMO–UNESCO hydrology programme (WCRP) and the COSPAR–IAMAP International Satellite Land Surface Climatology Project (ISLSCP). All of these may be dwarfed (or subsumed) by the plans for the International Geosphere Biosphere Program initiated by the International Council of Scientific Unions in 1986 and intended to be running through the 1990s. On a mere continental scale, schemes such as the EEC's CORINE programme to build an environmental data base for the whole of the European Community should be mentioned (Wiggins et al., 1987). A characteristic of most of the global schemes is the relative lack of geographers in central positions.

The Role and Attitudes of the U.K. Government

In so far as it is possible to speak of the U.K. government as having one attitude or policy on matters related to geographical data and its handling, this has until recently been one of neutrality and acceptance that sectoral interests (typically the interests of individual government departments) should prevail. However, several important developments have occurred which have initially seemed almost marginal to the theme of this chapter, but have progressively become central. The first of these is government's attitude to information which it has collected for its own purposes. A comprehensive review of government statistics carried out in 1981 under Lord Rayner's schemes to eliminate 'waste' led to the termination of many statistical series; see Hoinville and Smith (1982). This review was short-sighted in the extreme: it appears to have deliberately taken no account of the needs (and even the ability to pay) of non-government users, or of the benefits elsewhere in government outside the commissioning department.

Since then, a somewhat more utilitarian view has become evident, apparently directed from Prime Ministerial level. This has led to the creation of a Tradeable Information Initiative – a scheme set up by DTI under which departments are encouraged to sell whatever information they hold of value to the outside world. Sales are to be organized via commercial firms and, in the first two years, marginal costing may be employed if this is necessary to develop the market. Clearly this could well have major implications for academic geographers unless some very substantial research discount is made available, or unless collaborative ventures with the data suppliers are effected.

The second way in which government has modified its attitudes has been as a result of a series of studies. The first of these was the Ordnance Survey Review (or Serpell) Committee which reported on the future of the National Mapping Agency in 1979 (Coppock, 1984). This was perhaps the first firm statement of the value of OS mapping not merely as a 'backcloth' to the display of other data but also as a data set in its own right: government accepted virtually all the recommendations of Serpell, albeit as late as 1984. In that same year the report of the House of Lords Select Committee on Science and Technology concerning remote sensing and digital mapping was published (Coppock, 1984; Rhind, 1986) and, on most counts, accepted by government. Thus government accepted the need for data dissemination mechanisms and catalogues, the need to increase R and D in the area, that OS should actively support and promote GIS and that OS data should be made available more rapidly; a recommendation that Britain should concentrate on the more applied ends of remote sensing research was warmly welcomed – as was the view that an expansion in education and training was required (but which government said was a matter for universities and polytechnics!). The final recommendation of the Select Committee was, however, that a Committee of Enquiry be set up, to advise the Secretary of State for the Environment within two years on the future handling of geographic information in the U.K., taking account of modern developments in information technology and of market need.

This Committee was formally constituted under the chairmanship of Lord Chorley in April 1985 and duly reported (Department of the Environment, 1987) on its extremely wide topic. Having taken evidence from 400 organizations, and visited research laboratories, government departments, local governments, utilities, etc., it concluded that a fundamental change towards the routine use of geographic data on computers had begun, and suggested that this was analogous to the introduction of computers into accountancy twenty-five years previously – hesitant and error-prone at first but subsequently irreversible. The Committee's report recommended significant changes in the rate at which digital topographic data were collected, emphasized the need for a detailed examination of copyright on data, urged government to make more of its voluminous holdings of geographical data available in disaggregated form and argued for the adoption of post-codes and National Grid coordinates as twin and (to some degree) interchangeable standard forms of spatial referencing. A series of recommendations on the most vital areas for new R and D and for new developments in education and training were proposed, as was a Centre for Geographic Information (CGI). The government response was published in February 1988 (Department of the Environment, 1988) and broadly accepted most of the Chorley proposals, other than that for the CGI.

Where Now?

The Diversity of Applications and Users

The growth of computer technology has allowed almost anyone to carry out statistical analyses, to produce maps and to manipulate geographical data. Today's first-year undergraduate has capacities unavailable to all except the best research laboratories of ten years ago; the same situation will shortly be true of many schools. Over and above this, we have already seen that the present range of applications of the functionality in table 12.1 is large, growing rapidly and limited only by imagination. Yet many of the fundamentals were not defined by geographers: examples include the initial work on topological data structures which was done by mathematicians in the US Bureau of Census (see Corbett, 1979). Given this, how do we ensure both that geographers play a significant role in these developments and that users from non-geographical backgrounds obtain appropriate training in the handling of geographic data? The answer to the first of these lies substantially with individuals – those such as Blakemore, Gardiner, Mather, Openshaw, Unwin, Walker and Waugh, who have gained respect in their substantive research areas and/or in computing circles. Where inter-disciplinary conflict occurs, the role of organizations such as the International Geographical Union, the IBG, RGS and GA becomes important, if only to add legitimacy and insist on some level of representation. So far as the second question is concerned, the answer is obvious – the construction of expert, low-cost and comprehensive on-line tutors, as argued by Green (1987).

What is Wrong with Present-day GIS?: the R and D Agenda

Consequent upon the previous sections, we can itemize those areas in which shortcomings exist. In some cases, IT developments in general will solve the problems, but in other cases we can itemize some GIS-specific R and D which is required. The most important and more fundamental of the latter areas are listed below.

1 Building some 'intelligence' into the systems. This is needed to:

- (a) improve the quality of description of the entities extracted from raster-scanned data;
- (b) facilitate different 'view modellings' of all data;
- (c) act with some understanding of the data characteristics (e.g. to render illegal and impossible any overlay of data from 1/50,000

scale maps on those derived from 1/625,000 scale maps and to repeat all data analyses and linkage at different levels of spatial resolution to define the level of stability in the results);

(d) enable consistently simplified results to be produced, which can be interpreted by humans (e.g. to provide name placement systems for mapping);

(e) facilitate appropriate data-dependent search strategies;

(f) make possible on-line tutors to teach individuals about the theory *and* practice of geographical data handling;

(g) elicit structural knowledge from human 'experts' as guidance from future systems.

2 The development of spatial analytic theory and its algorithmic implementation – particularly to provide estimates of the quality of any analyses.

3 The creation of modelling and spatial analytic superstructures on top of, and integrated with, existing data retrieval tools.

4 The development and implementation of new data structures and mechanisms for decanting data from one structure to another with minimal and predictable loss of quality.

Conclusions and Recommendations

1 Much of the foregoing can be summarized in figure 12.1. It suggests that hardware will progressively diminish as a problem for geographical applications. Software will similarly decrease as a problem once reliable geographical programs become readily available and cheap – but we may well see a resurgence of concern over software as the lack of 'intelligence' characteristic of existing software is more widely appreciated: this should only be temporary if more 'intelligent' systems can be made to function. On this basis, human skills will remain critical over a longer time period until such time as 'expert systems' become valuable and widely available – since the geographers' expert knowledge is only partially reducible to a set of explicit rules, this may be a substantial way ahead. The single item of continuing importance is the availability of appropriate data at an acceptable price to the user. This, of course, may be more of a problem in the U.K. than in the U.S.A. and some other countries: the crux of the matter is that, whilst U.K. government requires a return on its data investment, it is impossible to measure use successfully and no simple proxy exists for it. Given this, ad hoc and costly proposals for outright purchase or renting of data are to be expected in the future, at least until such time as the half-life of the data renders it useless for anything other than historical analysis. The only circumvention of this looming problem to date – and possibly the last – is the BBC Domesday system (Goddard and Armstrong, 1986; Rhind, Armstrong and Openshaw 1988). It should be stressed that data suppliers view the academic world with

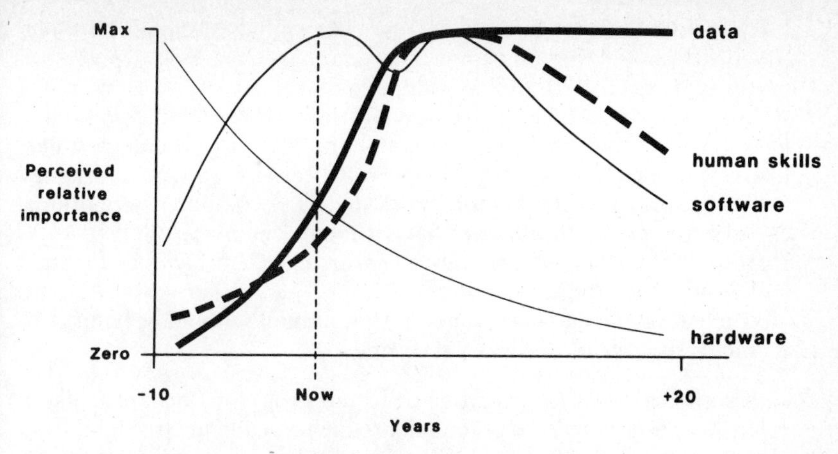

Figure 12.1 A forecast of the perceived relative importance of data, human skills, software and hardware over the next twenty years

suspicion, believing that data provided at a discount for research frequently 'leak' out into consulting applications.

2 As data collection becomes progressively more de-skilled, members of disciplines as disparate as land surveying and physics are becoming enamoured by the concept of managing data bases: the key to this thinking is simply that acting as a 'gate-keeper' to knowledge enables one to stay 'central to the action'.

3 Despite the enormous diversity of tasks to which geographical data are put, there is – at the functionality level – considerable commonality. This ensures that well-designed software can be applied in many application areas and, to that extent, the typical geography department can make use of the same tools in many parts of its research.

4 Despite the growing 'data mountains', remarkably little of the present work of the many thousands of individuals now working in this field world-wide is on modelling in any sophisticated sense. A number of commercial firms (often staffed by highly intelligent and motivated individuals) have seen this, and commercial superstructures for user-specific modelling will appear. Nonetheless, this area still represents an opportunity for professional geographers.

Given all this, the following action points suggest themselves:

1 Geographers should seek to obtain systems from commercial vendors, preferably (and often easily) at preferential rates for educational organizations;

2 these tool boxes should be re-configured into systems suitable for particular research projects;

3 close relationships should be struck with commercial exploiters of data bases, with government and with system suppliers;

4 we must recognize this is a 'team game'; the solitary academic working in this field – as either an exploiter of data bases, a GIS specialist or a spatial statistician – is at a considerable disadvantage;
5 we must recognize that we are now part of a near-global community linked by telecommunications; collaborating with friends of like interests in the U.S.A., Europe, etc. is often easier than with colleagues across campus and much more rewarding, i.e. computer networking ensures that the team does not have to be in one place;
6 Heads of Departments must seek to ensure excellent facilities for staff and ensure appropriate training;
7 all conventional language computer programming should be banned as an ineffective use of academic staff time.

Such a scenario is not for everyone or for all geography departments; it also has manifest dangers in potentially reducing academic freedom. It is, however, a necessary approach if we are to live with and exploit the opportunities for geographical research and teaching wrought by developments in computing technology. If we do not, others will!

Postcript

Between the Models conference and the production of proofs for this chapter, enormous changes had already occurred in the situation described. All, however, reinforce the contentions made herein. Thus the speed of development of technology has actually increased still further: what Dangermond and Morehouse (1987) predicted for 1992 will have been exceeded by late 1989 or early 1990. The advent of CD ROMs – derived directly from digital audio compact disk technology – looks likely to revolutionize the use of large data sets since each disk can store about 600 million bytes of archival data and can be reproduced for a few pounds. Institutionally, much has also changed; in the main, governments world-wide are coming to see the geographical data they collect as a form of real estate from which revenue may be generated. The greatest difference, however, is in the growth of awareness of what is possible. This is manifested in the setting up of research programmes by ESRC and NERC (notably the former's Regional Research Laboratory initiative) and by the dramatic success of such new publications as *Mapping Awareness* and the *International Journal of GIS*. In the UK, perhaps the most significant development has been the founding of the Association for Geographical Information, an umbrella body as recommended by Chorley and consisting of equipment vendors, central and local government, utilities, professional organizations like the land surveyors, and learned societies. A modest number of professional geographers are involved. It is, however, also significant that a 1988 survey by the Heads of Geography Departments in the UK indicated that GIS was then the second or third most widespread research topic in about half of British university departments.

13 Geography and Spatial Statistics: Current Positions, Future Developments

Bob Haining

Introduction

Progress in the use of statistical methods in geography over the past twenty years or so can be classified in terms of three main areas: methods for the analysis of space–time data, methods for the analysis of qualitative data (especially categorical data analysis) and spatial statistics. The development of each area reflects awareness in the discipline of the distinctive types of analytical problems to be faced and the types of data available for tackling them. In this chapter I examine some of the important developments that are current in spatial statistics and of relevance to research in both physical and human geography. Spatial statistics, as its name implies, is that branch of statistics concerned with the analysis of spatial data where the classical assumption of independence no longer holds. It is concerned with problems that arise in the application of various standard statistical methods (for example correlation and regression) to spatial data, but also with distinctive problems that are unique to spatial (and other 'ordered') data sets (for example interpolation, pattern description and image reconstruction).

It seems quite evident that much spatial cross-sectional data in geography (particularly data relating to small areas) cannot be independent. During the 1960s a variety of methods were developed for testing for statistical non-independence in geographical data. This work was summarized by Andrew Cliff and Keith Ord (1973) in their book on spatial autocorrelation, in which they applied these tests to problems in regression analysis and diffusion modelling. In addition to these areas of research, spatial statistics has to date linked up with geographical research in two important ways. First it offers rigorous procedures for the descriptive and inferential analysis of geographical data. The sorts of problems addressed by spatial statistics occur in geography where data are collected for an area for a single time period or where space–time data are collected but are to be analysed as a series of cross-sections. Second it is of relevance in certain areas of geographical (especially spatial econometric) modelling where underlying generating mechanisms operate in terms of spatial interactions and spatial transfers.

In the next section there is a broad overview of spatial statistics in geography and certain significant current developments are examined in detail. The final section consists of speculative comments particularly on the links between spatial statistics and two 'growth areas' of the discipline – remote sensing and geographical information systems.

Spatial Statistics in Geography: Present Position

Variation in spatial data arises from three components: a *deterministic* structured element, a *stochastic* structured element and a local random element or noise. The second element is usually represented in terms of the second-order properties of probability distributions while the first element is given a functional representation that can be equated with the mean (not necessarily spatially constant) of that probability distribution.

An implicit and untestable assumption that is made in using the methods of spatial statistics in geography is that geographical data consist of one or more of the three elements and represent a single realization of a probability model. This 'conceptualization of reality' in terms of probability models (particularly the first- and second-order properties of those models), Matheron (1971) argues, is justified if it allows us 'to solve effectively, practical problems which would otherwise be unsolvable' (p. 6).[1] This section examines these broad problem areas but first reviews two ways of characterizing certain second-order properties (the stochastic structured element) that are consistent with (but need not necessarily be dependent upon) the idea of an underlying probability model: spatial covariances and semi-variograms. These two ways of representing structure have both found their way into the geographical literature as methods for providing quantitative descriptions of spatial variation at varying scales. Spatial covariances and correlations are widely used, the semi-variogram approach seems to be gathering support mainly in physical geography and in areas such as remote sensing.

To set the scene, suppose the underlying probability model can be assumed to be second-order stationary (for definitions see for example Journel and Huijbregts, 1978). If $Z(\mathbf{x})$ denotes the random variable at location $\mathbf{x} = (x_1, x_2)$ and E denotes mathematical expectation then

$$E[Z(\mathbf{x})] = m \text{ for all } \mathbf{x}$$

where m is a constant. Furthermore for each pair of variables $\{Z(\mathbf{x}),$

[1] This definition of spatial variation in terms of probability models could be extended to the 'randomization hypothesis' which conceptualizes observed spatial structure as one possible ordering out of all possible orderings of the given set of data.

$Z(\mathbf{x}+\mathbf{h})\}$ the spatial covariance exists and depends only on the separation distance $\mathbf{h}$, that is

$$C(\mathbf{h}) = E[(Z(\mathbf{x}) - m)\,(Z(\mathbf{x}+\mathbf{h}) - m)] \quad \text{for all } \mathbf{x}.$$

It follows that the variance of $Z(\mathbf{x})$ exists and is equal to $C(\mathbf{o})$. The semi-variogram which is half the variance of the increment $[Z(\mathbf{x}+\mathbf{h}) - Z(\mathbf{x})]$ is given by

$$\gamma(\mathbf{h}) = \tfrac{1}{2}\,\text{Var}\,[Z(\mathbf{x}+\mathbf{h}) - Z(\mathbf{x})] = C(\mathbf{o}) = C(\mathbf{h})$$

In geostatistics the assumption of a finite variance is often felt to be inappropriate since variances are empirically observed to increase with increases in the size of the study region. In such a situation there can be no finite covariance either. However, it is possible to replace the assumption of second-order stationarity with the assumption that the *increments* have finite variance, that is the semi-variogram exists. This would be consistent with the condition of no finite variance for the original random variables but would allow the analysis of second-order data properties through the study of the semi-variogram the form of which reflects various attributes of the surface (see for example Journel and Huijbregts, 1978). This is known in geostatistics as the intrinsic hypothesis. As before $E[Z(\mathbf{x})] = m$.

The observation that variance increases with the size of the study region does not, however, necessitate discarding the covariance function, $\{C(h)\}$ for purposes of analysing second-order properties of the data. For example, variance will increase if the variable in the study region shows evidence of trend (linear or higher order). In such cases the source of the non-stationarity is seen to lie in the nature of the deterministic element which is no longer constant over the surface. However, the residuals from this mean may be second-order stationary (about a constant mean of zero) and so may be analysed in terms of spatial covariance plots after the non-constant mean has been removed, or allowed for. I shall consider this approach in more detail in a later section. The point is raised here, however, because in many geographical situations it may be useful to construct surface descriptions which consist of two components (a spatially varying mean and a covariance plot) rather than subsuming them both into a single descriptor, the semi-variogram.

There are several areas where the probability conceptualization and prior knowledge of $\{C(h)\}_h$ or $\{\gamma(h)\}_h$ or estimates of these functions is of practical use. A knowledge of the dependency structure in a spatial surface can be used to identify an appropriate spatial sampling framework (Quenouille, 1949) and in particular the distance between sample points in order to maximize the information content of observations for parameter estimation. If the distance over which inter-site correlation declines to zero is known, this can be chosen as the minimum sampling interval so that resulting observations can be treated as statistically independent (Webster, 1985).

The existence of correlation between sites implies that some of the statistical 'information' carried by an observation at one site is duplicated by observations at other sites. Such 'information loss' has implications for the sampling distributions (especially standard errors) of certain statistics and, as will be shown, usually needs to be allowed for. However, there are clearly situations where such duplication may be used to advantage, in particular where it is necessary to interpolate between observations on a map or where there are missing values in an otherwise complete spatial record such as sometimes occurs in the analysis of census data (Tobler and Kennedy, 1983), meteorological data (Gandin, 1963), geological data (David, 1977) and remotely sensed data (Barringer, Robinson, Coiner and Bruce, 1980). In such instances the information content carried by the observed data can be used to construct estimates of missing values and to place probability intervals around those estimates (Martin, 1984; Haining, Griffith and Bennett 1984).

Remote sensing includes the collection of data from satellites and aircraft. Such data, particularly multi-spectral data, can be used for the construction of land-cover maps (see Harris, 1981 for a review of research problems in this area). This involves the classification of pixel images, via their spectral signatures, into cover types. Spatial correlation is a commonly encountered property of such data. This can arise from instrument effects (Slater, 1979) and also when the alternation of surface types is on a scale that is larger than that of a single pixel. Even data processing algorithms, used to 'clean up' (or pre-smooth) images may add spatial correlation if based on a local smoothing operator. Campbell (1981) has shown evidence of persistent misclassification when supervised classification procedures are used which treat each pixel in isolation. Such misclassification occurs often in contiguous blocks (and is therefore more pernicious) and because it often looks 'plausible' may also be that much more difficult to spot than the occasional isolated misclassified pixel. A knowledge of spatial correlation structures is of use in devising textual measures as classification criteria to supplement the use of spectral signatures. In these, the information content of neighbouring pixels is used to aid the classification of pixels (Switzer, 1980, 1983). I discuss some of the approaches and problems in this area in the final section.

I conclude this section with a closer examination of two areas: the description and analysis of spatial surfaces (means and variances) and the description and analysis of relationships between two or more variables across the same surface (correlation and regression). These are important areas because they constitute some of the most basic tools used by geographers in all areas of the discipline to tackle a wide range of questions. I shall assume throughout that $C(\mathbf{h}) \geq 0$, for all $\mathbf{h}$. This seems to be the most frequently encountered situation in geographical (and, more generally, environmental) data analysis.

Univariate Statistical Analysis: Estimation and Inference on Means and Variances

Suppose n observations $(z_1, \ldots, z_n)$ come from a set of independent and identically distributed random variables with mean μ and variance σ_z^2, then the sample mean

$$\bar{Z} = (1/n)\sum_{i=1}^{n} z_i \tag{1}$$

is an unbiased estimator of μ and the variance of $\bar{Z}$

$$\mathrm{Var}(\bar{Z}) = \sigma_z^2/n. \tag{2}$$

By the central limit theorem, $\bar{Z}$ is, for large n, normally distributed, that is $\bar{Z}$ is distributed $N(\mu, \sigma_z^2/n)$. From these results confidence intervals for μ can be constructed from $\bar{Z}$ and the estimate of σ_z^2, and these results can also be used to suggest appropriate sample sizes in order to achieve a given level of precision in fixing the value of μ.

Suppose the observations are not independent, what is the effect on $\mathrm{Var}(\bar{Z})$ and on the estimator for σ_z^2, and is there still a central limit theorem to justify using the normal distribution? As regards the last question, Bolthausen (1982) has proven a central limit theorem (on $\bar{Z}$) for spatial sampling which covers some situations. As regards the first two questions, analysis must proceed with caution.

In the case of dependent variables:

$$\mathrm{Var}(\bar{Z}) = (1/n^2)\sum_{i=1}^{n} \mathrm{Var}(Z_i) + (2/n^2)\sum_{i<j}\sum \mathrm{Cov}(Z_i, Z_j) \tag{3}$$

The second term in (3) identifies the effect of spatial covariation on estimator variance. The true variance of $\bar{Z}$ is now larger than (σ_z^2/n). The magnitude of this effect can be judged by looking at $\mathrm{Var}(\bar{Z})$ under different spatial population models. Suppose the observations have come from a MVN $(\mu\mathbf{1}, \sigma^2\mathbf{V})$ model where the $n \times n$ matrix $\underline{V}$ describes the variance-covariance structure in the data. For discussion purposes suppose

$$\mathbf{V} = (\mathbf{I} - \tau\mathbf{W})^{-1} \tag{4}$$

where $\mathbf{W} = \{w_{ij}\}$ with

$$w_{ij} = \begin{cases} 1 \text{ if site } j \text{ is a neighbour of site } i \quad (j \neq i) \\ 0 \text{ otherwise} \end{cases}$$

This is the conditional autoregressive (CAR) model where τ is a constant

Table 13.1 The standard error of $\bar{z}$ (equation 3) as a percentage of the value given by (equation 2) for the CAR model

	τ		
Lattice size	*0.075*	*0.15*	*0.225*
5 × 5	114.5	140.5	203.0
7 × 7	116.1	145.7	229.6
9 × 9	116.2	147.7	245.9
11 × 11	117.7	151.1	260.0

such that $|\tau| < \eta_{max}$ and η_{max} is the largest eigenvalue of **W** (see Ripley, 1981).

Table 13.1 compares the true standard error of $\bar{Z}$ (using (3)), as a percentage of the value given by (2) for the CAR model with $\sigma^2 = 1$ (and hence $\sigma_z^2 = 1$ in (2)). The larger the value of τ, and hence the stronger the inter-site correlation, the greater the error introduced by using (2) rather than (3).

In practice σ_z^2 will not be known, and so must be estimated from the data. Again standard theory would suggest the estimator:

$$S^2 = (1/(n-1))\sum_{i=1}^{n} (z_i - \bar{z})^2 \qquad (5)$$

which in the independent case is unbiased. However, in the spatial case there are two problems. First, $\mathrm{Var}(Z_i)$ is not constant for all sites for the model specified by (4) above. For example if we compute the variances for the following simple (n=3) situation:

<table>
<tr><td></td><td>1</td></tr>
<tr><td>2</td><td>3</td></tr>
</table>

then with $\sigma^2 = 1$

$$\mathrm{Var}(Z_1) = \mathrm{Var}(Z_2) = (1 - \tau^2)/(1-2\tau^2),$$

but:

$$\mathrm{Var}(Z_3) = 1/(1-2\tau^2)$$

At best, therefore, S^2 can only provide an estimate of the 'average' variance. This is a general problem with spatial population models for finite lattices. Furthermore, in the dependent case,

$$E(S^2) = (1/n)\sum_{i=1}^{n} \mathrm{Var}(Z_i) - (2/n(n-1)) \sum_{i<j}\sum \mathrm{Cov}(Z_i, Z_j) \qquad (6)$$

Table 13.2 The bias in S^2 as a percentage of the 'average' variance on the lattice for the CAR model

Lattice size	τ		
	0.075	*0.15*	*0.225*
5 × 5	1.23	3.40	9.23
7 × 7	0.67	1.93	6.15
9 × 9	0.42	1.24	4.34
11 × 11	0.29	0.86	3.20

so that (5) tends to underestimate this 'average' variance by the second term in (6). Table 13.2 identifies this bias as a percentage of the first term in (6), the problem becoming less serious as n increases.

So far we have assumed a constant mean. In the last section the need to allow for variation in the mean was indicated. This is often approached by specifying a trend surface model where

$$\mathbf{Z} = \mathbf{A}\,\boldsymbol{\beta} + \mathbf{u}. \tag{7}$$

$\mathbf{A}$ is an $n \times k$ matrix of location coordinates for the n sites where k depends on the order (α) of the trend surface. (If $\alpha=0$ the constant mean case, $k=1$ and $\mathbf{A}=\mathbf{1}$ (a column vector of 1's). The $\alpha=1$ case is a linear trend surface with $k=3$). The vector $\boldsymbol{\beta}$ (which is $k\times 1$) is the set of unknown trend surface parameters. The element $\mathbf{A}\boldsymbol{\beta}$ describes the deterministic or functional component in the surface (it is the mean of the surface), $\mathbf{u}$ then describes the stochastic component. If $\mathbf{u}$ is an independent (white-noise) process this implies no structured stochastic (second-order) variation. This represents the standard trend surface model described, for example in Krumbein and Graybill (1965). Suppose, however, we allow $\mathbf{E}[\mathbf{u}\,\mathbf{u}^{\mathrm{T}}] = \Omega$ where Ω, the variance covariance matrix, is an $n \times n$ unknown, symmetric (not necessarily diagonal) positive definite matrix (For example, we could set $\Omega = \sigma^2\mathbf{V} = \sigma^2(\mathbf{I}-\tau\mathbf{W})^{-1}$ so that the errors follow the CAR model.) Model (7) now captures the three elements of a spatial surface indicated above. Analysis proceeds by specifying the order (α), the form of Ω, estimating $\boldsymbol{\beta}$ and the parameters in Ω and evaluating the goodness of fit.

Currently a model such as (7) can be fitted as follows. The vector $\boldsymbol{\beta}$ is estimated by ordinary least squares and the residuals computed. These are used to provide an initial estimate of Ω. The vector $\boldsymbol{\beta}$ is now re-estimated using the estimated variance–covariance matrix, $\hat{\Omega}$, in a generalized least squares procedure. The new residuals are used to provide a new estimate of Ω and the estimation procedure is iterated until convergence occurs in the parameter estimates. An additional complication is added when the order of the trend surface is not known (as is usually the case). Then a sequence of such models may need to be fitted and compared starting at $\alpha=0$. To avoid unnecessary computation it is

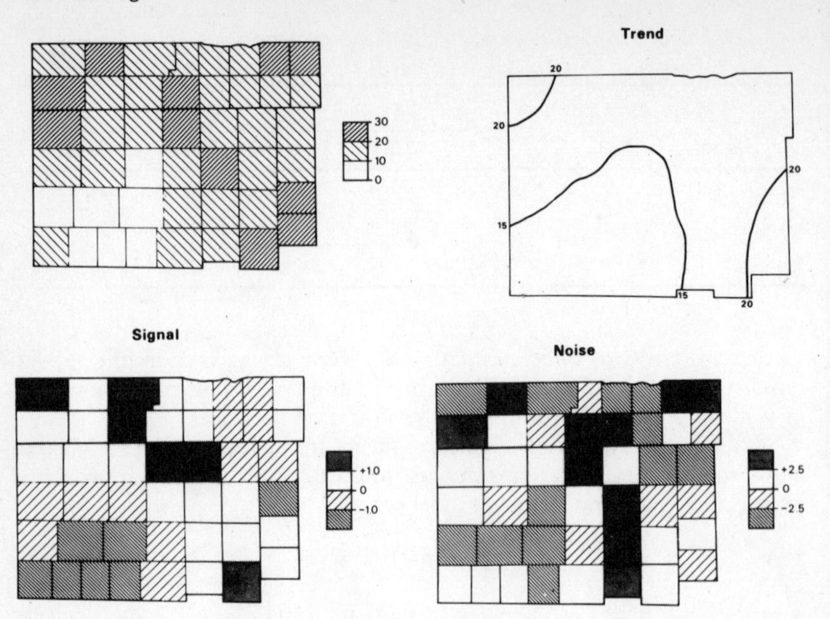

Figure 13.1 Decomposition of a crop yield map into trend, signal and noise components.

useful to test for residual autocorrelation in the ordinary least squares residuals using one of the methods in Cliff and Ord (1981) before proceeding to fit the full model. Approximate confidence intervals and hypothesis tests on parameter estimates can be devised using asymptotic theory (see Ord, 1975) and recently Mardia and Marshall (1984) have given sufficient conditions for the asymptotic normality of the estimators in those situations where the data can be assumed to have been drawn from a gaussian process. Further discussion, together with an empirical application using remotely sensed data, can be found in Haining (1987).

Figure 13.1 shows the results of applying this procedure to a set of county corn yield data taken from an area of several thousand square kilometres in the North American High Plains. The main objective was to identify changes in the crop yield surface over a period of nearly 100 years starting from 1879. This required good estimates of broad regional trends at each date for which information existed (Haining, 1978). The model fitted was that specified by (7) with

$$E[\mathbf{u}\ \mathbf{u}^{\mathrm{T}}] = \Omega = \sigma^2\ [(\mathbf{I}-\varrho\mathbf{W})\ (I-\varrho\mathbf{W})]^{-1}$$

which specifies a simultaneous autoregressive (SAR) model (Ripley, 1981). The top map of figure 13.1 shows the original data for one of the years (1929) and the other three maps are the three components that are identified by the procedure:

trend: $\mathbf{A}\,\hat{\beta}$
'noise': $\hat{e} = \hat{\mathbf{L}}^{-1}\,[\mathbf{Z} - \mathbf{A}\,\hat{\beta}]$
'signal': $\mathbf{Z} - \mathbf{A}\,\hat{\beta} - \hat{e}$
where $\hat{\mathbf{L}}^{-1} = (\mathbf{I} - \hat{\varrho}\mathbf{W})$ and '^' denotes an estimated value.

The approach outlined above, in the case of a constant mean (α=0) provides a better estimator than $\bar{Z}$ for the spatial mean (μ or, equivalently, $\mathbf{A}\beta$) in the case of spatial dependency in the data. The estimator ($\hat{\mu}$) is given by:

$$\hat{\mu} = (\mathbf{1}^{\mathrm{T}}\,\hat{\Omega}^{-1}\,\mathbf{1})^{-1}\,\mathbf{1}^{\mathrm{T}}\,\hat{\Omega}^{-1}\,\mathbf{Z}$$

and

$$\mathrm{Var}(\hat{\mu}) = (\mathbf{1}^{\mathrm{T}}\,\hat{\Omega}^{-1}\,\mathbf{1})^{-1}$$

(Note that if the variables are independent $\Omega = \sigma^2\mathbf{I}$ and $\hat{\mu}$ and Var($\hat{\mu}$) equal (1) and (2) respectively). One of the benefits of this procedure is that the standard error of $\hat{\mu}$ is less than that for $\bar{Z}$ in the situation where the data are dependent. Figure 13.2 compares the standard error of $\bar{Z}$ for the CAR model with that for $\hat{\mu}$ (which is a maximum likelihood estimator) on various-sized lattices (assuming σ^2=1). This has the practical advantage of giving tighter confidence intervals for μ. An empirical application is discussed in Haining (1988) using remotely sensed data.

The Analysis of Geographical Relationships: Correlation and Regression

The Pearson product moment correlation coefficient r is a measure of association for two gaussian variables (Y, Z) where

$$r = \mathrm{Cov}(Y, Z)/(\mathrm{Var}(Y)\ \mathrm{Var}(Z))^{1/2}$$

For estimation purposes ($\hat{r}$)

$$\mathrm{Cov}(Y, Z) = (1/n)\sum_{i=1}^{n}(y_i - \bar{y})\,(z_i - \bar{z})$$

$$\mathrm{Var}(Y) = (1/n)\sum_{i=1}^{n}(y_i - \bar{y})^2 = S_y^2$$

$$\mathrm{Var}(Z) = (1/n)\sum_{i=1}^{n}(z_i - \bar{z})^2 = S_z^2.$$

If Y and Z are spatially dependent then we have already seen the

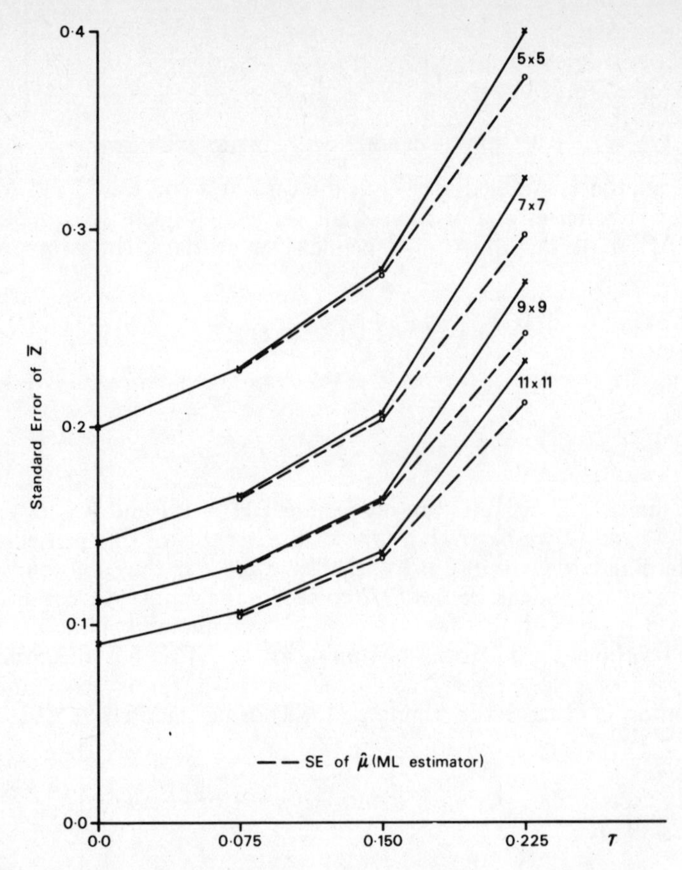

Figure 13.2 Standard error of $\bar{Z}$ for the CAR model on various-sized lattice.

problems associated with S^2 for estimating variances and the sample mean for estimating μ. However, here we concentrate on the effect of this spatial dependence on the sampling distribution of $\hat{r}$ as specified above. Results by Bivand (1980) indicated that the effects would be severe if both Y and Z were strongly spatially dependent seriously increasing the probability of a type I error.

In the case of independent gaussian variables (which are uncorrelated, $r=0$) then

$$(n-2)^{1/2}\ \hat{r}\ (1-\hat{r}^2)^{-1/2}$$

has the Student's distribution t with $n-2$ degrees of freedom (see for example Wolf, 1962). Recently Clifford and Richardson (1985) have suggested adjusting this statistic in the case of correlated data, replacing n with n' where n' is dependent on the spatial correlation structure in the two variables Y and Z. (Note that the adjustment also applies to the degrees of freedom.) The changes to the standard formula reflect the 'information loss' that (positive) spatial dependency in data implies because $n' < n$ so that the method appears to be evaluating the number of independent observations that contain the equivalent level of statistical information for estimating r. They report provisional simulation results which indicate that whereas making the adjustment keeps the range of type I errors for a 5 per cent test to between 3.2 and 7 per cent, without the adjustment, errors for a 5 per cent test range between 8.2 and 52 per cent depending on the level of spatial dependency.

The problems that arise in regression modelling have been frequently discussed and have been well summarized recently by Miron (1984). As in the trend surface discussion an iterative generalized least squares procedure is appropriate if the residuals from a spatial regression show evidence of spatial correlation. Perhaps the two most serious problems arising from failure to account for spatial correlation in any practical situation is that apart from giving an inflated measure of goodness-of-fit (r^2) there is again a risk of inflated type I errors, with non-significant variables being retained in the final model. Hepple (1976) gives a nice example of this problem arising. A full treatment of this question is beyond the scope of this review but is, in any case, well documented in Miron (1984) and Cliff and Ord (1981). Recently, Mardia and Marshall (1984) have derived asymptotic results for the sampling distribution of maximum likelihood estimators in the case of gaussian variables. They consider numerical algorithms and discuss procedures for specifying and estimating the parameters of models for Ω. This raises an important final point. In many of the situations discussed in this section, it is important to model spatial covariance, that is identify the best functional form and estimate the unknown parameters (see for example Ripley (1981), chapters 2 and 4 for a discussion of some permissible spatial correlation functions). Geographers have tended to examine only a fairly limited range of models. If these methods are to significantly improve the quality of geographical data analysis more attention will need to be given to properly modelling this element of the data.

Spatial Statistics in Geography: Prospects

The comments in this section are divided into three areas: remote sensing, the development of geographical information systems and future statistical training in geography. An underlying issue, at least with respect to the first two areas, is the need to develop computer software to facilitate rigorous spatial analysis. Much standard software does not allow the analyst to explore the sort of problems raised in the previous section. If every econometrician had to write his own time-series programs this would no doubt have a marked effect on the analysis of such data, and yet that, for the most part, is the situation facing geographers handling spatial data. It seems that a major initiative is required in this area using the expertise of statisticians, computer scientists and representatives of the main applied fields.

Spatial statistical theory and methods are already, and will continue to be, of considerable importance in the field of remote sensing given the earlier comments on the nature of much of this data and the types of problems associated with it. The raster based form of the data makes it particularly suited to rapid analysis, including the estimation of second-order properties, avoiding many of the technical problems encountered with non-raster data. The volume of such data need not be a serious problem since spectral analysis may be relevant for purposes of estimation and inference (see for example Larimore, 1977). Whether methods based on the semi-variogram or methods based on the covariance function should be preferred will to some extent depend on the nature of the data and the particular problem. At this time both avenues of approach should be kept open and explored.

One area of research in remote sensing concerns scene restoration (or image reconstruction). This involves identifying the 'true' scene from the image (or record) provided by the sensor. Methods such as ICM (Iterated Conditional Modes) for scene reconstruction exploit the dependency properties of the record that arise from the nature of the true scene. A provisional estimate of the true scene in each pixel is obtained using, for example, a maximum likelihood classifier where each pixel is classified based on its own record (without reference to any set of neighbours). This provisional estimate is then updated using a model for the scene in which the update value is dependent on the estimated neighbouring values at the previous round. This is continued until convergence.[2] The method described in Besag (1986) is a model-based approach in which dependency relationships between pixels are expressed in terms of *variate* interrelationships rather than the *correlation* structure. This focuses attention on the analysis of spatial data in terms of variate relationships recalling earlier work by Whittle (1954), for example, and also Besag (1974).

The remote sensing application of this approach must recognize the multivariate nature of the record. In addition environmental images frequently contain linear features (roads, railways, hedges) and there are, presently, difficulties in specifying appropriate models that can generate or retain such features in a scene. Finally the record for any pixel depends not only on the true scene covered by that pixel but also on the *record* at neighbouring pixels. This arises from the effects of pixel overlap. This requires some modification of the basic approach in its application to remote sensing problems.

A Geographical Information System (GIS) is a computer-based system for the storage, manipulation, retrieval and display of geographical data

[2] It has been noted that whilst this procedure could be computationally expensive the latest generation of 'super computers', that use parallel processing, are ideally suited to this sort of analysis, converting a computational problem that would have been sequential (each pixel in turn) to one that can be handled 'in parallel' (updating each pixel at the same time).

(see, for example, Rhind's chapter). Much importance attaches to developing efficient methods for accessing and retrieving the spatial data stored in the system. Increasingly attention is also moving towards manipulative capabilities. At present these appear to centre on overlay operations (in which two or more of the stored data sets are related) and 'buffer' operations so that composite maps can be examined for specific fields or sub-areas (see Green, Finch and Wiggins, 1985). In the future one may anticipate the development of analytical facilities, as for example in SPANS, but these will need to allow the analyst to handle (amongst other things) the sorts of problems reviewed in the second part of this chapter. It will not do to integrate these systems with 'standard' statistical packages for purposes of correlation, regression or indeed any other techniques for multivariate spatial analysis.

In conclusion, some of the comments in this chapter have pointed to the importance of statistical training for geographers that goes beyond standard statistical methods. This implies a change in the emphasis of quantitative textbooks that are written for geographers. The point has been made forcibly quite recently: 'Textbooks on quantitative geography still frequently contain large sections on standard statistical methods assuming independent observations. . . . These are unlikely ever to be appropriate in geography. . . . A revolution is needed' (Ripley, 1984: 342).

14 Modelling, Data Analysis and Pygmalion's Problem

Nicholas J. Cox

The relationship between 'modelling' and 'data analysis' provides one broad theme for discussing the chapters by Haining and Rhind. Although these expressions are often used rather loosely, a serviceable distinction can be made. In modelling, the construction of a model, preferably as a formal representation of a serious theory, tends to precede any detailed investigation of a data set. Much of the effort and interest is likely to be devoted to working out the consequences of model assumptions, either by analytical derivation or by numerical simulation. In data analysis, exploration of a data set frequently leads to the development of quantitative summaries intended to capture general patterns and structures. If these summaries are presented as 'models', they nevertheless are usually empirical generalizations for which theoretical interpretations are sought retrospectively. For modellers, therefore, the model is produced before the data analysis, while for data analysts it is produced by the data analysis.

A stark distinction like this will strike many readers as a caricature, and qualifications are immediately in order. Often in research, modelling and data analysis are combined intimately, and investigators may iterate back and forth between model and data, making any idea of a single logical path quite inapplicable. It is, I hope, a truism that model-based and data-based approaches should be complementary and mutually indispensable in many projects, and certainly in geography as a whole. In any case they intergrade continuously. A linear regression can serve either as a low-level empirical generalization or as a means of testing a serious theory which leads to linear relationships, either directly or through transformation (e.g. by taking logarithms of power or exponential functions). Hence the term 'regression model', while common statistical usage, may occur in very different situations where the amount of serious formal theory varies considerably, from zero upwards.

The qualifier 'formal' is important here, meaning, as is standard, expressible in some mathematical language, as a series of equations or an algorithm for example. Certainly data analysis never proceeds in a theoretical vacuum: there is always a context of ideas which influences the kinds of questions asked, the kinds of data thought relevant, and so forth. In practice, however, it is often the case that most or all of the theory is informal, i.e. verbal or even unconscious, telling us that variables are likely to be related, but giving us no clear guidance on the form of their relationship. Hence there is no implication here that data analysis is a purely inductive or empirical approach. The main point to carry forward is that of a distinction between modelling and data analysis defining endpoints of a continuum, along which there is variation in the character and role of models and the degree to which theory is developed formally.

Taking a broad view, I think it is correct to claim that the majority of quantitative work in geography until about 1970 was nearer the data analysis end than the modelling end of this continuum. There was a preponderance of statistical applications, including standard descriptive statistics, correlation and regression and various multivariate and spatial techniques. The period was one of exuberant experiment about which it is all too easy to be extremely critical in retrospect. Most of this work was (simultaneously) over-concerned with the use of what were to geographers fairly new and unfamiliar techniques and unaware or neglectful of specific technical problems which they posed. It was usually heavily descriptive in a manner which showed more continuity with the style of 'traditional' (e.g. regional) geographers than was generally appreciated. Effort devoted to producing or testing geographical theories was correspondingly limited.

Greater concern with modelling and theoretical work became especially evident in the 1960s. *Models in Geography* was naturally one of the founding documents of this movement. One important strand in geographical modelling is very well represented by Haining's chapter. It has developed from the realization that spatial data pose special problems, particularly because they are almost certainly mutually dependent, contrary to one common simplifying assumption in statistical theory. Negatively, this messes up many estimation and hypothesis-testing procedures based on mutual independence (although these are still included in most introductory textbooks available for geographers). Positively, it has led to a research programme of identifying and developing stochastic models and statistical methods that are appropriate for spatial data. This programme has been undertaken by a small but highly productive group of geographers who entered research about 1970, including Bennett, Cliff, Griffith and Hepple, as well as Haining himself. They have built upon the research of an earlier generation, especially Curry and Dacey. Naturally, geographers are not alone in their interest in spatial data. In statistics itself Besag, Cressie, Diggle, Ripley and Switzer have been particularly active since about 1970, following the earlier work of Bartlett, Geary, Matérn, Moran and Whittle.

While widely respected for technical expertise, I think it is true to say that this programme has had a very limited impact on geographical research as a whole. The most obvious reason is that even now strongly quantitative work is only a minority pursuit, especially in human geography (cf. Flowerdew, 1986). More specific reasons can also be identified.

First, the problems posed by spatial data have generally turned out to be messy, and lacking in simple, neat solutions. Recourse to simulation is usually necessary, and appropriate software is frequently developed *ad hoc* by researchers, and not widely available through standard packages, as Haining points out. These intellectual and practical difficulties have limited widespread understanding and application of the more correct procedures that have been produced.

Second, viewed as a whole the programme has often focused on models which lack fundamental theoretical interest and have not been applied to many substantive problems. There are important exceptions to this brash assertion, such as the work of Cliff, Haggett and Ord in epidemiology, but it can be developed in a little more detail. The models discussed in most cases have little or no basis in terms of physical or human processes or behaviour, the most usual hallmark of theoretical argument in geography. Thus autoregressive or moving-average processes do not seem to have a direct geographical basis, whatever the more or less plausible interpretations that can be suggested for them. This need not be a fatal criticism even when it is totally justified, because the mathematical structures might still be useful in data analysis. How far this is the case can only be judged from applications to a large variety of data sets.

In research of the kind reported by Haining empirical examples are often very brief and based on very small data sets, and may come together with a disclaimer that the example is not to be taken as a serious substantive application. This is legitimate for many individual papers – only so much can be done in any one paper – but it is disappointing to see it done frequently. Naturally, models can be valuable even if unapplied, so long as they are a tool for theoretical reasoning, and enlarge or revise our ideas. Otherwise the only role left for models is as structures used for the analysis of real data, and in the investigation of interesting substantive problems.

Pygmalion's problem was that he fell in love with his model because it (she) was so beautiful. With divine assistance the model became real. This wonderful solution is beyond our dreams, and while we may suffer Pygmalion's temptation we must strive to avoid it. Much of the work on spatial statistics has appeared to focus excessively on the models themselves, with disappointingly little theoretical or substantive impact. Any criticism must be shared by data analysts who have not taken up the ideas of this programme and, to give a more specific example, by agricultural and settlement geographers who appear to have thought rather little about the implications of Haining's work on the Great Plains for their analyses (Haining, 1978).

Few geographers can be unaware of the extremely rapid developments which have taken place in the availability and performance of computer hardware and software, and of the extent to which very large data sets are increasingly accessible via computerized data bases and information systems. Rhind's chapter outlines these changes and challenges us to face up to their implications, warning that others outside the discipline will seize the opportunities if geographers fail to do so. It is clear that geographers specializing in geographical information systems will survive only by combining considerable technical expertise with a taste and talent for the *Realpolitik* of grants, contracts, committees, administrators and entrepreneurs. A canny awareness of commercial and political realities will be as necessary as any particular intellectual qualities.

For the past fifteen years or more, Rhind has stood out as a singularly well-informed guide on where geographical computing is – and should be – going. I do not dispute his predictions, but guess myself that only a few geographers will follow in the directions he indicates and be significant developers in fields like GIS. Almost everyone else will be either users or non-users of geographical computing.

Actual or potential users would do well to think very hard whether the research questions that interest them can be answered by data accessible via information systems. Obviously it is difficult to overestimate the impact that modern computing facilities have had, and will have, on work in fields like employment and population, in which the geographer can analyse very extensive data sets, national and international, in a way hardly dreamt of quite recently. Even more, work in remote sensing is totally inconceivable without computers. But there are examples with the opposite interpretation. In my own field of geomorphology, most of the great advances have stemmed from sharp, intensive thinking followed by a collection of a truly relevant data set, which did not exist previously, and which was often extremely small by Rhind's standards. Several were the basis of a Ph.D. thesis (e.g. Schumm, Melton, Dunne, Church). Successes like these should still be possible for many years.

The advantages of very large data sets need little elaboration, but a few qualifications deserve mention. Sampling variation, as measured for example by the standard error of the sample mean, is inversely proportional to the square root of sample size, and increases more and more slowly with sample size. Multiplying sample size by 10,000 reduces standard error by 99 per cent, but of that 90 per cent comes from multiplying by 100 only, i.e. 1 per cent of the effort of multiplying by 10,000. Long before we reach very large sample sizes it is often the case that imprecision arising from sampling variation is swamped by biases which do not depend on sample size. Hence the greater precision of larger sample sizes is a phenomenon of diminishing returns and may not affect the main problem of data quality. In fact a small or moderate data set collected by a dedicated individual is frequently of much higher quality than a large data set, collected by several people who may be untrained, unscrupulous or simply a group of competent professionals who are working from different operational definitions in circumstances which are not genuinely comparable. This last situation is very common in fields like geomorphology.

The greatest worry, however, possibly trite but nevertheless arising from a genuine issue, is that too much research in the Brave New World will be guided by whatever data happen to be readily and copiously available. The danger will be not Pygmalion's problem, but the opposite, that preoccupation with data will limit imagination and creativity. The difficulty is that arranging access to data bases is far easier than stimulating people to ask the right questions in research and come up with good ideas. I am sure that Rhind would agree that this stimulation is vital. If GIS were to contribute to a new empiricism, which neglected the role of ideas, models and theories, then in some ways we would be back where we were twenty years ago.

PART V
The Critical View of Modelling

15 From Models to Marx: Notes on the Project to 'Remodel' Contemporary Geography

David Harvey

The publication of *Models in Geography* in 1967 marked a watershed in the development of geographic thought. It symbolized the consolidation of a whole new approach to the subject matter of geography by an influential and soon to become hegemonic group of the discipline's practitioners. My own essay in that volume, under the title 'Models of the evolution of spatial patterns in human geography', sought to address a number of very specific concerns. Far from dissipating, these concerns have remained a constant in my thought over the last twenty years. What has changed is the way of seeking answers.

The introduction to my essay put the question bluntly and succinctly enough. I argued that all geography is necessarily historical geography because it is only through a study of the genesis of social forms and an analysis of the process of change over space and time that we can understand our contemporary geography. I bewailed what I then saw (and still see) as an 'unfortunate gap' between the scholarly studies of historical geographers and the burgeoning analytical techniques of human geographers concerned with contemporary distributions. The focus had, therefore, to be on process, genesis and transformation. I rejected Hartshorne's continuing insistence on a purely chorological approach to geography that diminished the significance of time and history. I appealed to those geographical works (by authors as diverse as Ratzel, Whittlesey and Sauer) that dealt directly with social transformations over time and space.

The problem, as I then saw it, was the seeming inability to define a rigorous method for investigating spatio-temporal processes. Lacking that, geographers were either imprisoned within the idiosyncrasies of idiographic reconstructions or forced to resort to unruly metaphors or outrageous analogies (of which the organic analogy was by far the most pernicious) as substitutes for rigorous analysis and argumentation. I judged it imperative, therefore, that the geographer 'become a scientist and attempt, by the normal procedures of scientific investigation, to verify, reject, or modify, the stimulating and exciting ideas' handed down from earlier generations.

So what has changed? I still insist, even more so now than then, that all geography is necessarily historical geography, that a concern with genesis and spatio-temporal processes of social transformation must dominate over pure description of spatial pattern, and that we must search out ways to construct scientifically rigorous theory that can shed light on the often tortuous twists in the historical geography of human occupancy of the earth's surface. What has changed is my conception of the science and of the method that can properly be brought to bear on such a subject matter. For the overwhelming presupposition of *Models in Geography* was that a particular conception of scientific method, largely drawn from the experimental and natural sciences, could be brought to bear on all geographical problems, regardless of their form and content. It is in terms of the various challenges to that basic presupposition that the subsequent fragmentation and evolution of geographical thought has to be understood.

There was, I suspect, more than a little naiveté as to what constituted 'natural scientific method' and an even greater naiveté as to how it might be applied to the kind of subject matter that I, at least, had in mind. On the one hand, the positivist search for a 'neutral' language with which to render objective descriptions of a world too long viewed through subjective lenses appeared to be a thoroughly praiseworthy objective. The definition of such a language suited to understanding the rich variety of historical geography proved difficult if not intractable. In practice, that rich variety ended up being crammed into markov chains, simple learning and diffusion models, and the like, simply because these were the mathematical process languages that lay most easily to hand. It proved, in practice, very difficult to evolve any real theoretical argumentation with such mathematical languages and even more difficult to apply positivist (Popperian) standards of verification and falsification under conditions where the phenomena being studied looked more like a continuous stream of historical–geographical evolution. While it might be possible to build models of some elements within that general process, such as the spatial diffusion of country banking or automobile ownership, this did not necessarily do more than add a slightly different dimension to our understanding of more general historical–geographical processes. Questions of the transition from feudalism to capitalism, or the evolutionary path of capitalism itself, appeared to be uncapturable by that kind of modelling. Indeed when put in the context of these grander questions, the modelling effort appeared both puny and not particularly revealing even when it might reasonably be undertaken and reasonably verified by techniques that broadly rested on the conception of independent and replicable events.

Those who have stuck with modelling since those heady days have largely been able to do so, I suspect, by restricting the nature of the questions they ask. I accept that we can now model spatial behaviours like journey-to-work, retail activity, the spread of measles epidemics, the atmospheric dispersion of pollutants, and the like, with much greater security and precision than once was the case. And I accept that this represents no mean achievement. But what can we say about the sudden explosion of third world debt in the 1970s, the remarkable push into new and seemingly quite different modes of flexible accumulation, the rise of geopolitical tensions, even the definition of key ecological problems? What more do we know about major historical–geographical transitions (the rise of capitalism, world wars, socialist revolutions, and the like)? Furthermore, pursuit of knowledge by the positivist route did not necessarily generate usable configurations of concepts and theories. There must be thousands of hypotheses proven correct at some appropriate level of significance in the geographical literature by now, and I am left with the impression that *in toto* this adds up to little more than the proverbial hill of beans. This may in part be attributed, of course, to the sad degeneration and routinization of the modelling exercise into mere data crunching, numerical analysis, and statistical inference instead of careful theory-building. The latter presumes, however, that we know (or agree) what we are theorizing about. Perhaps it was because I was trying to theorize about the historical geography of capitalism that I broke with most of my colleagues and sought for a different though equally rigorous path for building theoretical understandings. But where could we turn for a properly constituted theory and method that could cut through to a new plateau of understandings of the world's historical geography?

One of the virtues of traditional geography, often dubiously capitalized upon to be sure, was its resolutely materialist orientation. Environmental or geographical determinism – perhaps the only real theory that geographers did develop with some coherence – came perilously close to a purely materialist theory of historical determination. And even the so-called 'possibilist' response, while it concentrated upon human agency, voluntarism, and even idealism as human motivation, still exhibited a strong appreciation of the constraining powers of environmental conditions and an attachment to the material landscape as an artefact to be observed and understood. The 'region' formed the typical 'theoretical object' of geographical enquiry (and, as such, came under increasing epistemological scrutiny of the sort that David Grigg offered up in his chapter in the *Models* volume). Only recently have geographers concerned themselves with the *terra incognitae* of the mind, and sought to integrate understandings of mental maps, geographical perceptions, and interpretative experience of the 'geographical life world' into their more traditional concerns.

It seems extraordinary in retrospect that a discipline so deeply imbued with materialism should have for so long remained untouched (some would doubtless say 'unscathed') by Marx's historical materialism. I am at a loss to explain the absence of any explicit Marxian tradition in 'Western' geography. But in 1967, Marx and Marxism were scarcely to be found in geography (even though André Gunder Frank published his seminal *Development of Underdevelopment in Latin America* in that same year). The only reference in the *Models* volume is in the essay by Pahl on 'Sociological models in geography' and there as a passing mention in the context of debates over sources and patterns of social change in sociology. The absence of any explicit Marxian tradition is all the more odd, given that many geographers had certainly flirted with left ideas in the 1930s (Cambridge was a hot-bed of that, for example) while in France a whole raft of geographers, with Pierre Georges at the helm, were Communist Party members with no noticeable impact upon their geographical work. It was almost as if the work of geographers was sufficiently materialist in its own right to obviate any debt, direct or indirect, to Marx. Lattimore certainly argued so when he insisted that his understanding of commodities came from tracking them on the ground across the trade routes of inner Asia, rather than from any familiarity with Marx's *Capital*. Wittfogel, who saw the connection between geography and Marx's historical materialism very clearly in the 1920s, muddled matters most horribly in the post-war period with his strange brew of geographical determinism, abstract historical materialism (I say 'abstract' because Wittfogel was invariably shown to be wrong on the ground) and virile anti-communism. McCarthyism took its toll in the United States, of course, and the cold war had its victims in Britain too (an episode in the history of British geography that has yet to be unravelled – it certainly had many leftist sympathizers marginalized or 'exiled' to posts in the far-flung niches of a crumbling British Empire). The problem in geography seems to have been that there was no-one of the stature of a Maurice Dobb or the guttsiness of an E. P. Thompson to keep the flame of Marxian scholarship (if it had indeed ever really been kindled) alive through the dark days of the cold war and into an era of heightened social concern in the late 1960s. In the absence of a Marxian option, those who sought remedial action on social problems (and many of those involved in the *Models* volume were deeply concerned with mobilizing the power of knowledge for purposes of human emancipation) did so through the medium of an enhanced technical bureaucratic rationality. And this was, of course, the background ethos and ethic (in so far as there was one) to the *Models* volume.

The confrontation between geography and historical materialism, when it finally came, opened up entirely new paths to understanding the historical geography of human occupancy of the earth's surface. It also exposed what Marx called 'the weak points in the abstract materialism of natural science, a materialism that excludes history and its processes', and which inevitably led those who saw the unity of geography as a unity of method into 'abstract and ideological conceptions' of the world. The alternative conception of 'science' that Marx proposed suggested a path towards rigorous theory construction and of cogent historico-geographical interpretation. It also indicated that the unity of human scientific endeavour had to rest upon the 'humanizing' and historical understanding of natural science rather than upon the shaping of the social sciences, geography, history, and the humanities according to the 'abstract materialism' of natural science.

But Marxian method is not understood overnight and its practice, like any craft, requires training, critical evaluation and, above all, experience. Lacking any Marxian tradition, those who took up the Marxian challenge were forced to be auto-didacts. That meant a willingness to experiment and make mistakes. It also produced an atmosphere of vigorous confrontation and excitement. Geographers were not alone in this. A vast wave of Marxian insights broke over all the social sciences in the late 1960s, reinforcing the hands of those political economists, historians and philosophers who had clung to Marx all along. That wave, in part drawing its energy from the social turmoil of the late 1960s (the inner-city riots in the United States, the Vietnam War, the student movement, the struggle for empowerment on the part of many oppressed minorities) surged into academia, transforming the intellectual landscape in seemingly irreversible ways. Geographers who turned to Marx were swept up in that, and many were so submerged in it that they entirely forgot their own disciplinary identity. In my judgement such submersion was vital. To the degree that the academic and disciplinary division of labour is an invention of the bourgeois era, it, too, has to be understood historically. To the degree that Marx proposed an approach to the unity of human intellectual endeavour entirely antagonistic to both traditional 'bourgeois' divisions and the radical reformulations for unity of the sciences (so attractive to geographers beset by the human–physical divide) proposed by the positivists, it required a fundamental epistemological if not ontological re-tooling. To pretend that we can put ourselves 'outside of' or 'above' society through the construction of some entirely neutral or objective language was, in Marx's view, immediately to falsify any claim to 'true' understanding. How could we aspire to 'true' understanding when our first move was to deny who, what or where we were in society? The problem was to find a way to the production of objective knowledge from *within* society's confines, recognizing the sociality of our own endeavour from the very outset. This we could do, Marx insisted, only if we were to see the production of knowledge as a political project irreversibly implicated in the organizing of power relations. In a class-bound society that means that the production of knowledge is inevitably caught up in the dynamics of class struggle. The 'rationality' of scientific–technical understandings, together with their legitimacy of application, depends quite simply on class orientations. Only through a self-critical understanding of class position and class interests would it be possible to construct objective understandings of the tortuous twists and turns in our historical and geographical development.

There are, to be sure, innumberable pitfalls and difficulties with the Marxian endeavour, and I would not want for one minute to suggest that it contains the instantaneous resolution of all those problems that positivism and the modelling strategies that derived from it either ruled out of consideration or simply left unattended. For my own part I have found the path to Marxian understandings both difficult (politically as well as intellectually) and deeply rewarding (though at this juncture I have to say it is more so intellectually than politically). But if, as seems self-evident to me (though most seem to want to avoid the issue), capitalism is a transitory form of social, economic and political organization (just think what it has done in a couple of hundred years and imagine what it would have to do over the next hundred just to maintain those logistical rates of growth), then someone, somewhere, has to be thinking about both what will replace it and the when and how of its transformation. And if the world is beset by all manner of problems – spiralling indebtedness and financial insecurity, militarization, widespread unemployment and not a little social anomie – then it is vital to find a way to represent those phenomena and their geographical dynamics in a critical but objective way. It is this that Marxism at least attempts, even if there are good grounds for thinking Marxist formulations incomplete, still open-ended, and by no means omnipotent in confronting the realities of daily life.

Any project to 'remodel' contemporary geography must take the achievements of the Marxist thrust thoroughly into account at the same time as it recognizes the limits of positivism and the restricted domain of the modelling endeavours that derived therefrom. This is not to rule all forms of mathematical representation, data analysis and experimental design out of order, but to insist that those batteries of techniques and scientific languages be deployed within a much more powerful framework of historical–materialist analysis. But the confusion between positivism, mathematics and data analysis has, I fear, gone far too deep to allow any easy disentanglement of that particular knot. Nevertheless, as Scott and I try to argue in the chapter that follows, theory construction has to be central to our concerns and it is to historical–materialist theory that we must turn if we are to void the plague of abstracted, ideological representations of the world that have flowed from two generations of modelling endeavour.

The irony, of course, is that such a call for remodelling geography along these lines will be perceived by most as an ideological call, precisely because it is necessarily political and built upon some conception of our collective agency in history. So let me end by turning the tables and making what I believe to be the correct counterclaim: that those who stick to the abstact materialism of natural scientific method, and who locate their modelling efforts within that frame, are the true ideology producers in contemporary geography. Our presence as actors in history, though it can be evacuated as ideological by the back door, is omnipresent out front.

16 The Practice of Human Geography: Theory and Empirical Specificity in the Transition from Fordism to Flexible Accumulation

David Harvey and Allen Scott

This chapter is built around a central thesis: that in contemporary society commodity production for profit – i.e. capitalism – remains the basic organizing principle of economic life, despite the various perplexing changes in the concrete forms this principle has assumed these past few years. The production of geographical knowledge, therefore, has to hold fast to a vision of the forces at work under capitalism in general. At the same time we must grapple with the complex dynamics of new and emerging structures of accumulation and with their detailed spatial manifestations.

From Fordism to Flexible Accumulation

What many now call the Fordist and neo-Fordist regime of capitalist accumulation was put into place step by step over the first half of this century as private and public initiatives were brought to bear on the means of organizing commodity production and social life in the context of the dynamics of the class struggle and the crisis tendencies of capitalism. On the one side, the core elements of the production system were steadily transformed in conformity with the dictates of mass production, and out of this process emerged the major propulsive sectors of the inter-war and post-war decades (together with the great manufacturing regions they engendered). On the other side, Keynesian and welfare-statist policies were put gradually into place as a means of coordinating and regulating this regime of accumulation. The whole was sustained by a tripartite social contract involving the large corporations, organized labour, and the state. In this process the state mediated historical compromises between management and the unions in which, for the sake of guaranteed shares in productivity gains (including social consumption), the latter agreed to an uneasy but relatively durable industrial peace.

This Fordist regime of accumulation reached its apogee in the long post-war boom, but ran into increasing difficulties towards the end of the 1960s. By this time, competition from Japan and the newly industrializing countries was becoming persistent and severe, stagflation started to emerge as a central problem and, in a context of diminishing productivity, Keynesian welfare-statism became an increasing burden on the public treasury. The Vietnam War and the oil shocks of the 1970s added further to the problems of the advanced capitalist societies in general, and of the U.S.A. in particular. The old tripartite social contract could no longer hold in the face of these crises. The 1970s and early 1980s were thus a troubled period of economic restructuring and social readjustment in which Fordist mass production and the political arrangements which had sustained it were dramatically reorganized. In the social space created by these processes of flux and decline a series of novel experiments in the areas of production and political life began slowly to take shape. These experiments represent the early stirrings of what is increasingly coming to be referred to as a regime of flexible accumulation, and which now seems progressively to be moving to centre-stage in all the advanced capitalist societies. In contrast with the rigidities of mass production and Keynesian welfare-statism, the new regime is distinguished by a remarkable fluidity of production arrangements, labour markets, financial organization and consumption. It has at the same time engendered new rounds of what we might call time–space compression in the capitalist world – the time horizons of both private and public decision-making have shortened drastically, while electronic communications systems make it increasingly possible to spread the effects of those decisions immediately over an ever wider and more variegated space, generating major changes in patterns of geographical development. These trends have been accentuated by the emergence of entirely new sectors of production, new ways of providing financial and business services, new markets and, above all, greatly intensified rates of commercial and technical innovation. Organized labour, already weakened by the restructuring and high unemployment rates of the 1970s, has been further undercut by the reconstitution of foci of (flexible) accumulation in areas lacking in previous industrial traditions – and by the importation back into older centres of the regressive employment norms and practices established in these areas. In the new climate of entrepreneurialism, privatization and competition, the social wage has also been dramatically reduced, and public austerity is now the watchword in all the advanced capitalist societies, even those still governed by socialist parties.

All of this has been accompanied, and in part ushered in, by the rise of an aggressive neoconservatism in North America and Western Europe. The political and electoral successes of this neoconservatism have tended to throw left-wing parties and organizations into disarray, and this in turn has had disastrous consequences for their ability to put together coherent alternative analyses and programmes. In a world in which flexible accumulation is already significantly undermining established expectations about the meaning and rewards of life in capitalism, so the rise of neoconservatism has helped to accentuate perceptions of the world as being made up of fugitive, contingent and localized fragments, like disorganized shards of glass. This fragmentation, however, has also been accomplished through an increasing centralization of power within a deregulated world financial system through which entirely new constellations of financial power are being brought into existence through mergers, take-overs, and inter-linkage. The problem here is that these events have effectively rendered opaque the continued self-reproduction of capitalism through commodity production and exchange and the perpetuation of class relations. The surface ebullition of contemporary capitalist society and all of its shifting transitoriness conceal or reduce to mere 'naturalness' its stubborn underlying continuities.

The Theoretical Crisis of the Late 1970s and its Aftermath

As Marxian social theory gathered momentum over the 1960s and 1970s it came up with increasingly elaborate and widely ranging studies of the logic of Fordist industrialization and the Keynesian welfare-statist policies that helped sustain it. These studies can be divided into three broad and non-exclusive categories. First, there were a number of major attempts to theorize the state, and in particular to show that the conjuncture of class relations in Keynesian welfare-statist capitalism led to ever-deepening predicaments of public finances (at local as well as the national levels), thus provoking crises of governance, legitimation and political control of working-class and other popular movements. Secondly, the broad dynamics of the production system were intensively scrutinized to try and understand how the laws of motion of capitalism, technological change, transformations in labour processes, the formation of the rate of profit, and the like – were manifest under conditions of monopolization and imperialism. Thirdly, and most important in the realm of geography, came a proliferation of critical studies showing how Keynesian welfare-statist society was organized around phenomena such as the delivery of housing, health care, educational and other services, and how land rent, planning and local class relations played a role in the perpetuation of inter-regional and inter-urban inequalities and as mechanisms of social control.

These studies were, to be sure, far from monolithic in character, and there were vigorous debates and disagreements at virtually every turn. Nevertheless, two features seemed to be widely, though not universally, recurrent in this work. On the one side, structuralist notions of social order constituted a major philosophical/theoretical underpinning of much of the work, particularly in its early stages. On the other side, many investigations (especially those with an explicit or implicit Hegelian sense of history) seemed to fall by default or design into a teleological view of capitalism and to presume that some kind of movement towards socialism could already be discerned in the developmental course of capitalism. Ideas with respect to the transition from competitive through finance–monopoly–imperialist capitalism to state–monopoly capitalism seemed to point, via fiscal and political crisis, towards socialism as the only possible next step. State intervention, it was often presumed, was increasing down a one-way street as the tasks of managing an ever more complex society multiplied, as social movements grew and intensified their claims and as the state was forced to respond in order to maintain social and political legitimacy.

We do not wish to deny the many useful, and indeed enduring, insights that this work produced. But from the mid-1970s an increasing disparity arose between prevailing theory and the actual evolution of most of the advanced capitalist societies. These discrepancies began, slowly at first, and then at an accelerating rate, to undermine confidence in the general value of Marxian theory as both scientific synthesis and as a programme of political action. In particular, by the time that Thatcherism and Reaganism began their rise, it had become apparent that the state was now in full retreat from many of its earlier social commitments and, astonishingly, the result was not so much a resurgence of the political and legitimation predicaments that its intervention was supposed to have resolved in the first place, but a general acquiescence in the politics of neoconservatism.

The massive re-structuring of the capitalist economy that had been going on for some ten or fifteen years was now approaching a major watershed. By the end of the 1970s, and certainly by the early 1980s, the Fordist regime of accumulation was being displaced by flexible accumulation as the dominant way of doing capitalist business. The resurgence of economic competition and entrepreneurial activity at all levels in the economy undermined theories of monopoly or state–monopoly capitalism and certainly made any implied teleology in earlier transitions moot. The rise of high-technology, artisanal, and service sectors of production, the fragmentation of institutionalized working-class power (the unions in particular), the increasing dualism in labour markets and the appearance of thriving new industrial regions and spaces in areas formerly shunned by industrial capital, suggested that the traditional views of the ability of working-class power to shape the trajectories of capitalist development were open to question. There were, as it turned out, very few arenas of Marxian enquiry that were not called seriously into question by the crisis of Fordism and the rise of flexible accumulation and by its associated politics of neoconservatism.

The movement of events provoked deep cleavages within the field of Marxian theory. Just as the bloom was fading from structuralist interpretations, for example, E. P. Thompson launched a trenchant attack upon its conceptual underpinnings. Thompson isolated for especial critique its rigid sense of social relations, its totalizing closure, and its depreciation of the role of consciousness in historical eventuation. Many of his criticisms were well-founded and his attack certainly opened the way for a new respect for history among Marxian social scientists. Structuralism, particularly in its Althusserian version, was depicted by Thompson as a pathological aberration on a par with Stalinism, and indeed conducive to the latter. At the same time, those who focused rather exclusively on the role of working-class power in the supposed teleology of capitalist history also had to face up to the fact that the working class as a whole (in spite of continuing struggles on the part of some segments such as the air controllers in the United States or the miners in Britain), seemed not only powerless but even partially to accept a role of complicity in the neoconservative thrust.

In view of the increasing mismatch between much of grand theory as it was then conceived and these emerging realities, and in view also of the damaging cross-fire among the major figures of Marxism as to how to react to these new circumstances, it is hardly surprising that we have since experienced a decisive retreat from theoretical work and an increasing fragmentation of research concerns. Along the way, strong counter-movements such as deconstructionism, post-modernism and post-structuralism, pragmatism, and a sort of 'new naturalism' (reinforced by realist philosophy) in the social sciences have also taken their toll and helped intensify the withdrawal from attempts at theoretical synthesis.

In human geography we have, as a consequence, seen a re-assertion of the primacy of empirical research and a fixation on the specificity of the local, as opposed to a continuing concern for elucidating the generality of capitalism in its totality. Here again, we do not wish to deny the important and fruitful lines of research that have been opened up by these developments, and to some degree this retreat into the dense empirical details of specific historical and geographical situations can be seen as a sensible and necessary step in the struggle to come to terms with the puzzling realities of the current (transitional) conjuncture. At the same time, as at least some of this work has decisively turned its back on any theoretical engagement with the nature of capitalism as a whole, so has it also become increasingly sterile in both scientific and political terms.

Recent developments in the area of conceptual elaboration have also raised a number of problems. By the early 1980s much theoretical work on the left in human geography had been reduced a variety of pallid, depoliticized and ultimately vacuous versions of the theory of structure and agency. We say vacuous precisely because by this time, in many accounts, the very notions of structure and agency had been shorn of any very definite content in terms of political economy or the laws of motion of capitalism as a whole. The idea of human agency itself came to be seen as an implicit suggestion that there might not even be such laws of motion, and there was a tendency to ascribe unplanned, macroscopic social outcomes to the workings of 'unintended consequences'. This is, of course, in one sense correct since such outcomes (like flexible accumulation itself) are the overall result of myriad individual choices, decisions and behaviours. But the problem here is that the very language of 'unintended consequences' suggests that broad social change is somehow only a dissolved side-effect of the inchoate swirl of human agency. This swirl, real as it may be, is, we insist, bounded and shaped by the stubborn logic of capitalism as a whole, and rationally explicable as such. This, surely, is the only reasonable interpretation that can be put upon structuration theory once it is grounded in concrete historical referents. At its best (as in some of the more polished realist accounts) the tendencies we are criticizing have given birth to research that evinces a sensitive understanding of the palimpsests of local events, and strong insights into the interplay between the fragments and contingencies of capitalist development. At its worst it dissolves into descriptive recitations of regional characteristics, with occasional Marxist atmospherics as remote stage scenery. In any case, how all of this related to capitalism in its entirety (or, indeed, if capitalism in its entirety could even now be said to be a sustainable concept) remained far from clear.

We maintain that something decisive has been lost in much recent Marxist work in human geography, and that the disengagement from explicit theoretical work rooted in political economy has been an understandable but self-defeating retreat. In particular, we take a stand against the notion that the current phase of capitalism is marked by steadily disorganizing tendencies in which the coherence and unity of the capitalist system is dissolving away. We are also opposed to the slippage that begins with the notion that history and geography are constructed by human agents in unique places with an open-ended future, and ends with the notion that anything goes, especially at the micro-scale. Indeed, we want to suggest that the issue of the local is considerably less potent than some of its protagonists seem to imply. Above all, we argue that a re-assessment of current analytical priorities dominated by empiricist lines of inquiry (infused with notions of contingency, the open-endedness of outcomes, and idealist views of agency) is long overdue.

We recognize the potential for misunderstanding in these remarks, and we hasten to add two provisos: (1) that the detailed realities of historical and geographical transformation are not immediately deducible from general theorizations, and (2) that we in no sense wish to deny the significance of freedom of action, of imagination, and of will in the construction of alternative realities. But these provisos also imply a third point, viz.: in order to comprehend the complexity of real historical and geographical processes in capitalism we must be armed with, and be prepared to operationalize, theoretical ideas about the workings of capitalism as a total system. Only in that way can we get behind the fragmentary, the contingent and the ephemeral characteristics of the modern (post-modern?) world, and tackle its underlying systematicities.

The Theoretical Imperative

We still live in a world dominated by capitalism. There are, therefore, scientific and political imperatives to build theory of sufficient power to keep the totalizing behaviour of this mode of production clearly in view, particularly when the surface confusions of contingency and ephemeral change mask underlying dynamics. We use 'theory' here in its usual Marxist sense, to mean the creation of the intellectual preconditions for self-consciousness of the structures of capitalist domination coupled with the construction of coherent representations and analytical tools to facilitate the struggle for human emancipation. Our ability to know the world, and to represent it truthfully, is essential to this emancipatory enterprise.

All representations of the world, including those put forward under the aegis of pure empiricism, carry implicit theoretical presuppositions and codings. It is therefore particularly important to continue the task of clarifying the presuppositions underlying research in human geography at a time when the increasing flexibility of contemporary practices of accumulation obscures underlying realities, when there is little agreement as to what constitutes a valid conceptual framework, when meta-theory has in any case been largely reduced to background atmospherics, and when the analytical foreground to research is occupied by a multiplicity of competing and fragmented discourses focused on the local and the empirical.

What we are, in effect, calling for here, is a major theoretical effort that transcends the disconnected plethora of approaches and findings that have been generated these past few years in the course of trying to come to grips with the surface appearances of flexible accumulation. These approaches and findings, we submit, must either be synthesized and integrated into some more general theory of the spatial dynamics of capitalism, or be rejected as mere surface gestures and representations that mystify rather than clarify underlying meanings. This task must be pursued in the full recognition that regimes of accumulation do change over time, and that the arenas of accumulation shift around in geographical space. There are, nevertheless, certain durable aspects of capitalism that we wish to insist upon in striving to build a holistic theory able to capture its totalizing behaviour. And that means dealing with basic concepts of class relations, capital accumulation, commodity exchange, money forms, finance capital, state formation and the various manifestations of oppression endemic to capitalism. We need, then, to keep the theory of the totality of capitalism very much in the foreground of all analysis.

We should be explicit, however, as to what we mean by 'totality' and the way in which a holistic theory might be produced. This is a crucial matter since there has been much criticism these past few years of attempts to build any kind of totalizing discourse or holistic meta-theory. We wish first to acknowledge the force of some of this criticism. For example, we reject the idea, derived mainly from Lukacs, that an understanding of the social totality has some ontological priority over an understanding of internal relations, subsystems, individuals and the like. This kind of argument has had particularly nefarious effects when researchers fall into the lazy habit of hypostatizing current theories of the totality. It also subsumes away the different logics and relationships that might properly be identified at meso- and micro-levels of analysis. We do not accept, either, that kind of totalizing discourse, often found among the leftist avant-garde in the inter-war years (and which was revived in some Marxist circles in the 1960s) that presumes either some dialectical necessity (teleology) in the transition to socialism or some total dominion over the future through the powers of reason mobilized in conjunction with the productive force of modern technology. Such discourses are quite antithetical to our conception of the role of individuals and other agents in processes of historical–geographical transformation.

We think of the production of holistic theory as a *project* of understanding the totalizing behaviours of capitalism. By the totalizing behaviours of capitalism we mean the way in which, e.g. (in addition to such familiar questions as the logic of commodity production or the operation of labour markets), the production of information, the marketing of that information through the media, the organization of pleasure and entertainment, the production of new knowledge, the division of labour within the household, and so on, are now mediated by capitalist social relations. At the same time, a reading of the financial press indicates immediately the far-flung global extension of capitalism and its imbrication in the social and political life of peoples scattered across the entire surface of the earth. It is this totalizing activity that holistic theory seeks to grasp; and plainly it will take major efforts of theoretical and empirical analysis to do so. How, then, are we to come to grips with this totality in the midst of all manner of fragmentations and empirical specificities?

Detailed investigation of the particularities of historical geography provides one gateway to comprehension of a totality that is always differentiated and multi-layered, always the product of human action, no matter how much individuals are held captive in their own material and ideological constructions under capitalist relations of production. Passage through these particularities can lead us to an understanding of the real universal qualities of capitalism. It is in terms of these universal qualities that we think of the totality, rather than in terms of a 'thing' that can be understood abstractly. Our task is to identify these universals within the fleeting, the ephemeral, the contingent and the fragmented aspects of daily life under conditions of flexible accumulation.

Recent inquiries provide us with abundant raw materials for such a theoretical project. But there seems to have been little effort devoted so far to the tasks of bringing these hard-won insights – sometimes cast in pure empirical form and in other instances incompletely theorized as part of some realist endeavour – back into the fold of a general theory of the political economy of capitalism. Which brings us to the problem of how that might be done.

Abstractions, we insist, should always in the first instance be rooted in the analysis of daily life. Their formation depends, therefore, upon the detailed appropriation of historical–geographical materials in a manner that respects the integrity and variety of human experience. The materialist method, however, entails a search for those 'concrete abstractions' through which the capitalist mode of production (or any other mode of production for that matter) is bound together into a working whole. Theory construction means the conceptual representation of such concrete abstractions and their linkage into a coherent analysis through a careful reconstruction of the necessary relationships that connect them together and ensure the reproduction of capitalism as a viable social system. Money is a prime example of a concrete abstraction. We deal with it daily, spend much of our life either getting it or disposing of it, and in many respects find ourselves bound to it, even ruled by concern for it. Analysis reveals how some form of money is inherent in commodity exchange and we can also see how money becomes the way in which the value of social labour is represented under capitalism to the degree that production for exchange becomes generalized. Careful study of actual processes permits us to identify other concrete abstractions such as 'commodity', 'division of labour', 'profit', 'labour power' and 'capital', at the same time that we gain information on the relationships between them. The very identification of such concrete abstractions opens up all kinds of possibilities for speculative theorizing. To the degree that money becomes a vital source of social power, for example, so the monetization of daily life, including those aspects that humanists are wont to treat under the rubric of the life world, becomes a real possibility (with many compelling examples that lie readily to hand), thus suggesting the totalizing hold that the sociality of money, and by extension the circulation of capital, can have over even intimate aspects of experience. These are the kinds of speculative engagements that can arise out of an imaginative consideration of a single concrete abstraction like money.

The task of theorizing here is three-fold. First, we strive to show how the various concrete abstractions that we can identify through historical materialist inquiry are *necessarily* linked. Analysis of relationships allows us to show, for example, how the concrete abstraction 'commodity' with its two faces of use and exchange value produces of necessity a money form that is itself divided between its roles as a measure of the value of social labour and a pure facilitator of circulation. It may then be possible to show how the tensions latent within the money form foreshadow the circulation of capital and the buying and selling of labour power. Theory construction here means the representation of the binding relationships that give the capitalist mode of production its contradictory coherence. Theoretical argument of this sort always has the capacity to produce new insights and findings, and this provides a second gateway to the creation of new understanding.

Secondly, then, we can seek out underlying concepts that have the power to synthesize and explain the links between such concrete abstractions but which are not themselves directly identified through the appropriation of historical and geographical materials. Such notions as the hidden hand of the market, the coercive laws of competition, class relations, equalization of the rate of profit, the necessity of accumulation for accumulation's sake, the annihilation of space by time, and the like, can be imputed as essential to the dynamics of a capitalist mode of production. What we are looking for here are, as it were, certain generative principles that help to explain the linkage between concrete abstractions just as they explain the dynamics of any capitalist mode of production.

The third task is to set this whole apparatus to work, now built into a synthetic but incomplete statement of the necessary laws of motion of capitalism, so as to interpret the historical geography of capitalism. There is, here, a moment in the research process of trying to interpret the actual dynamic theoretically. This can either be done retroactively (in order to explain past events) or through political practice (where the role of theory is to provide a better-informed basis for fighting for emancipation from oppression). In either case, the gap between the theoretical representations and historical geographical events, depicted in terms of concrete abstractions, forms a space out of which new concrete abstractions can grow. Here lies a third gateway to the creation of new knowledge. The reintegration of these into the theory allows the project of holistic theory construction to move forward to the point where we might hope to reflect the totalizing dynamics of capitalism, as in a mirror.

The building of such a theoretical structure through historical materialist research is, of course, a collective endeavour and there is much room for debate over the status and form of linkage between different concrete abstractions as well as over the interpretation of underlying terms and their projection as laws of motion of capitalism. Out of this there always arises a danger of arcane theoretical debate about theoretical issues (a kind of closure around 'theory for theory's sake' or some notion of 'pure theoretical practice'). That danger can be avoided if we scrupulously adhere to a basic methodological precept: namely that (1) the unity of theorizing and (2) the constant revision of theory by historical materialist inquiry, are both fundamental to the project. This means, then, that our current task is to appropriate as much historical–geographical material as we can, to search this material for new concrete abstractions and to match that process by an extension of the framework of theoretical argument which we already possess concerning the basic laws of motion of capitalism. Only through such procedures can we hope to build a general theory of the space-economy of capitalism under condition where flexible accumulation is becoming more and more the dominant form of economic organization.

Towards an Understanding of the Space-economy of Flexible Accumulation

We have already suggested that the rhythms and tempos of capitalist accumulation seem to be in a process of rapid transformation. The main thrust of this transition can be captured by the proposition that forms of accumulation in the advanced capitalist societies are less dependent than they once were on Fordist mass-production, and are now being modified on a significant scale by the turn to flexible forms of production, the emergence of new industrial ensembles, and the crystallizing out of increasingly custom-based patterns of consumption. One result, among others, has been a series of profound changes in the geography of the advanced capitalist societies. In terms of urban processes, for example, we witness a massive shift of employment into service sectors, a steady dualization of local labour markets, a new glorification of the entrepreneur (and spirit of privatization) in business and urban governance, and the appearance of new lifestyle patterns that carry over into the creation of new kinds of residential structures (such as those created through gentrification). In terms of regional development we observe an accelerating movement of industrial capital out of old production areas and into new regions, as in the cases of the third Italy, the U.S. Sunbelt, the various new high-technology growth regions of Western Europe or the burgeoning electronics complexes of Southeast Asia. International, inter-regional, and inter-urban competition increases as a result of these same processes and creates profound instabilities – new regions of spectacular growth now seem capable of suddenly becoming wastelands of decline, just as earlier thriving regions have also dramatically decayed.

Let us note immediately that here is an exceptionally rich terrain for significant empirical inquiries, informed by historical materialist ideas, into local communities and their economic bases, particular industries and their new forms of industrial organization, and the wide range of new urban and regional phenomena to be observed in the realms of consumption, culture and governance. Two further propositions must also be immediately advanced.

In the first place, as flexible accumulation has risen to prominence as a new means of organizing production and work, so it has encountered innumerable contingent circumstances. Its developmental course has been deeply affected by the local availability of special resources (including venture capital, entrepreneurial ability, labour power, as well as those material resources found in the physical environment), pre-existing patterns of urban and regional development, local cultures and traditions (such as those found in older artisanal production systems), and the like. It is our contention, however, that one of the crucial analytical tasks we face is to demonstrate how such contingent circumstances become internalized within, and recomposed by, the advancing development of a regime of flexible accumulation. We need to show, in short, how particular contingencies that on first sight appear as external and arbitrary phenomena are transformed into structured internal elements of the encompassing social logic of capitalism. This means that we reject the idea that there is some kind of absolute opposition between theory and contingency in historical materialist analysis.

We are led, in the second place, to focus more specifically upon the role of space in understanding processes of flexible accumulation. A century or two ago, space functioned for the most part as a wall, surrounding human action with a dense and seemingly impenetrable envelope from the standpoint of the daily processes of reproduction of civil society. With the steady evolution of the forces of production within capitalism, that wall has all but crumbled as the power to overcome and command space has increased immeasurably, producing the increasing 'time–space compression' that we have already noted. This has led some to argue that space is decreasing in significance for understanding the processes at work under capitalism. We insist that the converse is the case. The problem of space is not eliminated but intensified by the crumbling of spatial barriers. Command over space now becomes the vehicle for increasingly subtle intermediations and differentiations. So far from becoming uniform and homogeneous, space becomes ever more variegated, heterogeneous and finely textured, ever more complex in the manner of its usage. The nuances of differential command over space become crucial issues in the dynamics of class struggle. In the regime of flexible accumulation now unfolding before us, this complexity is evidently raised to a new and more perplexing level.

But in order to confront the changing role of space and the meaning of the contingencies which that space contains, we need to have at our command the sharpest possible theoretical tools. While it may be true that the 1970s was characterized by an excess of theoreticism, we believe the pendulum has now swung far too far in the opposite direction. Only by a resuscitation of the will to deal with, and to take seriously, the larger theoretical issues posed by capitalist development as a whole, can we begin to make sense of the profound changes currently sweeping the space-economy of capitalism. Only in this way, too, can we begin to re-build a coherent programme of political critique and action with respect to capitalist society as presently constituted in its neoconservative phase.

We want to end with one final but very crucial point. The reasons why Fordist accumulation has been disintegrating into flexible accumulation still remain obscure. We do not as yet have any good understanding of the logic of this apparent transition, or of the struggles that have led to this, as opposed to some other turn in the historical and geographical trajectory of capitalism. The crisis of Fordism was, of course, palpable enough, but the rapid and seemingly simultaneous proliferation of many of the elements of flexible accumulation throughout the world remains largely unexplained. The curious parallel movements of production into areas like the Third Italy, Flanders, the Sunbelt, and the like, the sudden proliferation of new consumption habits (such as those of mass distraction) worldwide, and the almost universal shift into neoconservative politics and post-modern cultural attitudes, suggest a sea-change in the logic of capitalist historical geography of monumental proportions. Detailed work by geographers may have much to contribute to the global understanding of these changes. To the degree, for example, that the Fordist regime ran into acute difficulties of labour control, it may well be that the very existence of a highly differentiated landscape of labour reserves of varying qualities and dispositions may have had a major role in enhancing geographical mobility and flexibility of production capital. This is, of course, a tentative hypothesis, but its very plausibility suggests something of the importance of space and geography in the turn to flexible accumulation. The ability of geographers to respond to the challenge, and thereby to participate in the production of a politics of resistance and emancipatory social change depends, however, upon the collective acceptance that our objective is to understand and confront the origins and impacts of an as yet incomplete transition to flexible accumulation through an intellectual project that accepts nothing less for its goal than the pursuit of a rigorously formulated and holistic theory of capitalist development in its current phase.

17 Models, Description and Imagination in Geography

Denis Cosgrove

There was a measure of poignancy for me in addressing the issues raised by *Models in Geography* twenty years after its publication, particularly doing so in the School of Geography at Oxford. In 1967 I was reading geography as a second-year undergraduate in that department. I was aware, as were my contemporaries, of a deep divergence between the discipline of geography as presented in the Oxford School and the subject as it was being promoted elsewhere – in the quantitative schools of North America such as Pennsylvania State and Northwestern for example, and in Cambridge and Bristol here in England. At Oxford we were following an undergraduate geography programme still essentially based on the ideas of the founders of British geography, A. J. Herbertson and Halford Mackinder. It was constructed on the study of regions coordinated into a hierarchy of scales from the global, defined by climate and topography, to a comparative national study of France and Great Britain, and finally, as our own individual dissertation, a geographical *description* of an area no more than 150 square miles in extent. This description had to encompass the physical and human character of the area. Oxford was defensively proud of its regional tradition. In the first year, we had read Herbertson's 'Natural regions of the world' (1905) and Mackinder's 'Geographical pivot of history' (1904). Strangely, I do not recall having been asked to study the latter's classic statement, itself 100 years old in the conference year, 'The scope and methods of geography' (Mackinder, 1887) which derived from the lecture Mackinder gave in justification for the founding of the first undergraduate course in geographical study in a British university, at Oxford in 1887. I shall return to that paper below.

It would be an exaggeration to claim that this course of study was intellectually stimulating, or that we as students were entirely contented with Oxford regionalism. We were aware that away in the Fens or over the Avon Gorge a very different conception of geography was being presented to our peers. 'Geographical models' was the phrase which captured this new geography, far more than 'quantitative revolution'. *Models in Geography* appeared in the departmental library and on the booklist for Paper IV, the history and philosophy of geography paper. That huge volume, with its formulae, diagrams and statistics, seemed impenetrable, at least to me, and I should be less than honest if I were not to admit that I later engaged the polemic between humanism and positivism on the side of the former in part because of my own inability to comprehend or manipulate with any facility the language of statistics and mathematics the new geography entailed. But I do remember the excitement of reading Christaller's *Central Places in Southern Germany* (1966) – itself only just published in English – for the elegance of his interlocking hexagons and hierarchical point patterns stretching in symmetrical geometry across the undifferentiated plain. With William Bunge's *Theoretical Geography* (1966), the only other example of the genre I recall seeing, matters were different. We were given the text for an essay on the uniqueness or otherwise of places and locations, and I found it deeply disturbing. I had elected to study geography rather than languages or literature, my other alternatives at eighteen, precisely because of a childhood sense of wonder and excitement at maps and descriptions of different places and peoples fostered by school geography texts. To suggest, as Bunge appeared to do, that New York and Paris, both of which were at that time for me imagined rather than experienced, or Iowa and Bavaria, were mere locations and spaces, knowable by invariant processes of reduction to spatial models, seemed to strip my geography of its central romance, its very soul. It was a profoundly depressing prospect. Indeed, at the end of my time at Oxford in 1969, the future for geography in general appeared dismal. It was difficult to nurture any hope of breathing life into the enumerative regionalism into which the Oxford project had lapsed, yet David Harvey's *Explanation in Geography* (1969) offered no alternative to a strictly deductive, hypothetical and model-based geography whose language henceforth would be the grey neutrality of mathematics.

Geographical Language

It is always through the lenses of the present that we read and represent the past, and when I claim that it was the loss of evocative description in a modelling geography that alienated me from it I am aware that I am recalling my response across fifteen years of a humanistic geography that has stressed the significance of language and literary style in geographical writing. Yi-Fu Tuan has celebrated the capacity of metaphor and the ambiguity of common language to capture the meaning of places. Gunnar Olsson has experimented with the play of words and their tense, slippery dialectic with things to reveal the complexity of geographical understanding. The self-styled 'literary geographers', Douglas Pocock and Douglas Porteous, have indicated the expression of a geographical imagination in novels and poetry, and I have myself explored aspects of visual and graphic languages in the representation of geographical landscape. Indeed language and its use has become a focus of considerable discussion in our discipline. Yet despite the dangers of reading my own past through my present, I do believe that, in common with some others, it was above all a distrust of the claim attached to theoretical and statistical modelling in the 1960s, that the truth of geographical relations could be captured in the cold language of figures, that formed the core of my rejection of this approach. The world of topographical maps, of imagined and increasingly experienced places and landscapes, seemed infinitely richer and more excitingly complex than these models suggested. I did not want the simplification of reality which they offered. Indeed, that claimed simplicity appeared increasingly bogus as the simple elegance of Christaller, Weber and Burgess gave place to rows of correlation coefficients and simultaneous equations, or pages of systems diagrams in which blocks of text, drained of verbs, adjectives or comprehensible syntax, were crammed into boxes or circles and joined with lines that obscured rather than revealed their meaning. The one thing these diagrams seemed not to do was *flow*, despite their collective name. Rather they 'fed back', constantly, more like those tail-eating grotesques that surround the portals of the gothic cathedral than the soaring columns of its interior, branching into arches and vaults with a clarity that renders the entire edifice a comprehensible and satisfying whole. Such constipated grotesques were originally intended to represent the temptations of a sinful, material world refused entrance to the complex yet comprehensible and transforming world of the spirit captured by the interior. In 1969 the cathedral of geography seemed similarly arrested at the portal.

Geographical description, the use of a rich and evocative language attending to the diversity of the physical and human worlds, was an object of derision in those early years of a modelling geography. While quantitative geographers read and were clearly moved by the fashionable topographical novels of the time – Durrell's *Alexandria Quartets*, Lee's *Cider with Rosie* or Thompson's *Larkrise to Candleford*, for example, which evoked a powerfully specific and highly individualized sense of place with all its colour and mood, the qualifier then applied to description within geography itself was 'mere'. This was the great weakness, so we understood, of the Oxford School: description could not be analytical, could not reveal patterns and coherence. It is sad in this respect that we were not given Mackinder's 'Scope and aims' paper to read, for it contains a memorable example of geographical description which is in fact model-based. Mackinder's task was to justify the teaching of geography as a distinct discipline in the face of hostility from existing university studies, most particularly geology, then the most threatened cognate discipline. The dialogue with geology, together with Mackinder's own acceptance of a certain environmentalism, accounts for the significance he gives to describing the physical environment in his example of how geography can act as a synthetic study, bridging the arts and sciences. Mackinder introduces his chosen example of South-East England by outlining the region's topography using the very concrete model of a white cloth laid carelessly over a table and creased into folds:

> Let us try to construct a geography of South-eastern England which shall exhibit a continuous series of causal relations. Imagine thrown over the land like a white tablecloth over a table, a great sheet of chalk. Let the sheet be creased with a few simple folds, like a tablecloth laid by a careless hand. A line of furrow runs down the Kennet to Reading, and then follows the Thames out to the sea. A line of ridge passes eastward through Salisbury Plain and then down the centre of the Weald. A second line of furrow follows the valley of the Frome and its submarine continuations, the Solent and Spithead. Finally, yet a second line of ridge is carried through the Isle of Purbeck and its now detached member, the Isle of Wight. Imagine these ridges and furrows untouched by the erosive forces. The curves of the strata would be parallel with the curves of the surface. The ridges would be flat topped and broad. The furrows would be flat bottomed and broad. The Kennet–Thames furrow would be characterised by increasing width as it advanced eastward. The slopes joining the furrow-bottom to the ridge-top would vary in steepness. It is not pretended that the land ever exhibited such a picture. The upheaving and the erosive forces have always acted simultaneously. As with the Houses of Parliament, the process of ruin commenced before the building was complete. The elimination of erosion is merely an expedient to show the simple arrangement of the rocks, which simplicity is masked by the apparent confusion of the ruin. Add one more fact, that above and below the hard chalk lie strata of soft clay, and we have drawn on geology for all we require.
>
> The moulder's work is complete; the chisel must now be applied. The powers of air and sea tear our cloth to tatters. But as though the cloth had been stiffened with starch as it lay creased on the table, the furrows and ridges we have described have not fallen in. Their ruined edges and ends project stiffly as hill ranges and capes . . . . (Mackinder, 1887; 1962: 223–4)

Mackinder's opening metaphor reminds us of those plaster scale models of physiographic regions – the Ridge and Valley province of Appalachia, the exposed dome of the Lake District or the scarplands of the Isle de France – which once graced the entrace halls of geography departments. Such models allow us to grasp the unity of a part of the earth's surface, to see at a glance the complex but coherent network of streams and rivers relating to it, to comprehend the logic of vegetation and human settlement. Unity in diversity, Humboldt's grand claim, becomes more than merely rhetorical. We can actually see and grasp this geographical marvel. Today we can achieve this from an aeroplane, for example travelling between Northern Italy and London: the great ribbon of the Po and the Alp-fed rivers flowing to it, banked and regulated within the geometry of centuriation and reclamation with the vast, unregulated spread of modern Milan superimposed; the mountains themselves, their *adret* and *ubac* caught clearly by the April sun and the great central valley of Switzerland which alone gives the possibility of unity to what is otherwise so logically a federation of independent valley cantons; then the Vosges and the deep greens of eastern France where still today we can observe the logic of nucleated village and radiating *sentiers* stretching to the belt of woodland that separates the neighbouring communes. Here too the larger logic of the Marne and Seine, and today the autoroutes that also lead to Paris, stretches across Vidal's still recognizable *pays*. Finally the Channel and Mackinder's own frayed tablecloth, overwhelmingly green and hedged, with social power inscribed in the remnants of country parks and the modern specks of blue than indicate the swimming pools of the stockbroker belt. The model, translated to the text, becomes metaphor and serves to remind us that geography, which in the words of Giuseppe Dematteis, can never be anything more than 'an analogical–metaphorical description of reality representing socially relevant facts in the forms of physical, terrestrial space' (Dematteis, 1985), has always relied on models. If there is a critique here of *Models in Geography* it is of the poverty of language acceptable to model-builders in the 1960s and of the imperialistic claims of quantification as the guarantor of objective truth.

Objects and Subjects

This claim of objectivity was of course itself a stick with which to beat geographical description in the 1960s, and the celebration of subjectively experienced geographies later became, in the writings of many humanistic geographers, the central plank of their polemic against a model-building quantification of the discipline. One reason advanced by Mackinder for the synthetic, descriptive focus of geography on the region was that the world of 1887 was objectively known: 'we are now near the end of the role of great discoveries. The Polar regions are the only large blanks remaining on our maps', he declared (Mackinder, 1962: 211). Like F. J. Turner, a decade later declaring the closing of the American frontier and with it the first stage of American development, Mackinder's was the conception of a closed, absolute space. He, and the regionalists who followed him, accepted as objective the geographical space defined by the coordinates of latitude and longitude decreed by Renaissance cartographers, coordinates which served to make the globe knowable and exploitable to European power and its corollary of economic, political, military and cultural hegemony. But the very concept of objectivity is a function of subjectivity. Only with the acceptance of the Cartesian Ego as the measure of existence can this concept of an objective, measurable and external world come fully into being. We owe to the theoretical geographers of the early quantitative revolution, to social physicists, central place theorists and others, the notion of relative geographical space. The absolute space of the regional tradition was shown to be but one of an infinite number of possible spaces, each of which is constructed by social necessity. Such a recognition of course allowed the making of a theoretical geography abstracted from the experience of the world: its experience for example as a revealed truth as understood by the medieval Christian, or as a lived experience as in the case of young children or many pre-modern societies (see Sack, 1980). Theoretical geography naturally reached for geometry as the language of space. In a classically platonic sense, theoretical geography saw in geometrical theorems the certainty which evades the world of direct experience in a secular society. Geometry guarantees this certainty and objectivity precisely and solely because it is a pure creation of mind, an ideal space because it is entirely subjective, a space of which experience geographical space with its ambiguity and contingency is but the mirrored distortion. Phrases such as these are exactly the ones we find in the commentaries of Renaissance translators of Euclid's *Elements*, men like Nicolo Tartaglia and John Dee. To be sure, there are more sophisticated geometries than Euclid's, although his remains that best fitted to deal with terrestrial space whose measurement was precisely the origin of *geo-metry*, but the claims of geometry and mathematics to certainty within their own logic remain indisputable.

Model-building, quantified geography, however, seemed to go beyond this, to the fallacy of proclaiming the objectivity of geometrical and mathematical models as themselves the guarantors of truth in representing geograhical reality. It was here that a geographical humanism, seeking to draw upon direct experience rather than abstract reflection, opened a fundamental critique of model-building and quantification.

For anyone who felt the dissatisfaction with the geography presented in *Models* there were in 1967 few texts within geography to which to turn for intellectual support. David Lowenthal's 'Geography, experience and imagination' (1961), with its detailed exemplification of different modes of geographical epistemology, offered certain possibilities for alternative perceptions of the world. But, taken with Julian Wolpert's critique of optimization and the concept of economic man (1964) and Kevin Lynch's *Image of the City* (1960), Lowenthal's project was rapidly appropriated by a 'behavioural' or perception geography which implicitly accepted the underlying truth of the world as represented in spatial analysis, merely recording human perceptions as 'distortions' from it – useful for understanding spatial behaviour but hardly offering a unitary geographical understanding. Indeed in William Kirk's terms it proposed a 'behavioural environment' as a subjective layer to be drawn against the template of a more real 'phenomenal environment' (Kirk, 1963). This sustained a form of determinism close to that produced in psychological stimulus–response theory, maintaining a dualist epistemology of subject and object. A similar reading could be made of J. K. Wright's 'Terrae incognitae', which had celebrated in 1947 the place of the imagination in geography and proposed 'geosophy' as the study of different social, cultural and individual experiences of the world. Of course, in all these writings much more was being claimed. Indeed Wright explicitly answered Mackinder's declaration that exploration was at an end, pointing out that we had not begun to explore the most fascinating unknown lands of all, those that lie in human minds and hearts.

I recall reading these essays after leaving Oxford, and they did appear to offer the possibility of an alternative geography to the cold geometries of spatial analysis and the abstractions of the model-builders. But they did not suggest how such a geography might be constructed. Two other works had a deeper impact because they seemed to do this, and to respond to a dissatisfaction that was less easy to articulate at that time, an aesthetic and moral dissatisfaction with the geographies of both Oxford regionalism and *Models*. The first of these was Yi-Fu Tuan's (1961) paper 'Topophilia, or sudden encounter with landscape', published in the journal *Landscape* alongside essays by its editor, J. B. Jackson, whose own provocative descriptions of American vernacular scenes captured precisely the excitement and romance in which my own geographical imagination was rooted and at a time when it was being fed by the first direct experience of living in an unfamiliar landscape, that of North America. Tuan described topophilia as the love of place, not necessarily that love which comes from rootedness in a specific place – the deep affinity of the patriot or immobile peasant for the landscapes of their lives and labours. Tuan wrote as a geographer, speaking of the excitement gained from poring over maps of tropical South America, from studying in inhospitable and foreign lands, from relaxing momentarily from a detailed survey of Fifth Avenue shops, and suddenly becoming overwhelmed by the noise, smells and sights of a palpable and insistently material New York. 'Topophilia' spoke to an aesthetic pleasure in experiencing geographical reality at once in its specificity as individual places and in the mystery of its difference from the known and the familiar. More recently David Stoddart has expressed the same sensibility, correctly recognizing it as the very foundation of geography as exploration (Stoddart, 1986). This is more than merely the excitement of the tourist or casual visitor, although it is not unrelated to that experience, and I have little sympathy with the strain of criticism within humanist geography that would regard tourism as 'inauthentic'. Indeed J. B. Jackson's (1980) defence of geographical education as 'informed tourism' strikes me as sound. But the intellectual and scientific investigation of a previously unfamiliar place or region introduces a specificity and rigour which heightens our aesthetic awareness of individuality and difference. I still pass on to students a piece of advice given to me at the same period of my geographical education by R. C. Harris: 'if you intend to become a geographer', he told me 'then select an area, any region but preferably not in your own country, and, regardless of the specific questions you ask of it or study you make of it – historical, economic, political, geomorphological – learn to know it intimately'. I take this as the foundation of my claim to be a geographer, it has been the basis of my research over seventeen years in the Veneto region of north-east Italy. The specific studies are architectural and cultural, they are set in the sixteenth century, but they are informed by days of walking, cycling and driving through the cities and countryside of the Veneto, arduously and inadequately learning its language, more enjoyably eating in its restaurants and drinking in its bars. The aesthetic pleasure of such a geography is far removed from the intellectual elegance of the theoretical models I could construct without ever setting foot in Italy. Only now could I essay a geographical description of that region. But in 'Topophilia' Tuan is curiously apologetic about such an aesthetic, concluding merely that it would improve our geographical writings if we allowed our affection for the places and regions we study to find expression, but implying that it is distinguished from, and subservient to, the scientific concerns of the studies themselves rather than integral to them.

Quite different in the role it attributed to the geographical imagination in geographical scholarship was a short work by the French geographer Eric Dardel: *L'Homme et la Terre*. Published in 1952 it remains largely unknown in Britain, although it has just been republished in an Italian translation accompanied by a series of commentaries (Copeta, 1986). It was brought to my attention in 1970 by Edward Relph, and introduced me to the wealth of literature that Dardel drew upon, particularly Bachelard's phenomenology of science and Mircea Eliade's studies of religious myth. Dardel uses these to construct an analysis of geographical space and a history of geographical knowledge quite different from any that I had then imagined. His concern is to reach below the formal science of geography as constituted in professional texts, university teaching programmes and formal cartographic representations, to what Carl Sauer had earlier called a naive or pre-scientific geographical consciousness. In his opening paragraphs Dardel describes it thus:

> but before geography and its preoccupations with exact science, history presents us with a geography in action, an unbounded desire to traverse the world, to cross the sea, to explore the continents, to know what was formerly unknown, reach that which is inaccessible: Geographical questioning precedes and produces the objective science, whether it be love of native land or the study of new countries, a concrete relationship ties humans to the earth, a human 'geographicity' as an aspect of our being and our destiny. (Copeta, 1986: 11)

Here was a geographer prepared not only to admit to such a basic compulsion but to subject it to detailed, critical and highly evocative analysis. He distinguishes geometrical space, abstract and manipulable, from geographical space 'with horizons, relief, colour and density', and says of the discipline which elects this latter space as its object of study:

> Geography, according to its etymology, is the description of Earth; more precisely the Greek term suggests that the Earth is a text to be deciphered, that the line of the coast, the profile of the mountains, the meanders of the rivers, form the marks upon the text. The objective of geographical understanding is to clarify these marks, to interpret that which the Earth reveals to mankind about its human condition and its destiny. Geography is not merely an atlas open before our eyes, it is a calling which rises from the soil, the wave, the forest, a chance or a denial, a power and a presence. (Copeta, 1986: 12)

The remainder of the text elaborates this theme, analysing the power of different kinds of space divided by Dardel according to the ancient elements of earth, air and water, as well as the human spaces of architecture and landscape (city and country). In the second part he deals with different expressions of geographical consciousness, constructing a history of geography which

> makes sense only if it is understood that the Earth is not a brute fact to be taken as given, but always inserted between Man and the Earth is an 'interpretation', a structure and a perspective on the world, an 'enlightenment' which reveals the real within the real, a point of departure from which understanding develops. (Copeta, 1986: 47)

Dardel's history of geography deals with mythical geographies of different cultures and historical periods; 'prophetic' geographies like religious and social utopias; 'heroic' geographies of epic literature and of journeys of discovery; and finally the 'scientific' geography of the laboratory, the cartographic studio and the library which seeks to inventorize and synthesize the elements of knowledge gained in other forms of geographical experience. While the last may be sedentary, while it may distance itself from the raw experience of the natural elements and from exploration, while it may seek explanations and literally de-scribe or interpret the world using models and hypotheses, for Dardel it can never be reduced to 'a pure and simple science of nature' located at the juncture of the human and physical sciences, for

> it is impossible to eliminate from its 'object' every moral and aesthetic value; impossible on the part of the observer to overcome completely the point of view or perspective from which geographical reality is embraced, to cancel out the subjectivity of the person for whom geographical reality becomes reality. (Copeta, 1986: 85)

Even physical geography is in this sense human. Mountains and seas are only such because humans name them, in themselves they remain un-described.

Dardel seems to offer a justification at least of the geographical urge which the geometries of geographical models appeared to suppress, and which Oxford regionalism seemed to recognize but failed to answer. Furthermore his writings explicitly acknowledged that moral and aesthetic concerns could not be excluded from geographical writing. In this his book spoke indirectly to two of the most fundamental objections that I, and it became clear in the 1970s, many others, had to the geography of *Models*.

Morality

It is a truism to state that the geography of the 1960s, of statistical modelling and a quantified spatial science, sought to develop paradigms of spatial efficiency closely allied to the modernist faith in technological progress and planning. These values of rational social order expressed in the spatial organization of transport systems, or urban land-use patterns, in regional equilibrium or hierarchical central place networks, imperialistically declared themselves as value-free. Moral questions were systematically excluded from geography by the simple act of denying scientific status to alternative value systems to that ordained by the liberal democratic ideology of the post-war western consensus. A similar denial of pluralism was later to be sustained by 'scientific' Marxist geography with its equivalent elevation of economic motivation as the foundation of human interaction with the earth. It is a matter thus of little surprise that many of the most enthusiastic radical geographers of the 1970s had formerly been equally seduced by spatial science. Moral discourse concerns itself as much with what ought to be as with what is. A 'human' geography by contrast requires an intersection of the personal and social beliefs and values of the geographer and the empirical evidence obtained according to the accepted rules of scholarship: refusal to suppress awkward or countervailing evidence to theories, public acknowledgement of sources and methods, logical argument and so on. A degree of self-examination, of personal questioning about motives and preconceptions, is therefore a prerequisite of such geographical investigation. This can become, as in the early writings of some humanistic geographers, overly self-indulgent and at times moralistic. Nor is it true to claim that humanistic geographers have been the only ones to recognize the significance of moral discourse in our representations of the earth as the home of human life. It was a feature of the Zeitgeist of the early 1970s, as the 'relevance debate' and particularly David Harvey's 'What kind of geography for what kind of public policy?' make clear (Harvey, 1974b). But it is difficult to coordinate such discourse and its inevitable emphasis on debate, interpretation and critical self-reflection with the imperatives of a mathematical or statistical formulation and their requirements of precision and simple true/false distinctions. I recognize that this claim may reveal merely my own inadequate grasp of the sophistication of a quantitative methodology. Perhaps such concepts as 'fuzzy sets' allow for the nuances and hermeneutics that moral discourse involves, but I am unaware of the case being argued within quantitative geography. In my own geography I am forced constantly against the conflicts within my own values and beliefs.

Aesthetics

Dardel's impact comes as much from his literary style as from the intellectual case he sustains. Geographical language, he claims, is by nature poetic, and is so because our response to the earth and the human patterns inscribed thereon is as strongly aesthetic as it is utilitarian. I recognize that there is an aesthetic within mathematical models and elegant geometrical constructions, but this is formally distinct from the poetics of geography as such, that which arises from the intersection of human consciousness and the patterns and processes that differentiate and sculpt the earth's surface. Once we allow for this dimension we have three possible ways of incorporating it into formal geographical scholarship: we can focus attention on the aesthetic qualities of the environment itself, both physical and as altered by human agency; we can seek to represent directly the sense of environmental beauty through the media adopted for geographical description; and we can focus upon the different ways in which humans have inscribed their aesthetic response in various representations of landscape. I will comment briefly on each of these.

The first part of Dardel's book, which focuses on different kinds of space, is essentially an exploration of the aesthetics of elemental nature. Drawing heavily on Bachelard's phenomenology, Dardel seeks to penetrate the essence of direct, unmediated experience of earth, to the foundation from which more sophisticated geographical sensibilities develop. This is a radically subjective geography where he explores such matters as the primary – perhaps primal – experience of entering the earth's crust, in a cave, a mine, the depths of a gorge or ankle-deep in a sodden plough furrow. Telluric space touches us well below the level of intellectual reflection. Similarly primary is the experience of being on water, out of sight of land, particularly under a clear, star-filled sky on a cloudless night; or that of feeling the wind move through tall trees as we stand at the heart of the forest. At the core of our being we are a part of the earth. Our elemental composition is that of the earth – dust from dust – and the organic processes; physiological, alimentary, sexual, that make and sustain human life are shared with the rest of the created world. Of course these are mediated in infinite ways, but, as most serious students of art have recognized, this unity of being and nature is a powerful aspect of human aesthesis. Once having acknowledged this aspect of the geographical imagination, however, there seems little that the formal study of geography can say of it or do to explore and develop it (see Eliade, 1978). But it does inform our work, it is one of the foundations of the field tradition. Indeed I suspect that it unconsciously informs particularly the sensibility of those who concentrate on physical geography, although it may be difficult to admit it. I cannot help but suspect that there is a measure of loss felt among many of my colleagues in geomorphology constrained to model stream flows or slope angles and to write their reports in a language which formally denies expression to this sensibility. A hint of this may be found in Linton's (1968) essay on landscape classification written in that period when geographers did attempt aesthetic measurement of landscape quality within the modelling tradition of perception studies. Linton's attribution of aesthetic values to Scotland's scenery tells us more about his own love of mountains, glens and high moorland than of any objective landscape quality. The whole project of quantitative landscape classification perfectly captures the inability of the quantitative paradigm to gain a purchase on this aspect of our geographical sensibility.

The idea of writing an 'aesthetic geography' long predates the era of modelling and spatial science. Sir George Younghusband, whose own sense of mystery in the face of landscape beauty was sharpened by his experience of a solitary trek on foot across the Gobi to Himalaya, made a plea for such a geography in his presidential address to the Royal Geographical Society in 1919 (Younghusband, 1920). It was taken up during the 1920s by Vaughan Cornish in a series of essays and books. Studies of the amplitude of sea waves, of beach forms, of light and shade effects of distance in landscape, as well as of human transformation of nature – from churchyard yews to the sublimity of the Panama Canal during construction – all these formed the raw material of his 'aesthetic geography'. Cornish sought a heightened literary style to express his response. Reading these essays today is rather embarrassing, for they are self-consciously 'literary': the landscapes described lack edge, too often they founder in a sea of saccharine. I fear the same would result were we to follow more recent pleas for geography as an art, from Donald Meinig (1983) and J. Wreford-Watson (1983), for example. The desire seems almost to abandon the scientific and scholarly pretensions of geography to become painters and poets of geographical beauty. There are, to be sure, writers who have succeeded in capturing the quality of landscape in this way: for me John Ruskin is the most obvious example; his prose and painting are at once accurate morphological studies of natural forms and moving evocations of natural beauty. But Ruskin was a rare stylist and a highly trained sketcher and watercolourist. Rather than seeking to emulate his success in writing an aesthetic geography we should perhaps pay more attention to his words to neophyte drawing students, that he would rather teach them to draw so that they could better observe nature than to observe nature so that they could draw better. There is in Ruskin's aphorism a justification for reviving the near-dead geographical practice of field sketching where a geographer produces a graphic model of landscape primarily to focus attention on topographical forms and morphological relations. The topographical map, whose history is so closely aligned to that of landscape painting, is both a synthetic model of geographical relations and our most authentically geographical expression of landscape beauty. One of the unintended consequences of a model-based, quantitative geography has been to speed the disappearance of the topographical map from the geographical text and the geography curriculum. In the Oxford course I referred to earlier the appreciation and interpretation of topographical maps, British and European, was one of the more satisfying experiences, and the 150 square mile geographical description required for honours was based initially on the One Inch Ordnance Survey sheet.

The third way in which geography may incorporate aesthetic sensibility is in studying the manner in which human groups and individuals have themselves represented their affection for the earth, for its variation and specificity, and have transformed it according to the dictates of taste and beauty. It is in this direction that I have been most fully able to satisfy my own sense of landscape aesthetics. In the same years that *Models in Geography* was being prepared David Lowenthal and Hugh Prince published a pair of essays on 'English landscape' (1964) and 'English landscape tastes' (1965), exploring the ways in which urban and rural England had been shaped and designed to express the aesthetic predilections of English culture. To be sure, and as they pointed out, the English culture which dominated this landscape aesthetic – and still does – was aristocratic, elitist and conservative. It does not begin to exhaust the range of possible meanings of 'English culture', but these essays revealed to me the possibility for a geographical interpretation of such landscape representations as parks, urban design, paintings, poetry and novels, any of which could open up questions of culture, society and human attitudes to nature.

There is a formal affinity between geographical models expressed in mathematical or statistical form and a landscape painting or a designed parkland: both represent an idealized world, a human vision of how the world might be. Both construct space according to formal rules and conventions, seeking to arrange objects within the space they create in order to clarify the relations between those objects. They are different forms of play between the real and the imagined, and both require us to learn a specialized language and a set of compositional rules if we are to interpret their geographical significance as comments on the world of experience, to gain purchase through them onto the complex relations of human life and the natural environment. In my own work it has been to the concept of landscape as a way of seeing and representing the world, especially in Europe since the Renaissance, that I have turned to answer the questions prompted by my geographical imagination. Landscape, developed as a morphological concept in geography, refers equally to a sophisticated mode of apprehending the world, one which has developed formal rules of space and content representation over the past six centuries; framing, perspective, selection, arrangement, chiaroscuro, for example (see Cosgrove, 1985). These rules and conventions are never purely aesthetic, they emerge equally from social and cultural relations, from the ways in which land has been owned and exploited, from the meanings attached to nature, to society, to city, to country, and from the structures of power and subordination in human relations (see Cosgrove, 1984). The analysis of a single picture, as Stephen Daniels has shown in the case of Turner's painting of Leeds (Daniels, 1986), can be equally revealing of the spatial organization and human experience of textile industrialization and the complex interrelations of country and city at the time as a spatial model or statistical representation of Leeds in 1830. Such interpretative studies of individual works explode the fallacy that generalization from samples is the only path to representativeness. The close and contextual analysis of a particular park, a single poem, a building or even – to return to the *Models* debate of the 1960s – a unique place or region, can be as revealing of geographical relations as the largest statistical sample (see Cosgrove and Daniels, 1988). And for many geographers, myself included, such study is in the end more satisfying because it allows us to retain a sense of real people moulding their lives and their geography in real places rather than abstract numbers patterned across an equally abstract space. The study of the unique also forces upon us a humility of theory in the face of an intractable materiality in the object of study. For all that we bring to bear upon the painting, the novel, the basilica, a sophisticated battery of theory and techniques they remain insistently themselves, whole and specific objects. We may represent them in prose and illustration but, as with every object of geographical study, we can never pretend to have constructed a fuller truth than existed within the object before our intervention.

Conclusion

Accepting the relativity and representational nature of geographical description leads us to the immediately contemporary debate about a post-modern geography. Earlier I referred to *Models of Geography* as a perfect statement of modernism. The faith in rational and unilinear social and environmental progress, in technology as means of achieving it, in the efficacy of abstract numerical relations to gain a purchase on 'objective truth', and the belief in planning as a way of overcoming the contingency and flux of unpredictable human life and its environmental relations are all characteristically modernist. All are central to the project envisaged by Chorley, Haggett and their contributors. In the post-industrial 1980s these beliefs stand shorn of their credibility like the Emperor's new clothes. Just as the urban superhighways, the megastructures and the housing tower blocks of 1967 now stand as monuments to a past age with dead hopes, so the grand models of urban social structure or urban systems have a similar air of decay. In a world of Covent Garden playspaces, Quinlan Terry's easy Palladianism, the gentrified terraces of Islington and the smoked-glass towers that house the City's Big Bang microscreens, fragmenting into a thousand refracted images the older urban landscape, geography today celebrates diversity and particularly contextual studies of locale: historical and cultural reconstructions and a revived description. This is true even among economic geographers, once the highest stars in the modelling firmament.

The post-modern perspective distrusts claims for a privileged path to truth or to accurate representation of a single reality. It takes intellectual stimulus from a playful celebration of difference, from the mirroring of multiple representations. In this geography it is perhaps the theatre with its frank construction of spatial illusion, its changing scenery to indicate different meanings and moods, its dialectic of audience and players, each refracting the world of the other, and above all its languages of symbol and ritual, which offers a better metaphor for the geography of our contemporary world than the cybernetic system of macrocomputing from which emerged the flow diagrams of *Models in Geography*. The relationship between words and things has always been a contingent, socially constructed and unstable one, and this is equally true of non-verbal representations of the world, including mathematics.

Such a perspective does not exclude numerical modelling as a mode of geographical description, it merely places it alongside a range of alternatives, none of which is accorded a privileged relationship with a reality which resists direct transformation into our representations of it. In the play of de-scriptions of the earth as the home of human life there is a lightness and a pleasure in the geography of 1987, rather absent in that of 1967. We no longer see our discipline in terms of competing paradigms, nor are we burdened by the puritanism of the 1970s which fed the polemics of humanism, positivism and structural Marxism. Of course with this liberation of meaning, method and description there is always potential danger. Dematteis has shown how the metaphors of the earth developed by past geographies – of Eurocentric Renaissance globes, of Victorian imperialist explorations, of racist determinism in our own century, the territorial systems of quantitative geography and the social spaces of economic Marxism – have all served the purposes of social and environmental domination, masking their ideological function behind their claims to science. The world remains a place of power, domination and subordination, its resources are amassed by some, denied to many, its very life processes seem fragile and threatened. All of these issues rightly bear upon the practice of geography and demand a place in its representations. But there is also beauty and goodness in this home of human life: variety, texture, colour and density, and these rest equally at the heart of geography. Geographical description of whatever kind, employing whatever models, can still use its metaphors to bring together human needs and the conditions which the earth offers to us, to bring together words and things in a dialogue that will always have its own poetic. Geography remains, in Dardel's words 'a chance, or a denial, a power and a presence'.

18 Some Critical Views of Modelling in Geography

Robin Flowerdew

Introduction

The term 'model' was popularized in geography in the mid-1960s by Haggett and Chorley (1965, 1967), and is associated by most geographers with the quantitative revolution and with the search for nomothetic generalization. Both then and later, the term was used in many different senses. In my view the dominant associations of the word today are with the large mathematical systems developed by Wilson (1974) and others to represent interactions in complex urban and regional systems. The term has been used little in the radical or humanist schools of geography, and is usually associated with a positivist approach.

The word 'model' has been used in very different ways. Harvey (1969) says that 'the general confusion surrounding the term is reflected in geographic research' and (as Haines-Young notes in his chapter) Haggett and Chorley (1967) quote Skilling (1964) to the effect that 'a model can be a theory or a law or an hypothesis or a structured idea. It can be a role, a relation or an equation. It can be a synthesis of data.' Although their discussion ascribes many features to models, the most fundamental is *selectivity*. Modelling is therefore a method of representing a complex state of affairs by reducing it to something simpler which embodies as many as possible of what the modeller sees as its most important characteristics.

The way in which the term has been used usually involves two further features, which are not necessary features of modelling. One is *generalization* – models are usually viewed nomothetically, as summing up the important characteristics of a class of phenomena, rather than just one particular state of affairs. Secondly, *mathematization* is frequently associated with modelling either through the formulation of a set of deterministic relations between parts of the system being modelled or through the calibration of a statistical model. Neither of these are necessary features of modelling – a specific event can be modelled, such as the diffusion of some specific innovation, and the term 'model' can be applied to non-mathematical constructions like hardware models (Morgan, 1967), flow diagrams (as discussed so entertainingly by Cosgrove) or even verbal descriptions (like Mackinder's tablecloth analogy).

Haggett and Chorley (1967) also consider the functions of models, suggesting nine possibilities. The distinction that seems most important today, however, is between models intended to *communicate* the essence of a state of affairs, often for teaching purposes, and models intended to *explain*, for the purposes of controlling, understanding or escaping from a particular state of affairs. The desirable features of a model may be different between these two situations.

Critical Views of the Role of Modelling in Geography

Given that modelling is an important activity within geography today, critical views can vary from those which accept modelling but criticize the way it is done to those which reject modelling as a worthwhile activity. This section contains a classification of four types of critical position which could be taken, ranging from more to less sympathetic.

Adherents of the first position would agree that modelling is a worthwhile enterprise for geographers, but would criticize it on the grounds that much, or perhaps all, has been done badly. If the key aim of the modeller is to represent complexity by something simpler, it is possible for models to simplify too much (to throw out the baby with the bathwater) or too little (leaving most of the bathwater in). The former case (oversimplification) may mislead students and generate misunderstanding and bad predictions; the latter (undersimplification) is of little use in teaching, does not explain and gives insufficient basis for prediction. This dichotomy is exemplified by the disaggregation question. Modellers of complex systems are usually aware of the problems of disregarding important distinctions between age groups, occupations or transport modes, but including such distinctions may lead to intractable mathematics, shortages of data for model calibration, and terror for the naive reader faced with strings of subscripts and summation signs.

Models may also concentrate on the wrong things (throwing out the baby and keeping the bathwater). For example, some geographers have been criticized for focusing on spatial factors when their critics regard these as unimportant in comparison to economic or social processes; behavioural models in housing have been criticized (Gray, 1975) for modelling choice and neglecting constraints. Sometimes models may neglect to fulfil the basic criterion of simplifying (throwing out the baby and filling the bath?); principal components analysis and more recently Q-analysis often seem to produce models more complicated than the original data. Lastly models may incorporate some of the salient points but omit others (some parts of the baby are thrown out and some kept! My baby–bathwater model is itself in this category).

This first set of criticisms does not question the appropriateness of modelling as a generally applicable strategy in geography. A second position would be that modelling is worthwhile but not the only valid strategy: all geographers should not be forced to apply modelling techniques to everything. This could take the form of a claim that modelling is not appropriate in some parts of geography: regional geography, cultural geography and historical geography are common candidates to be model-free zones. An alternative claim would be that there are important insights which cannot be gained through a modelling approach.

Thirdly, this position could be developed into an argument that modelling is a worthwhile activity in geography, but its role has been over-emphasized in the past two decades. Geographers should devote more effort to developing alternative methods. Proponents of this position might claim that modelling strategies have distorted the subject, by concentrating attention on some topics rather than others, by developing some types of insight more than others, or by putting undue emphasis on generalization at the expense of specific cases.

Fourthly, the most extreme position would be that modelling is not what geographers should be doing at all; that all geographers should abandon it as an approach. There may be variants even within this position. Some may argue that modelling is appropriate in the social sciences but that geography should be a humanity and not a social science; others may argue that modelling is appropriate in the physical sciences but not where the actions and feelings of people are under study; others may argue that it is never appropriate.

Criticisms of modelling may be based on objections to the generalizations that modelling usually involves. It may be thought futile to construct general models to apply to geographical events, especially where idiosyncratic human actions and free will are concerned. Or it may be that the geographer's purpose is to predict or understand specific events and situations; his or her interests may be in the unique case for which a general model is thought irrelevant.

Criticisms may be based on the mathematical form of many models. Despite the quantitative revolution, few geographers feel comfortable with mathematical symbolism and ideas, and are thus largely unconscious of the generality, clarity and elegance that mathematical modellers appreciate in a good model; even concepts like parsimony and goodness-of-fit, essential criteria in evaluating a mathematical model, are strange and unnatural ways of thinking about geography for the mathematically untrained.

A related criticism is that the people for whom the model is produced, whether students, policy-makers, clients or the public at large, may find mathematical models difficult to understand. A major point here is that such models may be used by one group for the purpose of mystification; a dubious claim or policy may be protected from criticism by justifying it in terms of a model which is hard (or impossible) for potential critics to understand. Mercer (1984) develops a more general argument in which mathematical modelling is seen as part of 'technocratic' geography serving the interests of existing power structures.

In the introduction to this chapter I argued that the fundamental attribute of models was their selectivity, and that generalization and mathematization were contingent attributes, applicable to many but not all models. Criticisms of modelling for its selectivity, or simplification of complexity, are less common. One form of such criticism might be a humanist position that people are full, rounded individuals and attempts to model them or their behaviour detract from their individuality and dignity (Rowles, 1978). Another criticism is that no model is adequate by itself; any model must be continually subject to reassessment, modification and replacement. In Feyerabend's words (1975):

> knowledge . . . is an ever-increasing ocean of mutually incompatible (and perhaps incommensurable) alternatives, each single theory, each fairy tale, each myth that is part of the collection forcing the other into greater articulation and all of them contributing, via this process of competition, to the development of consciousness. Nothing is ever settled, no view can ever be omitted from a comprehensive account.

Modelling and Philosophical Stances

The term 'modelling' is associated with empiricist philosophies such as positivism, which are dominant in geography. The term became fashionable along with the adoption of Hempelian and Popperian perspectives on the philosophy of science (Harvey, 1969), and is closely associated with positivism, critical rationalism and their variants. It was little used by empiricists in geography prior to the quantitative revolution, although some of their key concepts, such as the natural region, can be regarded as models, despite their lack of mathematization. Even treatments of specific regions, implicitly or explicitly, can be regarded as models in that they simplify complexity, perhaps by focusing on certain aspects of the physical environment and human activities to emphasize the relationship between the two, or even to convey a clear idea of what the region is like.

Humanist traditions in geography have made little use of the term 'model'. Most humanist geographers have not been interested in mathematization, and their work avoids mathematical or statistical models. Attitudes to generalization differ, though most are interested, at least for teaching purposes, in simplification of complex situations. Zelinsky's portrayal (1973) of the American national character, for example, has many of the attributes of a model. As Johnston (1986c) states, 'some humanistic approaches at least have scientific goals, in that they are seeking to uncover the nature of universal knowledge which governs all individual behaviour'. Smith (1981), for example, aims towards generalizations based on comparative analysis of the 'ostensibly unique'. Much humanistic geography, however, 'focuses on the particular and emphasises the singular' (Johnston, 1986c). Even this does not mean that modelling is irrelevant. Idealists are concerned 'to reconstruct the thought behind the actions that were taken' (Guelke, 1974); it is hard to see how this can be done without constructing some sort of model of the actor's thought. Similarly, if an existential geography 'endeavours to reconstruct a landscape in the eyes of its occupants, users, explorers and students' (Samuels, 1981), this requires something like a model to be developed.

Structuralist geographers also make little use of the term 'model'. It is tempting to regard the structures themselves as models; most empiricists would view them as generalizations and simplifications of a complex state of affairs; the class system, for example, is frequently seen as a simplification, one which may or may not be fruitful for understanding our society. The very process of dialectic seems to be a general model of how ideas and social forms develop. But there is an important difference – for an empiricist, a model is a way of trying to understand or to simplify reality (or whatever perception can tell us about reality); for a structuralist, the structures are real, though not observable, and hence cannot be thought of as models. Instead of a model being a simplification of observable complexity, for a structuralist the observable world is a transformation of the real structures. Nevertheless, there is an important tradition within geography which is influenced by Marxism but which uses models characterized by abstraction and mathematization as well as selectivity (e.g. Webber, 1987).

For a realist, explanation of observable phenomena involves the 'attempt to discover appropriate structures and mechanisms' (Keat and Urry, 1982) and in this endeavour the construction of models is of prime importance. A realist must attempt to discover appropriate structures as a basis for explanation of observable phenomena, and does so by constructing a model, whose adequacy as a description can be examined by working out and testing the model's consequences. For Sayer (1984) at least, generalization is not an important goal and mathematization is of little value. Sarre (1987), however, attempting to apply realism to a geographical research project, finds the statistical analysis of actions, characteristics and circumstances to show regularities and/or differences between groups to be 'surprisingly helpful', these regularities not being considered as ends in themselves but as clues to the existence of particular mechanisms.

Cosgrove on Models

Much of Cosgrove's chapter is expressed in highly personal terms; his criticisms are made in terms of why he himself found modelling unsatisfactory or unappealing, although there are certainly implications for geography as a whole. His attitude to modelling falls within my third category, with numerical modelling placed 'alongside a range of alternatives, none of which is accorded a privileged relationship with a reality which resists direct transformation into our representations of it'. The works of Tuan and Dardel perhaps represent these alternatives, and

Cosgrove is consistent in claiming little more than personal preference as his reasons for applauding their work.

Cosgrove clearly accepts that modelling is not necessarily quantitative, as shown by his discussion of Mackinder's account of the topography of South-East England as a model. Much of his critique however groups models and quantification together, and his explicit complaints about *Models in Geography*, 'the poverty of language acceptable to model-builders in the 1960s' and 'the imperialistic claims of quantification as the guarantor of absolute truth', are both specific to quantitative models.

One aspect of his critique can be identified with the first type of criticism outlined above – that modellers have tended to throw out some of the most significant aspects of reality. Morality and aesthetics are isolated as two extremely important areas of human concern that are commonly neglected in mathematical modelling. While this is generally true, the former at least is hardly surprising given the positivist concern to separate value from fact, and to develop analyses that can stand independently of political and ethical belief. There seems little reason why non-positivist modellers should not tackle these fields – much discourse on these topics is characterized by selectivity and generalization, if not mathematization. Cosgrove's main points are that he does not want the simplification of reality offered by models and that they involve the loss of evocative description. Most modellers would accept that the extreme generality of many 1960s models was a reaction against the refusal to generalize of their immediate predecessors, and that geographical studies of individual places and small social groups have much to tell us.

Models themselves have claims to be evocative descriptions; many are descriptive in intention and a consistent claim has been that they are theoretically fertile because they evoke similarities between apparently dissimilar phenomena – similarities which have generated interesting ideas although not all have stood up to close scrutiny. Evocativeness can add much to our enjoyment of geography, and may be particularly useful in communicating our findings, but we must think about what is evoked in different people and, as with models, whether the associations evoked really aid understanding. It is pleasant to be reminded of Cosgrove's 'tail-eating grotesques', for example, but the analogy is hard to elucidate further; he presumably does not intend the apparent inference that feedback loops are more sinful than the simpler but evocatively named 'cascading systems'. And Dardel's phrase, that geography remains 'a chance, or a denial, a power and a presence', which Cosgrove finds evocative enough to use as the culmination of his paper, evokes little in me but blank incomprehension of what that combination of words could possibly mean.

Harvey on Models

Like Cosgrove, Harvey is autobiographical, and his tracing of common themes between his *Models in Geography* chapter and his subsequent work gives interesting insights into both. His critique of the modelling tradition also falls into my third category, but he is definitely more negative about the relative importance of modelling. Thus he accepts that our ability to model 'spatial behaviours like journey-to-work, retail activity, the spread of measles epidemics, the atmospheric dispersion of pollutants, and the like' represents 'no mean achievement'. He argues, however, that modelling allows us to say little about more important questions like 'the sudden explosion of third world debt in the 1970s, the remarkable push into new and seemingly quite different modes of flexible accumulation, the rise of geopolitical tensions, even the definition of key ecological problems'. Marxian historical materialism offers an approach to understanding these questions.

Like Cosgrove, Harvey does not define 'models', and his discussion of modelling refers to a school unified by the presupposition that 'a particular conception of scientific method, drawn largely from the experimental and natural sciences, could be brought to bear on all geographical problems, regardless of their form and content'. His critique refers mainly to that presupposition and not to models in general. Of the three attributes I have ascribed to models, generalization is the one most relevant to this critique, because it ignores the specific historical conditions in which particular events are situated. Mathematization is explicitly not rejected: 'this is not to rule all forms of mathematical representation, data analysis, and experimental design out of order, but to insist that those batteries of techniques and scientific languages be deployed within a much more powerful framework of historical–materialist analysis'.

Harvey and Scott's Contribution

The prime attribute of models, their selectivity, is central to Harvey and Scott's concerns. Although their extensive methodological discussion avoids the term 'model', this may be more because of its positivistic connotations than because they reject selectivity as such. Indeed, much of the paper is a defence of a highly selective view of reality against the empirically specific tradition they criticize, with its emphasis on particular localities. I do not know whether Harvey and Scott would accept a description of their theoretical project as development of a model of the historical geography of capitalism; their concern 'to explain the linkage between concrete abstractions' sounds very like model construction.

Conclusion

Modelling as a term has been relatively little used by those who have challenged the hegemony of positivism in geography over the past two decades, but modelling as an activity remains important for most, though not all, geographers. The role of modelling and the types of model employed, however, are not always the same. The need for models to simplify and to communicate would be accepted by Harvey, although he would be very concerned that the models stressed the crucial features of reality. Their function in generalizing is less widely accepted, although not totally discarded by either humanists or realists. The mathematical or statistical model is criticized not in itself but for what it does not do: for Cosgrove it cannot replace evocative description, and for Harvey it cannot answer the important questions. Neither presents it as the useless, evil or counter-revolutionary bogey-man that some of our humanist- or Marxist-inspired students see it as.

For geographers like myself with a background and interests in modelling, two main lessons can be drawn from these criticisms. The first is that it is important to consider the philosophical basis of what we are doing. There have been many important critiques of the positivist approach associated with the modelling tradition in geography, and we should be aware of them, and how our own practice is or is not vulnerable to these critiques. There have been many interesting philosophical discussions in geography recently, and again it is worthwhile to assess our modelling activities in the light of these ideas.

Secondly, criticisms can serve directly to help us improve our work. Do our models do what they are supposed to do? Are we using appropriate methods? Do our methods, and our methodology, encourage us to ignore the really interesting and important questions? Do the models highlight the key features of the subject under study, or only those that are analytically tractable? Are they understandable to our clients and to the public at large? Do they embody hidden (or overt) values, and are we prepared to stand by those values?

PART VI
Research Priorities: Where do we Go from Here?

19 The Future of Modelling in Physical Geography

Mike Kirkby

Introduction

Although the leading edge of modelling has moved on, the vast majority of physical geographers continue to rely mainly on statistical models, primarily based on regression. This is both the strength and weakness of our subject. It is a strength because it continues to generate the direct field observations which remain one of our greatest assets and these feed more ambitious models. It is a weakness because it highlights the generally poor level of communications between those with a theoretical bent and those who prefer field work. Physical (and other) geographers must also ask what they have to contribute which is not better done in other science and engineering disciplines. The answer, to me, lies in the willingness to synthesize across conventional boundaries, and in a readiness to incorporate novel field insights into geographical models, even if the field/theory division means that these insights are rarely tested rigorously.

One of the most successful areas to illustrate these points has been in hydrology, where there has been a considerable geographical input over the past twenty years which is now significantly feeding back into engineering hydrology. The original impetus for that research lay in the recognition that overland flow was much less widespread than was implied in the then-current hydrological explanations. This led to a substantial amount of fieldwork and to the development of models in which geographers have been significantly involved, and both of which continue to provoke new ways of looking at the hydrological responses of hillslopes, channels and whole catchments.

A second area of significant contribution has been in understanding the evolution of hillslopes and drainage basins over long time periods. Here too the interaction with hydrological understanding has been significant. Initial work was able to reconcile empirical field measurements of slope process rates with the landforms they produce, but is only very slowly being refined towards a proper understanding of the process mechanics, currently, with most effect perhaps, for soil erosion. Increased detail in understanding is bringing with it an appreciation of the difficulties, and the evidence needed, to extrapolate from short-term distributions of storm frequencies to meaningful long-term process rates over periods which inevitably include climatic change.

A third area of growing importance is the development of better understanding and the development of explicit forecasting models, which span the full range of interests represented within physical geography, and generally involve partners from neighbouring disciplines. It stretches and tests the ability of geographers to synthesize effectively both across discipline boundaries and across a range of time and space scales. For example, there is a need for dynamic models of plant growth, evapotranspiration, runoff and erosion. These are required both for managing ecologically sensitive and drought-prone areas, and for coupling with General Circulation Models which may, in turn, improve our knowledge of climatic process and thence climatic change.

In tackling the challenge of these needs it becomes increasingly important for physical geographers, and others involved, to have an effective training in the physical, chemical and biological principles which underlie their environmental systems of interest. We are already in danger of losing our grasp on these difficult topics, and only improved scientific training and the recruitment of interested scientists can maintain our hitherto substantial contribution.

For most physical geographers the methodological issues posed by modelling are less severe than for many human geographers, because little stands in the way of adopting a simple positivist view. In common with other contributors from physical geography to this book, I feel little difficulty in adopting a traditional science paradigm.

Perhaps my most significant reservation is that many geographical problems of interest are posed at a rather coarse, landscape scale, and often related to changes over thousands if not millions of years, so that there are real difficulties in reconciling the scale of the problem with the underlying scientific principles, which commonly refer to much finer time and space scales. On reflection, however, this difficulty of spanning scales or of aggregation and disaggregation is just as severe a problem in physics and other fundamental sciences. Reconciliation between, say, atomic physics and Newtonian physics is a far from trivial exercise; and we must expect similar difficulties in our own, less well-developed, field. To date these problems have perhaps been faced more explicitly in some urban models. In physical geography, concepts like magnitude and frequency (Wolman and Miller, 1960) or the various time spans proposed by Schumm and Lichty (1965) show an awareness of the problem which falls far short of a solution.

Although we are aware of great, and so far unresolved, difficulties in relating the evolution of a drainage basin over geological time spans to, say, the detailed findings of soil physics, we should nevertheless beware of developing new principles which are exclusive to geomorphology, or other branches of the subject. Perhaps the best cautionary tale in the recent history of the subject may be seen in the original enunciation of the principle of 'landscape entropy' by Leopold and Langbein in 1962. Upon reflection, and with the benefit of hindsight, this may be seen in part as a re-statement of the established principle of least work; and in part as a falsely drawn analogy between elevation above base level and absolute temperature.

Forecasting the future is an invitation to disaster, even with the most successful of models and over the narrowest of ranges. How much more ignominious is the record of those who have, in the past, forecast the future direction of a whole science? The most that I will attempt is to look at some of the directions in which interesting models might change, and examine a few examples which currently seem to show promise for evolutionary development. This process must miss the revolutionary changes which periodically re-shape every science.

Possible Directions for Change

Changes, brought about by new developments and shifts in fashion, may occur in either the technical content of models or in their subject scope. Their content includes their formal mathematical or logical basis, the style of model intended and the forms of outcome sought. These aspects of a model are far from independent either from one another or from the subject matter: there have been a number of approaches to the modelling of at least some problems and there is generally at least some freedom of choice in the approach taken.

The formal basis for modelling has generally concentrated on functional relationships between variables. Whether obtained from multivariate analysis or by *a priori* reasoning, the form of a functional relationship contains no explicit direction of cause to effect, although different response times for connected variables may carry a causal implication, the more rapidly responding variable being influenced by the slower responding rather than vice-versa. These relationships may express geometrical constraints, statements of balance or continuity, or rates of action of physical or chemical processes. In many cases the rates of change over distance or time set up one or a system of usually partial differential equations which formally describe both static distributions and system evolution over time.

Although functional relationships provide the predominant basis for models, a number of novel approaches have been proposed since the 1960s. Although perhaps these approaches should be viewed with scepticism, there is no doubt that they have provided a stimulus towards new ways of looking at our material which have been beneficial, even when the novel methods are finally abandoned as inappropriate. Thus the concept of landscape entropy, referred to above, has led to a number of new ways of examining problems of hydraulic geometry, and has refocused attention on the analysis of river channels and channel patterns as an efficient response to their inputs of water and sediment (e.g. Parker, 1979).

More recently catastrophe theory (Thom, 1975) has been used to describe the morphology of some thresholds in geomorphology (e.g. Thornes, 1980 for the stream head), and the associated theory of bifurcations has been successfully used in urban models in the context of dynamical systems analysis. There have also, however, been a number of attempts to apply the theory which show no similar forecasting potential.

Another approach which has shown bursts of interest is the use of fractals (Mandelbrot and Wallis, 1968). There was a flurry of activity, mainly among hydrologists, in the 1970s, and more recently there has been growing attention from physical geographers (e.g. Culling and Datko, 1989) who have used fractals to describe the morphology of lines and surfaces, and are looking for genetic interpretations for the difference. For example, the difference in fractional dimension between transects downslope and across the slope is being interpreted in terms of the relative intensities of diffusive processes in the two directions. The Hurst effect, which describes the long persistence shown in many hydrological and geophysical records, is also associated with fractals. Although again largely a descriptive tool, this approach has provoked thought about the basis for extrapolating short-term measurements over periods involving climatic and other major changes in conditions.

Also linked with fractals is the concept of attractors and chaos in the solution of non-linear equations, including many which may arise in the physical environment. Together, these associated approaches have provided valuable insights into the structure of geographical variability and into the inherent limitations to obtaining unique outcomes even within conceptually deterministic systems of interest.

There are many possible ways of classifying relevant model styles. Perhaps the most widespread type remains the 'black-box' model, in which outcomes are forecast from inputs, usually on the basis of multivariate analysis over an available range of data. A second important category, and the one which shows greatest current development, is the 'physically based' model which incorporates some process understanding, often constrained within a mass or energy balance framework. The third style of model considered here, and one which is beginning to show rapid development in our field, is the stochastic model in which parameters, inputs and/or processes contain a random element. These three styles are certainly not mutually exclusive, and many current models contain elements of all three types.

Black-box models are unquestionably the most widespread in the recent literature of physical geography, and are the most widely accepted and applied in terms of practical forecasting. Among the most successful are the universal soil loss equation (Wischmeier and Smith, 1958) and the hydraulic geometry relationships of Leopold and Maddock (1953), both pre-dating the twenty-year period of model development considered in this book. These and similar examples show their age, perhaps most noticeably in being pre-computational in their simplicity and derivation. It is still worth noting their strength and practical forecasting ability, which derives partly from the vast data base they incorporate, and partly from their very simplicity, which may give the user greater confidence than a less transparent scheme. For these reasons it is unwise to dismiss black-box models as obsolescent, particularly when most current 'physically based' models still contain a substantial amount of purely empirical representation for the processes acting.

Physically based models contain some scientific rationale for relevant processes together with mass and/or energy budgets. In many such models the budget provides the most reliable foundation for the model, whereas the process components are more or less empirically derived. No current physically based models completely avoid empirical components, although, with the better physical understanding of atmospheric processes, meteorological models are generally better founded than those in geomorphology, hydrology or biogeography. A budget has a dramatic influence on the forecasting power of a model where the budget is central to the relevant outcome, even though the process understanding is relatively weak. Thus mass budgeting has proved successful in hillslope and hydrological models, and energy budgeting in evapotranspiration models. Because energy is used with low efficiency in transporting sediment, however, energy budgets, though relevant in principle, have generally contributed little to effective sediment budget models, although Bagnold's (1977) stream sediment transport function may be an exception.

Stochastic components have increasingly appeared in recent models, initially in the network models of Leopold and Langbein (1962) and Shreve (1966), and now more widely. Their inclusion is not generally used to represent an inherent indeterminacy in the physical environment, but rather to represent either processes which are below the scale of interest of the model or model inputs which are simulated via their frequency distribution. Stochastic components may also be used to represent errors in parameter estimation or spatial variability, again usually via a specified distribution. In most non-trivial cases, stochastic components are added into a deterministic model to provide a distribution of outcomes which may be compared with a field or laboratory prototype. Given the low cost of simulation runs, relative to the cost of experimental replication, even where that is possible, there is little doubt that simulation models should increasingly contain stochastic elements and that the widespread availability of computer power favours this growth.

The improvement of computer facilities has had an even greater impact on the types of model outcome which are usually sought. In the 1960s the relative inaccessibility of computers encouraged the formulation of simple models, which could either yield general analytical solutions, or which could give analytical solutions for suitable equilibrium or steady-state assumptions. Digital computers encourage the construction of more complex models, and force the user into particular solutions. This trend has the advantage of allowing greater realism, but it may be achieved at the cost of comprehensibility, and certainly at the cost of generality. Particularly for non-linear systems, the range of possible outcomes may not form a simple pattern and there is a severe danger that the modeller no longer fully understands the web of interactions within his model, even though each individual interaction is clear. To approach generality, the modeller is able to conduct the equivalent of controlled experiments, changing one parameter or input variable at a time to observe the sensitivity of the outcome to each change, but this process falls far short of full generality.

Physically Based Models

Physically based models use some insights into the scientific basis of the system described, to give a model which incorporates at least a partial theory of how the real-world prototype functions. Such a model is not necessarily better at forecasting within its data range than a regression model, but is less prone to making absurd forecasts outside that range. The range of application of any physically based model is still limited, however, by the range over which the processes included dominate the behaviour of the system. Over an appropriate range, a sound physically based model should be more readily transferable between sites, on the basis of site measurements which should be explicitly related to model parameters. The limitations of this approach in practice can, however, be seen by reference to flood hydrology, where no satisfactory model is yet available for an ungauged catchment.

The insights within a physically based model may refer either to process mechanisms or to the budgeting of mass or energy. At an early stage in understanding, process models commonly rely on regression equations related to independent variables which have been selected either *a priori* or on the basis of, say, a step-wise regression or partial correlation analysis. For example, most current models for slope evolution incorporate process rates which are empirically related to gradient and catchment area or overland flow discharge on the basis of erosion plot data. An alternative simple approach is to choose mathematical functions with one or two parameters, which behave appropriately at extreme values, and to use empirical data to give the parameter values. This method may give a less good fit to data over a limited range, but a more physically consistent response outside it.

Some current models have significantly improved on these rather crude methods, and there is very considerable scope for further improvements. For example the work of Rutter et al. (1971) and subsequently by Sellers and Lockwood (1981) on interception of rainfall by vegetation and its subsequent evaporation (figure 19.1) has been based on increasingly complete specifications of the physical processes of heat and vapour transfer within plant canopies. Another significant study is the work by Dunne and others (e.g Dunne and Aubry, 1986) on the mechanics of

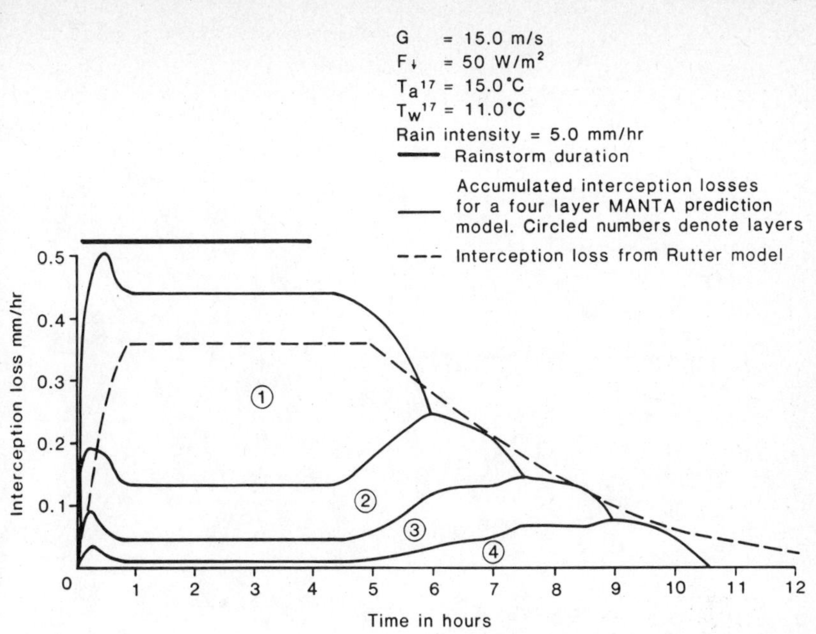

Figure 19.1 Estimated forest interception losses for single-layer (Rutter) model and multi-layer (MANTA) model (from Sellers and Lockwood, 1981).

overland flow on Kenyan savanna surfaces. This work has used field observations of surface irregularity and associated differences in infiltration capacity to forecast the pattern of overland flow generation. Flows have been routed downslope for the typically short-duration storms, and combined with sediment transport equations to forecast the pattern of rill incision, which compares well with that observed during a simulated rainfall event in the field (figure 19.2). These kinds of detailed study present difficulties in aggregation from very detailed short-term models to forecasts of change over significant longer periods. They do, however, offer an improvement in theoretical understanding, and help to identify physically measurable parameters as a basis for the transference between sites and environments which is essential to a deeper understanding of landscape evolution over substantial areas during time spans which include changes in climate.

Over the past twenty years, models have increasingly incorporated more detail of mass budgets in particular, although there have also been improvements in energy budgeting. Energy budgets have long been a crucial element in evapotranspiration estimates, but it is only in the past decade that serious attention has been paid to energy advection as a significant component of the budget (e.g. MacNaughton and Jarvis, 1983). Mass budgets for water have also been combined with energy

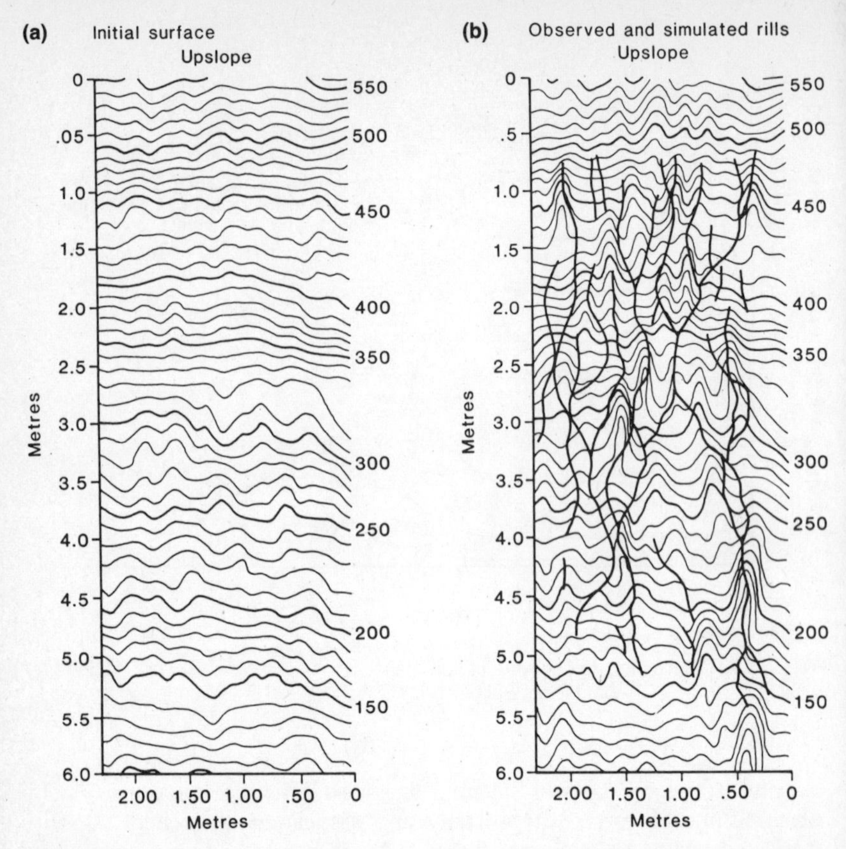

Figure 19.2 (a) Surveyed initial savanna surface. (b) Observed rill pattern after rainfall (heavy lines) and changes in contour form forecast by digital model of flow and sediment transport (from Dunne and Aubry, 1986).

budgets to provide one basis for estimating canopy composition in savanna areas (e.g. Eagleson and Segarra, 1985).

Mass budgeting has played a major part in the formulation of current models for slope evolution, which have gradually developed since the 1960s until they show enough realism to allow comparisons with surveyed landforms (e.g. Kirkby, 1984; Ahnert, 1987), for example for the South Wales cliff sequence examined by Savigear (1952) and shown in figure 19.3. Some recent work has extended the mass budgeting approach to long-term soil evolution both within a single profile (Kirkby, 1985a) and for a downslope catena (Kirkby, 1985b). Within the mass budget approach, the analysis of unstable hollow growth in response to small perturbations, based on the work of Smith and Bretherton (1972) is now being applied to the development both of rills and of the permanent valley heads which determine overall drainage texture.

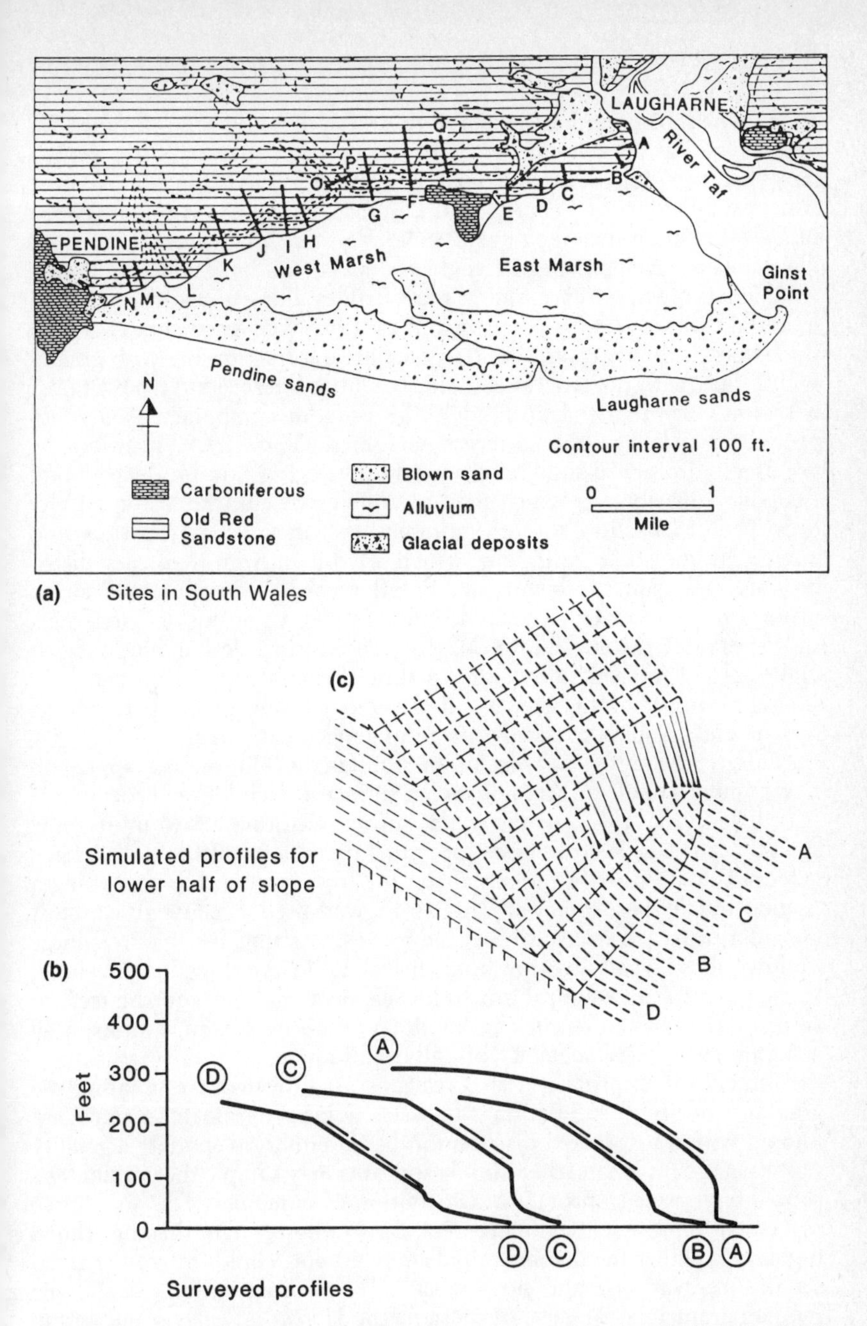

Figure 19.3 **(a)** Map showing locations of profiles surveyed by Savigear in South Wales. **(b)** Surveyed profiles A to D. **(c)** Computer simulation of change in slope form along coast, based on cliff attack followed by basal protection ((a) and (b) after Savigear, 1952; (c) from Kirkby, 1984).

Mike Kirkby

Stochastic Models

With increased computer availability and speed, the opportunities for adding stochastic components to physically based deterministic models have been exploited increasingly. Few models have the almost pure stochastic content of Leopold and Langbein's original (1962) network models, although such models still have some teaching value. Figure 19.4 illustrates a development of Langbein and Leopold's (1968) simulation of kinematic waves representing gravel riffles along a stream, in which material can move diagonally as well as directly downstream, except at the banks. It illustrates the tendency for greater bar spacing with greater width, but, with no process content, has little interpretative power.

In less trivial stochastic models the random component is used to produce a distribution of forecast outcomes. Figure 19.5 illustrates the use of a hydrological simulation model (Beven, 1987) to produce a flood frequency distribution which may be compared with the historical record. The stochastic elements represent the storm characteristics and antecedent moisture conditions, which are drawn from frequency distributions based on the historical rainfall record. This approach shows promise as a rational alternative to curve fitting as a means of extending a short-period hydrological record for engineering design purposes. It makes use of the generally better quality of rainfall records in comparison with runoff data, but is still limited by our inability to model hydrological response reliably for an ungauged catchment.

Figure 19.6 shows one stage in the simulated evolution of drainage on a two-dimensional surface. For such a simulation, either the process rate (Cordova et al., 1983) or the initial surface (Kirkby, 1986) must show spatial perturbations from a smooth function. In figure 19.6 this has been achieved by perturbing the initial surface from a smooth plane, using a pseudo-fractal perturbation. This form was used because it contains irregularities at all scales, so that the processes acting are able to enlarge hollows above a stability threshold, and fill hollows below this threshold. In figure 19.7 a temporal stochastic element has been introduced to simulate the growth of cut and fill along a hollow axis in response to a randomly generated sequence of daily rainfalls.

A stochastic approach is also relevant as a means of estimating the goodness-of-fit of simulation forecasts. Where parameter values are known within a specified distribution based either on spatial variability or measurement errors, then simulation runs may sample the parameters' probability space to provide a distribution of outcomes. The qualitative agreement represented by figure 19.3, for example, might then be refined to define whether the theory embodied in a model could lead to the given set of observations, and how sensitively the observations define the model parameters. The use of these methods is at an early stage within our field but, with the rising ratio of experimental to computational costs, is likely to show rapid growth, hopefully in association with more thoughtful design of critical experiments.

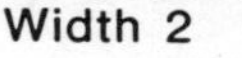

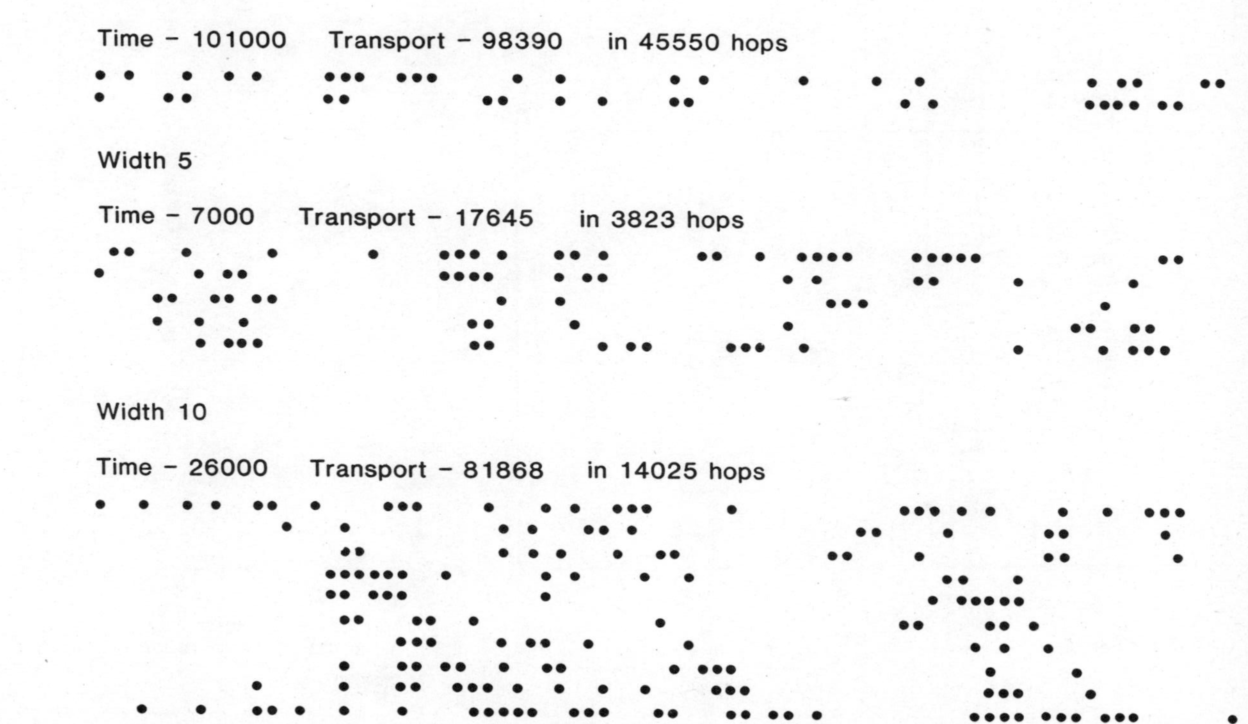

Figure 19.4 Stochastic model for gravel kinematic waves. Material may attempt to move down-channel or diagonally, but is impeded by material in its way or by the banks; **(a)** to **(c)** show the effect of increasing width on bar spacing.

(a) Inputs

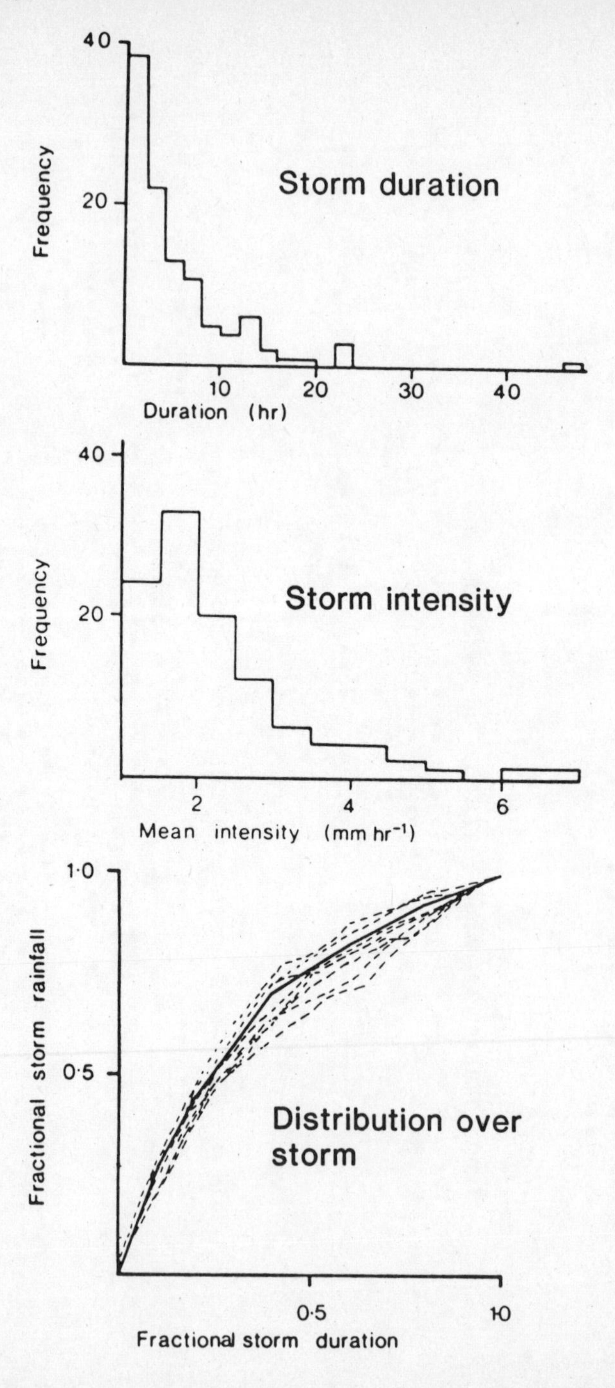

Storm duration
Frequency
40
20
10
20
30
40
Duration (hr)
Storm intensity
Frequency
40
20
2
4
6
Mean intensity (mm hr⁻¹)
Distribution over storm
Fractional storm rainfall
1·0
0·5
0·5
1·0
Fractional storm duration

(b) Outputs

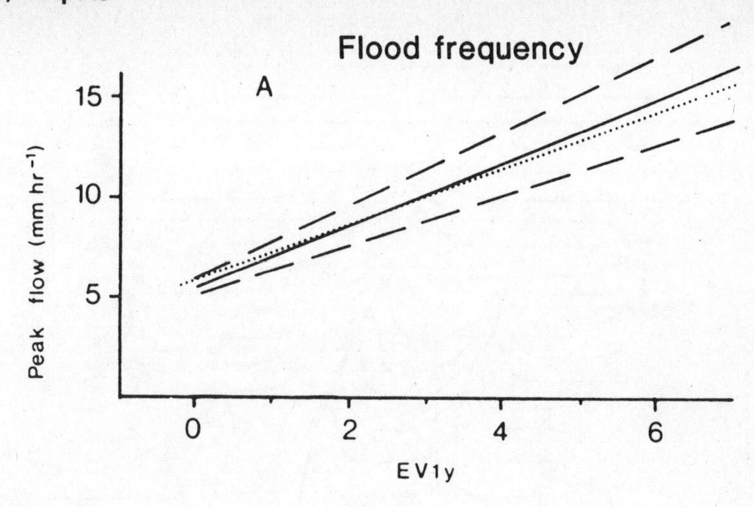

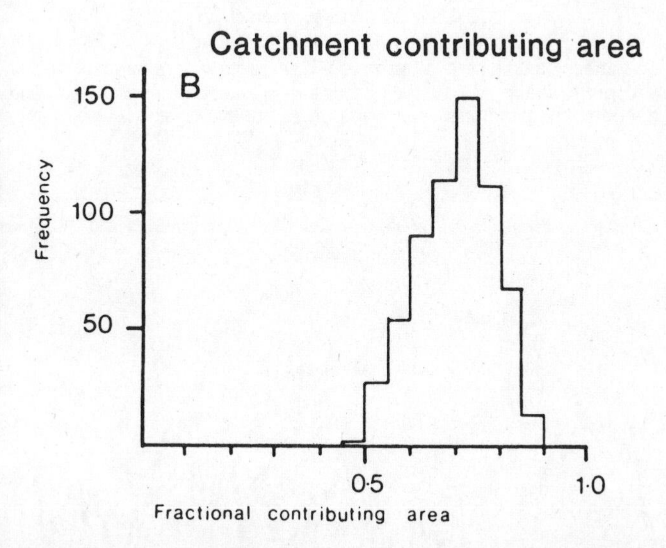

Figure 19.5 Forecasts of flood frequency distributions on the basis of repeated storm simulations: (A) shows assumed input distributions, based on available rainfall data; (B) shows forecast outputs (from Beven, 1987).

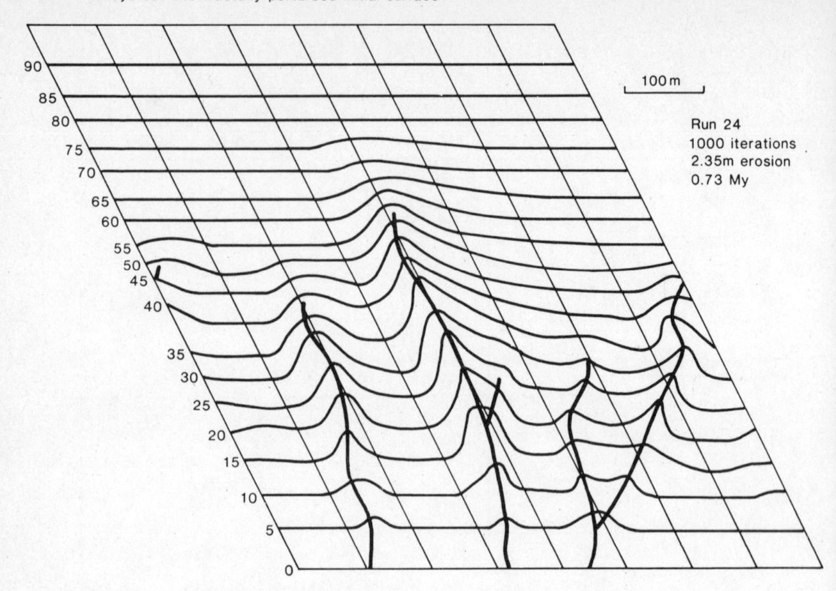

Figure 19.6 Two-dimensional slope evolution model, showing valleys forming from an initially uniformly sloping surface, with fractal perturbations. After a number of iterations, upslope areas become smoothed and downslope areas are carved into valleys.

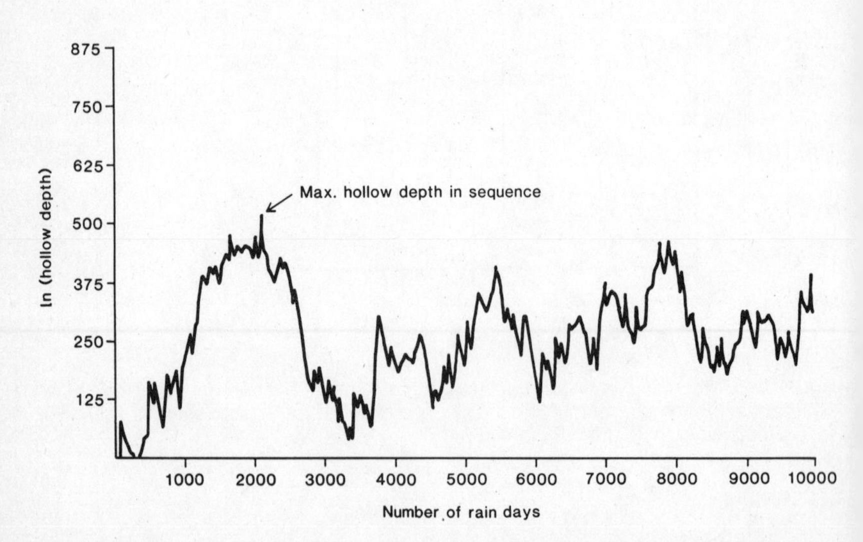

Figure 19.7 Simulated sequence of cut and fill in a stream head hollow, as fluvial processes incise alternately with slope process infilling. The sequence represents a point near the 'equilibrium' stream head, over a period of about 100 years.

Interactive Models

Although there are dangers, as pointed out by Thornes (this volume), in constructing bigger and more complex models by combining a series of component sub-models, nevertheless there is both scope and need for models which encompass a wide range of mutual interactions. Areas of currently rapid growth are in linking surface hydrology with atmospheric General Circulation Models, and in incorporating plant eco-physiological approaches into both hydrological and geomorphological models.

Where large models with many parameters are combined, problems tend to arise because of inherent complexity, through inconsistent definition of associated parameters in separate sub-models, and through ignoring all but the most direct linkages between the sub-models. The difficulty with complex models ultimately lies in the user's comprehension of exactly what his model is doing. One of the most critical steps in constructing a model lies in deciding which of the interlocking web of mutual effects are due in direct functional or causal links, and which are indirect, or at least negligible in relation to other, first-order effects. Finally this choice usually relies on intuition and *a priori* reasoning much more than on analysing the correlation structure of field data, though that can give some negative evidence. Figure 19.8 illustrates this choice for a program designed to estimate overland flow and soil erosion, using climatic inputs to estimate the differences in natural vegetation and soil properties in a still highly simplified way (Kirkby and Neale, 1987).

The resulting set of interlaced feedback loops tends to behave in ways which are no longer intuitively obvious, and it becomes increasingly difficult to know whether the effects produced are unexpected but relevant forecasts, or the result of incorrect identification of linkages, or even of incorrect programming. Tests with simple inputs and equilibrium cases can be very effective for checking some aspects of model performance, but tend to minimize the effects of unsteady conditions and transient states, some of which may be important in applications. Where model responses are non-linear, and computer modelling allows non-linearity to be introduced very simply, then the patterns even of equilibrium states can be extremely complex, as has been illustrated most widely in the context of Mandelbrot sets. It is argued that the cumulative effect of such non-intuitive responses, in models of even moderate complexity, may make them seem to the user almost as difficult to understand and refine as the aspect of reality which they represent. It is perhaps for this reason that the greatest progress in modelling complex realities has been made by building up from building blocks which have a very limited range of mutual interactions. Therein lies both the strength and weakness of the approach: strength because the whole remains manageable, and weakness because significant interactions tend to be ignored.

In defining sub-models to be combined, each clearly needs to meet the criterion of being a well-defined system: that is to have a large number of internal linkages and relatively few external links. The difficulty in this

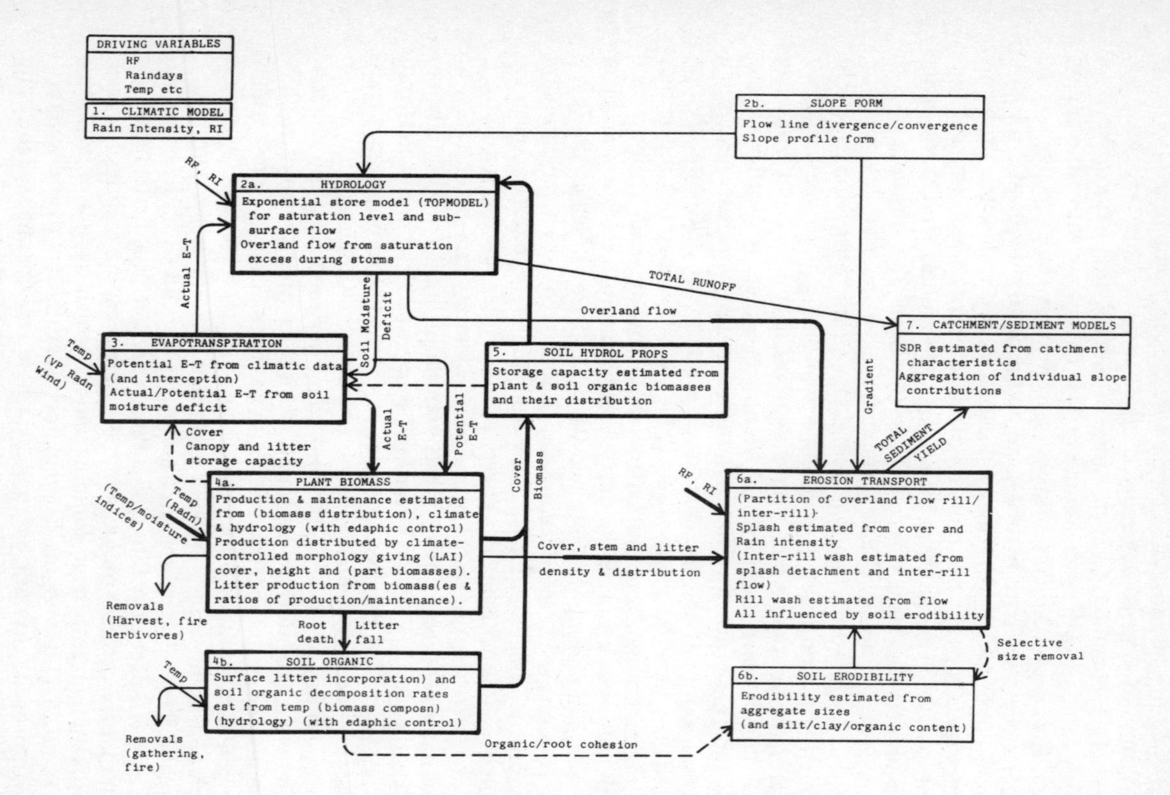

Figure 19.8 Flow diagram showing the main processes and interactions included in an erosion and vegetation model (from Kirkby and Neale, 1987).

approach is that, in for example combining sub-models for hydrological and vegetation sub-systems as sketched in figure 19.8, the two components, although conceptually distinct and with strong internal linkages, interact with each other almost everywhere, the vegetation influencing most, if not all, near-surface soil properties, and vice-versa. The way in which such a combined system behaves is very different from the behaviour of two separate models which are linked only by explicitly hydrological exchanges, for example as evapotranspiration.

Although a relatively simple example, this case illustrates the problem of combining component models, each with its own parameterization of the vegetation, but without explicit coupling of the parameters. If the vegetation is allowed to change dynamically, then the parameters may no longer forecast the effect of changed conditions on hydrological response. In modelling systems of increasing complexity, more and more first-order linkages tend to be ignored, so that forecasts can depart very severely from reality under rapidly changing conditions. Many of the first generation of complex computer models for the atmosphere, for crop growth and for flood hydrology suffer from this problem. Associated with it is the rapid growth in numbers of parameters, which mean that model optimization is both exceptionally time-consuming and very insensitive. Similar outcomes and qualities of fit to data sets can frequently be obtained for many sets of parameter values, and a great deal is usually left to judgement about the most qualitatively plausible process scenarios.

The nature of the problem can be summarized usefully by noting that the number of interactions which are needed in a model should increase roughly with the square of its size. In most early attempts at constructing large models, the building block approach has, however, resulted in only an approximately linear increase in this number, together with a disproportionate increase in the number of parameters. These difficulties are not reasons to avoid large and complex models, but only reasons to build with care and good craftsmanship.

Conclusions

The optimism expressed in *Models in Geography* has partly evaporated over the past twenty years, as geographers have continued to seek new methods and new goals. Among the debris of abandoned models there remain some partial successes, and some now-established approaches, even though modelling and theory development have once more become minority pursuits. Yet physical geography is in great danger for lack of a consistently rigorous approach to theory, to models and to experimental procedures. In many areas of research we are increasingly under threat from other scientists who are directing their interests to the physical environment, and we frequently suffer in comparison with them. Two important aspects of this problem are first the gap between geographical and scientific training; and second the need to convince other scientists of the complexity of the environment and our expertise in understanding that complexity.

As geographers, and perhaps particularly as physical geographers, we benefit in Britain from the continuing strength of geography as a school subject. It provides us with the students who do much to justify the existence of so many large university geography departments. At the same time, however, it provides students with a minimal level of scientific and mathematical training on which to build. This has been, and continues to be, reflected in our staff and our research, and our future must rest on the development of a more rigorous training.

At the same time, we do learn and teach about the complexity of our environment. We have much to teach traditional science about understanding a complex real world. So far, that message has been largely ignored, and our ability to model the environment is consequently doubted. If geography is to continue to offer an effective and distinctive view about environmental processes, we must improve the quality of our scientific training, but without sacrificing our breadth of understanding or our eye for relationships in the field. Otherwise models of the physical environment will no longer be *Models in Geography*, to the detriment of both our subject and of scientific understanding.

20 Whither Models and Geography in a Post-welfarist World?

Robert J. Bennett

Introduction

This chapter seeks to appraise the development of geographical models within the context of the 'culture of the times' in which they have been developed. It is suggested that the primary objective of models has been seen as seeking to improve the quality of public and private decision-making. In retrospect this objective can be seen as contexted within a framework of welfarism: of improving the quality of life and provision of needs through collective and governmental intervention. This naturally has led to a focus on relative deprivation and social, as well as spatial, inequality. It has been natural, therefore, for a rapprochement to emerge between social theorists and model-based analysis whereby the latter is contexted within the objectives of the former. The chapter argues that there is, however, a danger in this rapprochement and indeed for geography as a whole. The danger is that all modelling, and all geography, should be seen in *social* terms; and that social should have a singular interpretation. This is identified as an explicit or implicit *ontological shift*. It is argued that no theory has a privileged status, but must instead be supported by its ability to be confirmed by practice and by outcomes. In policy terms this means analysis of success or failure, and often of cost-effectiveness. It is argued that it is only by opening theory to challenge that geography can contribute to dialogues on policy or education in a world of changing political, social and economic structures – a world which is identified, loosely, as moving towards a post-welfarist culture of the times.

The Context of Model-building

Any intellectual activity, and a discipline such as geography can be no exception, is contexted and finds its raison d'etre within the culture of its times. Particular concepts and methodologies, such as quantitative approaches or the role of 'models' which are the subject of this book, are no less so. In order to understand the entry of 'models' into geography one naturally looks to the legacy of the 1960s Chorley–Haggett contributions through *Frontiers*, its related Madingley lectures, and to *Models*; from these it is natural to go on to the 1970s *Networks*, *Physical Geography: A Systems Approach*, *Geography: A Modern Synthesis* or *Environmental Systems*. The Chorley–Haggett contribution has been immense, highly significant in its influence on the discipline, and is rightly given a prime place in this volume. But it must be recognized, as Haggett and Chorley themselves have done, that as much as leading the field (which they certainly did often against stiff institutional resistance at the time), they also responded to what I will refer to as the 'culture of the times', or the policy culture. In the loosest and most general sense this indicated a clear path for disciplinary development through the acquisition of skills in data analysis and interpretation, statistical and quantitative generalizations, to operational models – quantitative or analogue – which allowed simulation, assessment, appraisal and, perhaps, more future-orientated forecasting and hence formulation of advice on policy. It was natural to take this argument forward into a managerial or control systems framework and to suggest feedback and feedforward control strategy. It was natural also to think of such approaches contexted within the general objective of 'improving' outcomes, satisfying quantitative or non-quantitative performance criteria, and hence linking models strongly to policy formulation for private sector or state sector planning in environmental as well as economic and social spheres. It was less natural to think of such approaches moving from descriptive generalization and understanding to the status of general or universal laws. Yet this status implicitly is claimed for models in 'Models, paradigms and the new geography' (Chorley and Haggett's opening chapters in *Models in Geography*, 1967); and certainly it became easy for Harvey (1969), working from the limited base mainly of Nagel and Braithwaite rather than Popper, to claim that this was the *necessary* base of explanation in geography: and even that if geographical phenomena did not behave according to general laws, they could be treated as if they did.

Here is not the place to rehearse again the issues of general laws, claims to objective understanding and the like. This has now been covered extensively in geography within the discussion, critique of, and rapprochement with, positivism (see e.g. Bennett, 1985). Here I wish to emphasize a different aspect. What I have referred to as the 'culture of the times' not only suggested a way forward in terms of progress in the technical apparatus for analysis, it also was contexted within the general aim of improvement of outcomes; i.e. it had a direct policy relevance and offered a means technically of *improving the quality of public and private decision-making*. If deleterious environmental or social outcomes could be foreseen, then they might be corrected. If our models allowed us to understand how certain stimuli produced certain system operations and hence particular outcomes, then it seemed as if we merely had to touch the right buttons, manipulate the correct controllable input stimuli, and a more desirable set of outcomes could be achieved. Almost all models became grounded in policy relevance and adopted some form of implicit or explicit rational decision model. In this sense the optimism which surrounded the technical possibilities for progress in environmental and socio-economic management and regulation within geography reflected the general optimism in the 1960s about both technical systems and more general concepts of social improvement. Major examples of successful control systems in industrial applications were being innovated and applied very rapidly in the 1960s, and there was still the memory of the achievements of technical change during the war effort in 1939–45. Indeed the whole field of cybernetics as a systems approach to social development is recognized by Wiener and others as being a direct spin-off of the control systems technology developed in the 1930s and 1940s (See Wiener (1949) and the discussions of the relations between science and geography recently developed by Livingstone (1985)).

The optimism was clearly too great. In retrospect it is now easy to see that not only were there immense technical difficulties in constructing credible models of the complex interacting systems with which geographers have to deal, but that the problems of achieving manipulation and control of system inputs were far from trivial, and indeed confronted policy-makers with a whole series of judgements which have rightly been revealed in subsequent discussions as having ideological bases of either an explicit or implicit kind.

The technical problems can be overcome; or perhaps more correctly at least most of them can be overcome for most of the time. Immense progress has been made in understanding how to construct, parameterize and test many of the models with which geographers are concerned. Particularly important advances have been made in understanding the problems presented by the *spatial* nature of the information and statistics with which geographers deal, and hence with the appropriate means of their treatment. I would highlight as key areas the understanding of the problems of spatial autocorrelation,[1] autocorrelation in space and time,[2] the interaction between spatial process and spatial structure,[3] the effect of data units and their modifiable scale and boundaries,[4] and the methods of simulating dynamic systems of spatial structure and interaction.[5] These developments also underpin the field of geographical information systems which is discussed elsewhere in this volume. At the same time, this understanding has also allowed geographers to make progress with 'simpler and dirtier' methods and to make major contributions in commercial areas, as argued by Openshaw in his chapter, and as applied, for example, by Beaumont. A good case is the 80:20 rule: that

[1] Chiefly, of course, the seminal work of Cliff and Ord (1973).

[2] In addition to the work of Cliff and Ord, also Bennett (1979) and Cliff et al. (1975), Haggett et al. (1977).

[3] See reviews by Cliff et al. (1975), Bennett and Haining (1985), Haining (1981).

[4] Openshaw and Taylor (1981), Arbia (1986, 1987).

[5] Wilson (1974), Wilson and Bennett (1985).

80 per cent of the answer can be obtained in 20 per cent of the time. For many applications this is enough; high levels of technical sophistication are not required.[6] But to overstate this case, which is a danger in some readings of Openshaw's chapter, ignores the fact that we can only have confidence in simpler approaches if we know from theory their technical strengths and drawbacks in the circumstances in which they are applied. Hence I would place very high emphasis on the importance of the recent developments of our theoretical as well as operational understanding which have been achieved as a consequence of the Chorley–Haggett stimulus to adopt model-based approaches in geography.

But this progress on technical issues disguises, and certainly has limited relevance to, many of the emerging questions to which answers are now required. It is a simplification, but not too unjust, to say that at a time when modelling approaches have achieved their highest degrees of technical success they are becoming less relevant to many of the most important questions of our time. This is recognized from different perspectives by both Batty and Harvey in this volume. In that modelling approaches, as any intellectual exercise, are driven in major part by relevance to current issues of policy, we are drawn inevitably to the influence I wish primarily to explore: in the broadest definition, the effect of the culture of our times. And it is at this point that I wish to address the nature of geography also in a wide sense, as now needing to rethink its overall strategy and priorities.

The Welfarist Legacy

If geography in the 1960s embarked on a technical path that could facilitate policy intervention and rational planning, this was only a step for the development of one discipline in one direction. Within geography, within other disciplines, and within society as a whole a wider set of changes were occurring in the post-war period which facilitated, encouraged and readily took up the model-based view and what it had to contribute. This 'culture of the times' can be viewed in retrospect, and with some injustices of simplification, as a consensual agreement on welfarism.[7] In the 1940s in Britain there was a remarkable series of legislative developments: to institute a system of full state education, a health and town planning system, a major emphasis on state housing provision, the acceptance of a full employment policy and the setting up of a social welfare structure. Many of these developments have their counterparts in Europe, America and other OECD countries. Britain is

[6] See Beaumont (1987) and other discussions in *Environment and Planning A*, vol. 19, no. 11.

[7] Clearly 'welfarism' is a simplification as a term, the use of which tends to suggest that other developments in this period are ignored. However, the term is used as an illustrative one, which seeks to capture the nebulous concept of culture of the times, and its dominant political consensus.

not particularly remarkable in its introduction of what we now recognize as the welfare state. The origins for the legislation in this period also clearly derive from much earlier, particularly in the Old Age Pensions Act 1908 and National Health Insurance Act 1911; and these in turn owe much to the nineteenth-century reforms of the Poor Law represented, particularly, in the Metropolitan Poor Act in 1867 and the rethinking kindled in the 1905–9 Royal Commission on the Poor Laws, particularly the minority report by Beatrice Webb.

The 1940s saw only a step along a trend which was well developed, but it was a major step in its recognition, and particularly its institutionalization of the responsibility of government; to ensure high and stable employment, to provide income support for the poor, to provide health care, education and housing, and to *plan* effective provision. Planning and rationality, not only in town planning, became by-words of the whole process and not least in economic management. The economy was to be manipulated through Keynesian mechanisms as the base for economic growth and stability. Indeed employment is recognized almost as a state 'service' in the 1944 Employment White Paper. It was, therefore, but a small step to think of intervention and control, and therefore to interface economic development with social needs, and hence to bring into play macroeconomic, microeconomic and planning models to drive economic systems towards social goals and targets. Intervention and social objectives therefore became the dominant concerns.

But the legislative and technical apparatus also reflected a wider change in the general culture of expectations. Welfare since the 1940s, in both the general sense, and in the specific sense of caring for the most disadvantaged, took on a now institutionalized status of 'entitlement'. Market theorists had argued that the outcomes of the exchange process cannot be morally appraised because they are *unintended consequences* of a multiplicity of actions for which no one individual or group is responsible,[8] hence welfare should be based on a principle of benevolence or altruism. This view became discredited, and it has now become 'normal' to accept a welfare state in which people have claims for welfare against society as a matter of *rights* or justice not merely as the outcome of altruism. In the case of the particular issue of income support this entails a whole lattice of subsidiary concepts. One is of '*relative deprivation*'. Even if the poor lived fairly well in terms of food and shelter, they lacked access to certain goods. Indeed the more highly developed an economy, the greater the level of general wealth, the more those who lacked certain items could be argued to be in relative poverty. As recognized by Galbraith (1958) and others, needs thence become easily confused with wants and desires and the moral judgement as to the distinction of 'want' and 'need' breaks down. Since economic development has allowed more or less continuous expansion of the definition of needs, so also the definition of relative deprivation or poverty has had to

[8] This view is particularly evident most recently in Rawls (1972).

be expanded; thus we have a theoretical basis for a welfare system based on '*rising expectations*'. This in turn leads to the conclusion that the lack of particular material goods is not only a misfortune, but a *social injustice*. If there is a social injustice then individuals have an obligation to the state to enable it to relieve injustice; and this moral duty, with the legislation of the welfare state, is transformed into a legal *right*, or 'principle of entitlement', for those who are relatively disadvantaged. In the 1940s, and since, such social goals became enmeshed in more general moral positions, particularly from the legacy of the Wesleyian and puritan traditions, in which entitlement should not be based on rank, or discrimination between deserving and undeserving, but on general Christian ideals of fellowship. This led to an emphasis on across-the-board provision. As summed up by William Temple, Archbishop of Canterbury 1942–4, commenting on the Beveridge Report: 'the first time anyone had set out to embody the whole spirit of the Christian ethic in an Act of Parliament'. Interventionism and social goals, therefore, took on a moral mission.

The consequence of this development in institutions and ideas is the establishment of a welfarist culture since the 1940s. The key aspects of this culture which I wish to emphasize are the dominance of interventionism, social objectives and moral mission. Indeed collectivism effectively became synonymous with socialism (cf. Pinker, 1979: 232). Let us take the specific focus, as an example, of welfare support for the poor. The idea of poverty has become displaced by welfare. It is true that new concepts of 'absolute' poverty have been rediscovered; for example pockets of poverty with special needs, not addressed by across-the-board services, have needed 'supplementary benefits' and targeting, and a 'culture of poverty' within those groups resistant to the welfare system has needed further targeted total support. But apart from these and other 'special' cases, welfare is often now not relief from problems, but the provision of services. Relative deprivation has become whatever is lacking. Recent interpretations for example describe this as 'restriction of choice' (e.g. Douglas, 1979). If it were not already clear that a transmutation has occurred, Townshend (1984) makes it vivid; within his indicators of poverty he includes the lack of hot breakfasts, of birthday parties, holidays and the habit of dining out. Such indicators are not measures of needs but of inequality in the fullfilment of wants, and those perhaps primarily of inequality in 'style of life'. By this argument deprivation can apply to any difference in habit, preference, capacity, even personal character. In this sense therefore it becomes merely a matter of judgement what is the norm from which deprivation is experienced. Hence it is but a short step to issues of morals, social ethics and, inescapably, to ideology.

These are general developments in the 'culture of the times'. For geography, as in many disciplines, there has been in some sense a consensus around the idea of research which seeks general improvement in the quality of public and private decision-making and hence in the level of living conditions and access to the 'stuff' that makes up the 'quality of life'. Models as an approach and geography as a discipline has much to contribute within this framework. Indeed frequently we see pleas for the discipline in terms of its 'social' or 'policy' 'relevance'. The model-based approach has been, and continues to be, one key aspect of this. I am sure that social geographers also believe themselves to be contributing to general improvement in society; physical geographers help to facilitate environmental understanding and management for social goals; and so on. But where does the suffusion by the idea of welfarism defined as 'relative deprivation' leave us? It seems to throw up a concern for inequalities of whatever form. Social class specifications are a focus which leads to studies of relative deprivation in place. The social context of geography leads naturally to regional inequalities being identified as a major focus: inner city versus suburbs; rural versus urban; north versus south in Britain; 'north' versus 'south' in international development; and so on. Different focuses of spatial inequality are also thrown up with the passage from 'Fordism' – mass production, division of labour, and class concentrations over space – to 'flexible accumulation' over space – interpreted by some as a means of creating a new form of working class. And there is 'spatial division of labour' which is not seen as an outcome of different lifestyles, but as a pattern of inequality. There are many other examples. My argument at this stage of the discussion is not to comment specifically on any one of these; merely I wish to identify the moralist position which has emerged. Differences exist because of geography. Geographical as well as other restrictions on choice, differences in lifestyles, differences in quality of life, access to certain goods and services can be identified. Because they can be identified, under a welfarist view, *morally* they should be overcome or at least ameliorated. But the alternative moral position is merely that they need to be better understood as distinctive regional, local or neighbourhood structures. Inasmuch as models help to identify the mechanisms which lead to the development of inequalities, and may even be used to help to reduce some of them, then the model paradigm is also framed and utilized within this welfarist framework. Certainly models were seen by Chorley and Haggett as a means of understanding which could be used to facilitate foresight and design an appropriate planning response. And now perhaps nowhere more than in the emerging field of geographical information systems is the capacity offered both to identify yet more forms of geographical inequality and to utilize simple models in offering policy advice. Recognition of this common cause, driven by an underpinning welfarism, perhaps more than anything goes to explain the emerging rapprochement between, on the one hand, the model-based quantitative/analytical approaches within geography, and on the other hand, the approaches of radicals, Marxists and critical social theorists as well as 'realism'. Indeed some of the chapters in this volume appear to reflect such a rapprochement rather fully. However desirable this development may be from many points of view, I wish to argue that it places both the discipline and model-based research in a potentially vulnerable position. I will explore this first by noting the ontological shift which we seem now to be collectively accepting.

The Ontological Shift

Because of the extent to which we have become suffused with welfarist notions it becomes difficult to disentangle research concerns from dominant 'social' goals and hence to avoid becoming driven by the single issue of the search for, and redress of, inequalities however identified. Clearly major advances can be made with such a social concern. However, within the welfarist culture there is a clear ontological shift which is not only dangerous but inherently methodologically circular. There are two components to this ontological shift: first, there is a generalist shift: the claim to the existence of beliefs or ideals independent of the world. This is achieved first by stating that all statements have an ideological base; from this position it is only a short step to argue that beliefs or ideals therefore have an equal status with that of fact: 'fact', as far as it exists at all, therefore derives from belief or ideology. Second, there is the specific shift: 'social' is defined solely, or primarily, as a concept of class within which relative deprivation is the operational indicator used as evidence of class difference. Let me illustrate this by two examples of recent lines of argument inside and outside the subject of geography.

First there is the discussion that has taken place about the relation between geography and social science. An important thrust of attention in recent years has been directed towards exploring this relation. Although it predates Harvey's (1973) *Social Justice in the City* this debate was given great stimulus by his importation of Marxist concepts and the notions of social inequality derived from the class exploitation theory of industrial capitalism. Although social science and its influence on geography is certainly not uniform, the influence of Harvey has allowed a rather general acquiescence in the idea that not only is there great relevance in introducing theories and methodologies from social science, but that geography *is* social science. From there it has been only a short step to argue that geography should be 'de-defined'; i.e. there is no subject of geography as such, merely an examination of the *spatial* aspects of general social processes.[9] The most extreme view of de-definition is not as frequently stated as the more limited presumption that geography

[9] Eliot Hurst is primarily responsible for this argument for de-definition, but in a more subtle form it is also implicit in most social theorists' contributions to geography, for example, Sayer (1984). Realists would argue that they overcome this outcome by confronting the theory with data. However, it is clear that although realism represents an important step forward, and is itself criticized by Marxists and others for stopping the theoretical project short, it is still trapped in a 'social conception' as a starting point. (See e.g. Massey and Allen, 1984; Saunders and Williams, 1986; and the critical discussion by Harvey and others in *Society and Space: Environment and Planning D*, vol. 5, no. 4, 1987.)

is social science. But the presumption of geography as 'social science' has no less underpinned a shift in ontology to a social ideal with a dangerous methodological circularity. For example we have the intensive debate around the issues of human agency and social structures within which both are seen as having only social contexts of space. Space as either an absolute or an environmental construct is ignored. The 'environment' is merely a social product. If the environment is acknowledged at all explicitly, as in Harvey's (1974a) paper, this is only to reject it as an ideological construction used to maintain a hegemony in class rule. Resource constraints and environmental limits are not interpreted as a factual reality, even in a limited relative sense, but only exist as constructs of society; e.g. scarcity or hazard is merely the product of social control or limitations of access to security. Indeed most socialist and Marxist accounts seem to lead to a common materialist position with the capitalist process itself, that the earth is a mere raw material to be used as required. The consensus between some researchers which seems now to be emerging around the philosophy of realism again must be understood within an ontology in which we are discussing 'social' realism from the starting point of a social theory, and the relation of socially constructed 'facts' to theory. The conclusion, therefore, is inevitable from the assumption; if only social relations are important then only social outcomes can be identified. Geography, within this conception of 'social science', loses its distinctive disciplinary focus either as the study of the relations of environment and society, or as the study of absolute and relative space; both of these traditional focuses include major 'non-social' aspects. The alternative social science definition de-defines significant parts of the discipline as it has been normally conceived.

Second, there is the reinterpretation of quantitative and statistical material as purely social constructs. This has particularly direct relevance to some of the issues of modelling in this volume. Under this view, facts do not exist; they are merely social constructs. A good example of the ontological shift that this involves is evident in a book used by many social scientists and now widely used by some social geographers.[10] In this Irvine and Miles start off by arguing that they wish to offer a critique in order to 'expose the political positions embedded in supposedly factual and technical arguments' (p. 2) – we can all agree on that. But later this is transmuted into revealing the 'single ideological framework underpinning the concepts and categories' (p. 113) such that 'official statistics present not a neutral picture of British Society, but one developed in support of the system of power and domination' (p. 114). Again we see the shift later where we might all accept that 'State institutions in which official statistics are produced . . . are themselves social products' (p.

[10] See Irvine and Miles, chapter 1; Miles and Irvine, chapter 10 in Irvine, Miles and Evans (1979). This position is implicitly and explicitly accepted in Harvey's chapter in this volume; other versions of the same argument can be found in Hindess (1973) and Triesmann (1974).

119), but where the ontological shift is completed by the statement (p. 127) that 'Only by understanding that statistics are produced as part of the administration and control of society around exploitative class relations can we grasp their full meaning'. In fact we should not have been deceived that there was any other possibility since the authors state quite clearly at the outset that they aim 'to assess possibilities for using statistics in developing radical social theory and practice' (p. 2). Hence the ontological shift closes the circle in that a pre-belief (in radical social theory) is confirmed. The outcome therefore cannot be expected to be other than an ideological statement since an act of faith was the starting point. We are all entitled to make acts of faith; the issue is the extent to which faith is presented as objectively confirmed or as a conclusion inevitable from assumption. The dangers are particularly worrisome when arguments such as these are accepted uncritically and repeated as objective conclusions: as for example in Sayer (1984) or Johnston (1983). Just as the critique of positivism made such an ontological shift, so does the social critique of statistical data. The danger in such an ontological shift is not just self-deception, since people are entitled to believe what they wish. The danger is that en masse the discipline accepts arguments as fact which are merely particular ideals or partial points of view, and erects them to the status of *the* theoretical base within which problems must be confronted. This in part relates to the issue of models compared to social theory accounts where, for example, Harvey in his chapter sees modelling efforts as a 'restricted domain' and as the true ideology producers in the discipline. But it is also much wider: as in Harvey and Scott's chapter that social theory has a claim to the status of being able (at least potentially) to provide a holistic theory of human behaviour. It is also rather fully reflected in the 'search' for the 'new' working class which now dominates a considerable field of research: 'class' assumes a status almost independent of reality.

Methodological Approaches to Improving the Quality of Public and Private Decision-making

I take this deliberately cumbersome subheading for the discussion since it seems to capture the post-war consensus position of the discipline, and indeed of the applied sciences as a whole, to the study of social and economic phenomena. If we seek to go beyond the presently dominant social theory approaches in geography as bases for an operational methodology, because of their false ontological status, how do we address the general issue of improving the quality of public and private decision-making? To make progress in the argument first requires answering the question of how we may approach a given research question of evaluation, effectiveness or analysis of process. Some authors have claimed much for post-modernist ideas (see e.g. Ley (1985) and Dear (1986), who follow Lyotard (1984) and Baudrillard (1983)). There is clearly merit in research on the visual and behavioural impacts of the rapidly emerging post-modern architecture and land-use plans as part of geography's general concern with urban form and structure. However, it is far from clear that post-modernism as a methodology offers any way out of the ontological circle discussed above. Indeed it seems to offer only arrival at the ultimate conclusion that, if all belief has potentially relatively equal status, such that merely whim or fashion or fun should be the basis of our research priorities, then no fact exists, no confirmation is possible, and social sciences as a whole ultimately can make no contribution – other than to entertain! The circle begun by Nietzsche is closed.[11]

Let me, however, argue an alternative position as analysts, particularly as academic commentators, sought as authors of advice and judgement, of how we should proceed to 'model' and to test our understanding by a reality which we take neither as purely social construct nor as a purely personal taste or fun. Let me distinguish in this discussion the role of the academic, the practitioner and the politician; realizing in so doing that I also therefore reject the concept that all actions are totally political (i.e. social) since this would lead us back to the ontological circularity highlighted earlier. For a practitioner some of these issues are much simpler than for an academic or a politician. An administrator or practitioner can be guided by pragmatism in which various approaches can be tried which are based on reasonable assumptions and values – no more than a loose conceptual scheme may be required. Evaluation is in terms of success or failure (do they work), although cost-effectiveness might be included. But the question of 'fact' or 'truth' is largely irrelevant. However, for a politician the issue of 'truth' is important: in the sense that his particular set of assumptions does not derive from pragmatism but from a guiding philosophy or ideology in which policy failure may indicate a failure of the ideology and hence challenge to his political party, his credibility and his position. But as we have stated earlier ideology as such can be no final guide or arbiter on policy since it leads to an ontological position that closes the question of confirmation. The academic is in yet another position: he or she should be concerned with truth, even in the most implicit or general way. Hence the academic should be concerned that outcomes confirm explanation. Outcome is, in some sense, part of an empirical confirmation. For most academics, in their positions as consultants, their input is in terms of advice on means of policy achievement, not ends (which are more the realm of politics). However success must be still governed essentially by criteria of adequacy related to observation of outcomes – or experiment. In this sense the challenge to policy and welfarism is: does it work? If there is evidence

[11] In another context, for example, Goodrich (1987) argues that although nihilism is the product of despair in a decayed culture, it also offers 'the hedonistic possibilities of displacing that culture and these values and of creating new forms of community and value'; in this sense post-modernism, concludes that nihilism is the only hope for progress in understanding. This approach, however, offers no scope for erecting relative values or seeking to prioritize.

that the basis of policy changes the incentives to behaviour in a manner which frustrates the desired outcomes, then this is a basis to challenge of the premises upon which the policy is based. It cannot be argued, as some analysts, particularly many socialists, do, that this is solely the work of other aspects of society which, if only they could be changed, would make the outcome different. To argue this is to lapse into idealism. It is also not dissimilar from the 1960s modellers' position: if only we could enter all variables we would have perfect capacity to forecast and hence to control.

In this statement I make a direct challenge to those conceptual frameworks which are not part of a validated and tested scheme. Unvalidated schemes inevitably produce circularity since the criterion of reference is the concept itself. Conceptual coherence alone is no criterion of validation since this only tests for internal consistency and the logical relation between propositions and outcomes. Such schemes are idealist and they offer no means for choosing between one scheme and another. This is particularly obvious for Marxist schemes which make claims to knowledge based on explicit acts of faith, such as labour theory of value, and a particular meaning attached to class structure. But although far from homogeneous it is also true of much social science theory. Modern realist notions do not provide escape from the trap of circularity since they also seek empirical confirmation of prespecified outcomes; failure of outcomes to meet theoretical specifications is still interpreted in terms of the conditions being not right. The enthusiasm for locality studies in geography, like the drift to micro-corporatism in political science, can be partly explained in these terms: as searches for the particular micro-conditions which allow the theory to be confirmed: not the alternative position that the theory was incorrect in some way in the first place. We must conclude, then, that such conceptual schemes offer no full or authoritative basis for academic judgement or policy advice.

In oversimplified terms the distinctions drawn here can be addressed through three levels of assessment. The first is, 'did the policy reach the target or people it was intended to reach in the intended form?'. The second level is, 'did the policy have the intended direct effects on behaviour or conditions of the target or people it reached?'. These should be the main focuses for our academic debate, I would argue. The third level, 'did the policy improve society as a whole?' is important to be aware of within holistic social research, but is not amenable to being answered unless we accept a welfare function, net 'happiness' index, or explicit moral position and hence an ideological belief. Where does this distinction lead us? The conclusion I have drawn has two chief consequences which I wish to emphasize. One is that recent ontological shifts must be exposed and recognized for what they are – attempts to make pejorative prescriptions for research and policy which have no more basis in reality, no more privileged status, and no more likely success, than any other. As such they are purely political beliefs. Second, welfarism also is similarly subject to confrontation by the reality of outcomes. Although this certainly does not mean an end of social policy, it does suggest a major agenda for research on the welfare state and its reappraisal in the light of outcomes.

Post-welfarism and Geography?

I have deliberately enmeshed the discussions of geography as a discipline, and of models, within the wider question of welfarism. This is not essential but it does serve to emphasize the role of models and the discipline of geography within the changed context of the times. Indeed even the move towards 'realist' theory must be seen as in part a response to this changing environment, even if it is only partial, as I argue. We are now entering a phase of development in western societies which some commentators, particularly of the 'new right',[12] see as post-welfarism. The concept of post-welfarism, I would argue, is perhaps the most critical question which currently confronts academics and their approaches to analysis – yet this volume hardly touches on the question. It is easy for supporters of social theory from the political 'left' to seek rejection of post-welfarism in terms of their own critical theory, but this ignores the reality that post-welfarism is not exclusively a 'new right' concept; indeed a good case can be made that the 'new right' theories are only receiving attention because of more general social changes, or what I have termed a shift in the dominant 'culture of the times' which is seeking a re-evaluation of welfare. This re-evaluation is challenging, in particular, the core concept of relative deprivation and its consequential focus on relative inequality and hence 'social' and 'spatial' justice. If every difference in outcome is translated into an entitlement to state intervention, then clearly the only end point is total state intervention in everything. This is certainly incompatible with a market economy, hence the 'new right' attack on welfare. But there is a wider problem. If relative deprivation, and entitlement as of right, cannot be a sole base for action, what is the alternative? And what is the role of regional and local inequalities which have focused much geographical research? Setting aside the 'new right' perspective, some would see the future as requiring both the abandonment of the concepts of socialism and social democracy for the alternative of a social market economy.[13]

[12] See e.g Minford (1983), Minford et al. (1983), Scruton (1980); also Minogue and Biddiss (1987). See also recent discussions in Bennett (1989 a,b).

[13] Perhaps I need to make a short comment on social democracy. Although like every other concept introduced in this chapter social democracy is far from uniform, it can be characterized, overlapping with liberalism and distinct from socialism, as tied to culture, constitutions and coalitions as a means of introducing collective disciplines into a society otherwise dominated by individualism. Some have identified this primarily with J.S. Mill and interpretations by Hobson, Hobhouse, Wallas, Hammond, as a means of balancing individualism with cooperative interest. But these writers only offer limited insight into contemporary problems. The similar concern of social democrats with class and

The post-war consensus underlying the welfare state was a triumph of social democracy. Consensus welfarism was consensus socialism: 'we are all socialists now', it was proclaimed by Sir William Harcourt even as early as the 1880s; with the result that socialism largely hijacked social democracy in the post-war period. The corporatist evolution of the state since the 1940s and its beneficent planning of life and work represented a compromise with capitalism: a 'welfare capitalism' or 'mixed economy' as it was often termed.

What has occurred in the period at least since the early 1970s has been a steady breakdown in this corporatist position and in the welfare compromise. It is in this sense that I refer to post-welfarism. But if the changes were occurring in the early 1970s, it was not until the Thatcher–Reagan era that it became fully evident and unavoidable as a theoretical issue. Thatcher, in particular, has sensed that in Britain a 'sea change' has occurred in which old structures no longer have their previous relevance or power. At the core of this change is the definition of social theory and hence of socialism and its critique of capitalism. Thus, it has been argued, Thatcher has been able to translate the image of capitalism from 'dark satanic mill' to shopping mall; capitalism has become seen as the means of creating and distributing the good things in life. Where social theory emphasizes the negative aspects of capitalism's capacity to create new wants and hence new 'relative deprivation', the emergent 'culture of the times' has been happier to see the market as both the creator and provider of new wants. Rather than markets being seen as an inhumane and exploitative system, socialism and even corporatist social democracy have come to be associated with the odious and paternalistic treatment of individuals. Indeed many socialists have recognized these problems: e.g. Williamson: 'Many of those who should benefit from what should be the greatest successes of British socialism to date – public ownership, the welfare state, municipal housing – actually regard these "successes" as part of the general force which is oppressing them'; or Ken Livingstone: 'services . . . are increasingly seen by those who receive them as instruments of control over their daily lives' (both quoted in Gyford, 1985: 4). Where the Thatcher era has heralded consumer choice and economic change, social theory and socialist politics has sought to defend the mode of production and to trap people in labour-intensive work practices and unattractive jobs vulnerable to technological change; the spirit of market freedom of individuals has heralded a consumer and service economy which has offered the release from the least attractive toils and labours, and has seemed to offer the potential to satisfy many of people's most avaricious dreams.

'interest' groups seems to undermine the scope for progressive thought. Certainly in Britain social democracy as yet seems to offer no clearer guidance on possible futures than socialism. Hence my simplistic lumping together of these two distinct and largely irreconcilable views. The meaning of a 'social market economy' is as yet even more problematic, but does seem to offer better possibilities for combining private rights and collective interest.

But, most important, the challenge of this period has been to ask the presently unanswered question of what are the alternative policies to market economies that can be shown to work. If there is a post-welfarism which has a clear 'social' base which justifies government action this will have to be defined, argued for, and proven. It is no longer credible intellectually or politically to argue for many previous forms of government management or intervention. Mass public housing, demand-inelastic systems of health care, education objectives which seek to equalize outcomes rather than potentials, 'full employment' Keynesianism, regional policy, even the existence of local government, are each examples where policies have been developed over the past forty years, but where there have been significant and recurrent failures. These policies are still cherished as idealist visions by many politicians and social theorists. However, I argue that it is now necessary to accept the scrutiny and assessment of effectiveness.[14]

It is here that I see the intellectual challenge for the discipline and for model-based approaches. It is clear from what I have argued that a welfare state based on social theory, particularly both Marxist and realist forms of idealist ontology, cannot be accepted as an authoritative base for policy: they cannot be proved to work, and can only be evaluated for internal consistency. Even social democracy offers no easy solution. It is equally clear that unfettered markets cannot be the guarantor of all social goals or human fulfilment. Markets do fail. Not only is there the issue of inequality, but there are also the wider questions of collective goods as well as 'human' needs. If markets could provide all our individual and quantitative needs, they certainly do not provide all our collective and qualitative needs. Policy and politics are not merely about which administrative procedure or party is the best guarantor of the highest level of individual want satisfaction. They also embrace the meaning of relationships, within and without the family, and in wider collective milieux. Welfarism has provided the approach of mediating the responsibilities of the well-off to the disadvantaged through the mechanism of government. The level of support may not be generous, nor sufficient, and it is being eroded, but it does give a level of support. The two key questions for a post-welfarist society are: first how can support be improved by practical policies that can be demonstrated to work and are reasonably cost-effective; and second at what point does governmental action end. Should government be concerned merely with facilitative policies which allow human needs to be satisfied; or does it involve wider 'service' needs of income, food, shelter, health care, and so on. It is in the second area that it is clear that neither social theory nor pure market systems provide fully satisfactory guidance. Markets frequently fail and we need a new

[14] Perhaps even more crucial is the need of many social theorists to accept comparison with the range of working Marxist regimes in the Eastern bloc. Despite claims of interest in socialist economies as a key research question, e.g. by Michael Dear and others, virtually no social theorist has had the courage to approach this question.

method of defining what are the appropriate collective and public goods in society, which allow the best conditions for human flourishing and fraternity, belonging, dignity, respect. These 'needs' cannot be codified as rights or entitlements in a welfare system. Indeed satisfaction of 'basic needs' and 'rights' through welfare often intensifies the problem of the need for solidarity: the satisfaction of basic needs seems to increase the scramble for scarce existing resources (see e.g. Hirsch (1977), Thurow (1980)), promulgates 'interest group' politics, and hence alienates individuals further from each other. Hence what is needed is a better definition of what is 'socially' possible through collective action and what is not. Social theory and social democracy has, perhaps, promised too much and hence has led to disillusionment in its own promises.

The grand objective of the discipline should be to contribute to the debates around these issues. But it must be a contribution to practice. It is a deeply embedded belief of idealist philosophies, such as Marxism and to a lesser extent realism, that needs can be satisfied, and indeed that needs can be ranked so as to avoid dispute. Yet how can we rank freedom compared to solidarity, the degree of individual rights compared to collective or interest group responsibilities, or fraternity/solidarity compared to equality? Social theory is purely ontology if it believes that these questions can be resolved in principle. Instead I would argue that they are contradictions which can only be resolved both by practice, through different circumstances, combined into practical policy solutions, and by better theorizing the distinction between politics and practice so as to avoid intrusion of the political into the area of the unattainable. A good discussion of this problem is given by Ignatieff (1984).

In a post-welfare society this leads us to ask if it is the government that is the most appropriate arbiter and provider of collective needs; and indeed whether governments should be the determiner of private fates. New practices of service delivery are only one aspect of the emerging question of *how* individual and collective rights and needs are rationalized and *which* practices can be demonstrated as successful. These seem the more urgent and important questions than those of theoretical debates on ends.

This would lead naturally at this point in the discussion to list future research priorities in the field of policy and practice and to outline the role of models within this field. Space here is not sufficient to do this, nor would I yet be that presumptuous, although some implicit indications have been given which are being developed in a later project. Also it must be recognized that the issue of interplay between market, individual, social and collective has been contentious over the whole of human history, and much has already been said or written which points our way forward. What I wish to emphasize here in conclusion is that geography, and both the social theory and model-based approaches within it, cannot continue with an accepted ontological status for intervention or welfarism. The welfare state, and its associated public decision-making, no longer has the privileged status that it can be justified by mere statements of belief in public or governmental goods: public goods and the policy that provides them have to be demonstrated to be *effective* in meeting social needs; policies have to work and be more effective than alternatives. Within the necessary reappraisals that must occur difficult academic and political judgements will be required, and particularly problematic will be attempts to redefine the concepts of 'rights' and collective action at an operational policy level.

The danger, and therefore the message for the discipline of geography, as I see it, is that a widespread ontological shift has allowed much of the moral high ground on policy issues to be captured by idealist frameworks with scant empirical base or practical outcome. I criticize particularly the attempt to assert their privileged ontological status. These approaches, just as any other, have no particular right to the moral high ground; nor are they any more immune than other theories to test and confirmation in practice merely because they have a 'moral mission' or claim a greater 'social' base of 'caring'. As practitioners move forwards to reassess and to seek new policy responses to old problems in the fields of welfare (e.g. of health care, education, help to the poor, regional policy, urban regeneration) there is a danger of the discipline, and indeed of a major part of social science as a whole, being left in isolation propounding idealist theories for policy issues which have already been proved to fail or are no longer relevant. Defence of welfarism cannot be justified merely by crede, but by sound policy which has succeeded, or has a chance of succeeding. Geographers must participate in this growing field of evaluation and innovation of policy and its outcomes. They must do this not through further conceptual refinements such as 'realism' but through analysis of the rationale for action based on assessment of success and failure in the realms of public and private policy. For welfare this means experiment with new structures in which are embedded new incentives. For the academic discipline of geography this means adaptation of its frameworks of teaching and research. I would argue that one aspect of this requires more intensive training in analytical methods including model-based approaches, information systems and elementary analytical skills; Openshaw sums up some of these needs rather well in his chapter. The key aspect, however, is a better understanding of the structure of economic incentives and rights, rather than class. As noted by Asa Briggs, we have become very much dominated by an early nineteenth-century concept of class – mechanistic and based on strata. We need to identify a new language not of class but of 'rights' or 'nature'. By which I mean 'choice' conceptions of rights which promote autonomy, freedom, self-determination and human development, and not 'interest' conceptions of rights which make people passive beneficiaries of the services of others.

In making steps forward in this direction we must recognize that new social relations and new methods of combining needs and resources will be required, and these must be based on practical capacities of collective action and individual choice fulfilment. A major research agenda for the discipline therefore lies in the field of reappraisal of the welfare state and its alternatives in the light of outcomes.

It is for this reason that this volume is so timely and important. Without changes in our teaching and research, and without a thorough reassessment of the technical infrastructure required for policy appraisal related to models and their uses, there is a real danger of the geographical priesthood remaining whilst the congregation departs. The subject will rightly no longer claim the share of students that it has received in the past, and it will have lost its capacity to contribute useful knowledge to the research process, to policy debate and to practice.

Modelling Through: An Afterword to *Remodelling Geography*

Bill Macmillan

Introduction

Editorial comments on collections of papers tend to be anodyne, even-handed and brief. Editors generally know their place – to be seen but not heard. Favourable references to individual contributions are normal (courtesy demands it) but criticism is rare (propriety forbids it). In closing this book I am going to find it hard to be brief because of the importance of the issues to be dealt with, and I am not going to attempt to be even-handed. Rather, I propose to explore certain themes, which means concentrating on the chapters in which those themes are articulated. Nor will I avoid criticism: much has been said in earlier chapters which is challenging and it is a strange kind of compliment to the contributors not to take up the challenge.

What are the themes of the book? Surprisingly, after more than two decades of modelling, we are still debating what a model is, we have done little more than begin to understand what systems are, and we are only just getting to grips with the basic problems and properties of spatial statistics. Predictably, we remain divided over the efficacy of model-building as a mainspring of the discipline's development: it seems to be agreed that modelling has its place but there is profound disagreement as to what that place should be. There is general interest in the confrontation of theory with evidence, and argument about the forms which that confrontation should take – some of it centred around the concept of the 'region'. There is a strong sense that changes in the socio-political and computing environments are rendering academic model-building anachronistic in one context and peripheral in another: there is a feeling that some modellers believe in a better yesterday and that others are being played out of the game. Criticisms abound and they flow from the pens of the 'modellers' as much as the 'critics' but the overall tone of the book is one of frustrated optimism rather than pessimism. The frustration comes from the belief that there are great challenges before us that we are ill-equipped to meet, but which others are willing and able to take up.

I propose to pursue these issues by looking first at the past and present state of modelling in geography. This will involve identifying different styles of modelling and the criticisms levelled at them by contributors. In so far as there are ready responses to these criticisms from the work of other contributors, I will present them as I go along. It will also involve a counter-criticism by modellers of geography at large and a review of the arguments on the implications for modelling of the changing culture of the times. Having looked at the past and present state of modelling, it will be possible to consider the task of remodelling geography. That will involve a reappraisal of the position of theories, facts and values in the light of the comments on the culture of the times. From this it will emerge that theory construction and validation are the central issues to be faced, so I will go on to look at what contributors think we should be theorizing about and how we should be doing it. The relationship between theories and applications will also be addressed, as will the crucial question of teaching. In conclusion I intend to suggest that the main failings of the modelling community have been failings of conviction and imagination rather than of method, and that the constraints on further progress are not what they are sometimes supposed to be.

Modelling Achievements

As noted in the Preface, one of the reasons for holding the Oxford Conference and producing this book was to disabuse students of the idea that modelling in particular, and the scientific method in general, belong more to the history of geography than to its present and future. Much of the recent literature on the philosophy, nature and practice of the subject encourages this view. It is, quite simply, wrong. On top of this, the history of modelling is sometimes thought of as being short on successes. It isn't, but by no means all of the achievements are mentioned in this book. None of the contributors has tried to catalogue the models in his or her field and some have taken most of the major developments as read. Thornes, Harvey and Godfray, and Henderson-Sellers, all point out some landmarks but do not attempt comprehensive reviews. Some of the other authors scarcely provide a review at all – they have their minds on other matters. As a consequence, there is no mention of a number of significant achievements: recent studies on the integration of environmental effects into economic input–output modelling and the development of sophisticated spatial demographic and demo-econometric modelling tools spring to mind as examples. In this context it needs to be said at the outset that modelling has helped geography to achieve widespread and significant advances.

Modelling's Feasible Region

Even the critics acknowledge that what has been done by modellers 'represents no mean achievement' (Harvey). What the critics argue is not that modelling has been unsuccessful but that its past and possible future successes are limited – a view from which the modellers themselves would not dissent. What the debate turns on is the nature and extent of the limitation.

Modelling in Geography

Some of the arguments about limitations are concerned with the scope of one particular style of modelling and others centre on the relative scope of competing styles. Both these types of argument tend to take place within the modelling community. External criticisms, on the other hand, often fail to distinguish between styles and tend to be the poorer for it: they may be cogent if applied to one form of modelling but not to another, so they are difficult to sustain in a general form. Some subclassification of styles is essential if the criticisms are to be addressed satisfactorily.

Different commentators classify modelling in different ways, but there is a common underlying distinction in most schemes between 'applied' and 'theoretical' models, a distinction which is central to many of the arguments in this book. Haines-Young sees it in terms of 'technological' and 'scientific' modelling while Kirkby employs a black-box/physically based/stochastic subdivision, where the last two of these categories are, roughly speaking, scientific in Haines-Young's terms and the former is technological. Some human geographical models also belong to this second category, especially those designed for planning purposes. The defining characteristic of technological models is that the demands of planning, or of 'application' in a more general sense, are paramount in their construction.

The distinction between the technological and scientific styles of modelling is, then, one of context. In addition, it is often supposed that the latter are theoretical and the former are not – witness Batty's comment that 'models in the Lowry (1964) tradition . . . are essentially mechanical artefacts' – but this dichotomy is essentially false, as Haines-Young makes clear. Technological models are *implicitly* theoretical and their major deficiency results from 'the failure of workers to make explicit what theoretical foundations exist and to expose them to critical evaluation'. The context of the constuction of technological or applied models will be discussed later. For the moment, something more needs to be said about the major sub-categories of scientific modelling.

Scientific Modelling

Scientific models are built for explanatory rather than pragmatic purposes. When *Models in Geography* was published it would have been reasonable to distinguish two types of scientific model or, rather, two approaches to scientific modelling: an inductive approach centred on the application of a particular statistical technique and a deductive approach involving speculative mathematical modelling. The past two decades have seen great strides made in both of these areas but they have also seen another development – the emergence of data-based modelling. Thus, for this volume, there are three areas of activity that demand attention.

Statistical modelling The statistical approach to modelling is both the most frequently criticized and the most popular: it is the object (or apparent object) of most anti-positivist criticisms; and as Mike Kirkby points out 'although the leading edge of modelling has moved on, the vast majority of physical geographers [and some human geographers] continue to rely mainly on statistical models, primarily based on regression. Indeed, in Thornes's words, 'regression analysis and modelling were regarded as essentially synonymous' in the early days of modelling because 'regression model concepts are much more readily grasped than differential calculus'. Those early days, in Cox's view, were ones of 'exuberant experiment about which it is all too easy to be extremely critical in retrospect.'

Over the intervening years 'geographers have tended to examine only a fairly limited range of models' (Haining), but new possibilities are beginning to open up. Haining points to the progress made in spatial statistics and the development of methods for the analysis of space–time and qualitative data. He notes that for geographical work, standard statistical techniques (and standard packages) are of limited value because of their presumption of independence of observations, but he shows how recent advances are making it possible to circumvent this problem.

Meanwhile, Thornes suggests that general statistical modelling has lost much of its earlier significance, and Unwin (echoing Haining) agrees that 'classical statistical methods are often inappropriate' and adds that 'their use tends to force investigators into the analysis and description of system morphology rather than process description'. Unwin sees a bright future for one statistical approach, though. He argues that 'the full potential of stochastic process modelling of time, space and space–time series does not seem to have been realized', a view supported by Kirkby's observation that stochastic modelling is 'beginning to show rapid development'. Whether we should treat this development as statistical or mathematical is a moot point.

Mathematical models Mathematical modelling has developed significantly over the past twenty years. In climate work 'the mathematical, more usually termed numerical, climate model is by far the dominant model type' (Henderson-Sellers) and much the same could be said about the fields of location theory, spatial interaction, and a number of other areas.

In the early days of modelling there was much talk about systems but little or no real understanding or application of systems theory. It has taken a long time for this to change and in Thornes's view it is still the case that 'systems analysis is a popular concept enjoyed by many but whose deeper ramifications are understood by relatively few'. Nevertheless, the application of modified Lotka–Volterra models in biogeography (noted by Harvey and Godfray), some of the developments in climate modelling described by Henderson-Sellers, and the major trend seen by Thornes in the development of models of evolutionary behaviour which 'shift the interest from equilibrium conditions *per se* to stability or otherwise of the equilibria', are all signs that systems theory is finally making its mark. Some progress is also being made in human geography thanks to the work of Wilson and others (see, for example, his book on catastrophe theory (1981a)) but much more could be done: generally speaking, there is still too much concentration on equilibria and too little on evolutionary behaviour.

To set against the rather slow progress in adopting system theoretic methods in human geography is the successful development of what Batty calls the optimization paradigm. He says that 'the most dramatic accomplishment [in urban modelling] . . . relates to a theoretical synthesis of model estimation, specification and application around the idea of optimization' and he observes that relatively recently, 'the paradigm was formally extended to embrace random utility theory . . . and the link between micro and macro-modelling, so long a source of difficulty and contention finally fell into place'.[1]

Data-based models Applied optimization and the (numerical) solution of systems of differential or difference equations in system theoretic studies both require computers, as does nearly all practical spatial statistical work. These points alone would be good grounds for Openshaw to argue that 'maybe the moment is opportune for a substitution of the term "mathematical" by "computer" when applied to models'. But he has more in mind than this. He wants to see a change in perspective 'to favour applications with data'. The focus in a modelling exercise, in his view, should not be on a mathematical model or even a particular statistical technique but on a set of data. For Openshaw, the way forward is to adopt strictly inductive methods. He believes that the 'emphasis on mathematical modelling as a largely data-free activity has established a style of modelling which has continued despite massive changes in both computer hardware and data environments', and to reinforce this point he adds that 'few of the texts on models or modelling methods in geography mention the word "data" or "computers"'.

The effects on modelling of changes in the computing environment are

[1] In my view, for what it is worth, this link is as yet incomplete. All the components are available but there is a problem of logical closure that remains to be resolved, centring on the consistency of equilibrium prices.

also noted by other contributors. Kirkby comments that 'digital computers encourage the construction of more complex models, and force the user into particular solutions. This trend has the advantage of allowing greater realism, but it may be achieved at the cost of comprehensibility, and certainly at the cost of generality.' Batty observes that in planning contexts (and the same can be said of human geography generally) the advent of computers 'presented a working environment or medium akin to the physical scientists' laboratory'. He also makes the important point that for the presentation and comprehension of large quantities of information, computer graphics is a highly effective medium and 'in spatial problems it is essential'.

Haines-Young takes a different tack and examines 'the way in which techniques from the fields of artificial intelligence and expert systems engineering provides a richer language for modelling than has been available in the past'. He argues that 'the significance of developments in AI in general and with expert systems in particular, lies in the fact that they show how our notion of modelling can be expanded to capture the more qualitative aspects of the world and our knowledge about it'.

Rhind and Openshaw would no doubt acknowledge these points but they are more concerned with the burgeoning information economy. As Rhind makes clear, 'despite the growing "data mountains", remarkably little of the present work of the many thousands of individuals now working in this field world-wide is on modelling in any sophisticated sense'. He believes, however, that 'commercial superstructures for user-specific modelling will appear' and argues that 'this area represents an opportunity for professional geographers'.

There is some implicit support for Openshaw's argument for data-based modelling in Rhind's work but other contributors are clearly unhappy with his call for a return to strictly inductive models. Haines-Young claims that preoccupation with data-based or data driven models 'would seem to be a recipe for directing attention away from developing a theoretical framework', whilst Cox says that it will 'limit imagination and creativity'. Batty sums up these sentiments in one of his section headings: 'data, data everywhere and not a thought to think'. In his opinion, 'the lack of data has never really been a central issue, and in any case it is open to question as to whether any of the appropriate data is now available. . . . The issue is as much one of better theory as it has always been'.

Criticisms of Modelling

Having made these simple distinctions between modelling styles we can now take a look at some of the criticisms of modelling that pepper the earlier chapters. At the start, it is worth reiterating Flowerdew's observation that there are different levels or degrees of criticism. In this book, they have ranged from calls for fine-tuning to wholesale objections to supposedly key aspects of the modelling enterprise. The latter have been directed primarily at work in human geography.

'For most physical geographers', Kirkby argues, 'the methodological issues posed by modelling are less severe than for many human geographers, because little stands in the way of adopting a simple positivist view. In common with the other contributors from physical geography to this book I feel little difficulty in adopting a traditional science paradigm.' It follows from the general acceptance of this view that the criticisms that are made about physical models tend to be concerned with the best way to achieve broadly agreed scientific objectives.

In human geography the criticisms are much more varied. For Openshaw they 'can be summarized under four headings: wrong objectives, insoluble problems in model construction, lack of empirical validation, and little evidence of success or of markets for the traditional [modelling] produce'. He reckons that the 'limits might already have been reached on the likely achievements of human model-building', and argues that 'widening the scope of models to incorporate social processes, however desirable it may be, is also impossible!' To claim that something is impossible requires a good deal of confidence: I am equally confident that he is wrong. Difficult it may be, impossible it is not. The associated argument that 'the search for understanding through process knowledge and causality has never really worked', seems to me to be equally flawed.

For Wilson, 'geography is failing in one of its main traditional functions: to synthesize (and contribute to the creation of) knowledge about regions'. Like Openshaw, Wilson recognizes a disjuncture between models and real-world systems but the two part company over the best strategy for rectifying this situation. Some comments will be made later on the Wilson strategy.

The above criticisms are concerned implicitly with quantitative models. Indeed, it is often assumed that all modelling is quantitative, since, to use Haines-Young's words, modellers have shown an 'almost complete preoccupation with things that can be quantified'. In his view, this is a deficiency which ought to be rectified and he thinks it can be as noted earlier.

What appears to be a preoccupation to Haines-Young looks more like an obsession to Cosgrove. Referring back to the 1967 Chorley and Haggett volume, he argues that 'if there is a critique here of *Models in Geography* it is of the poverty of language acceptable to model-builders in the 1960s and of the imperialistic claims to quantification as the guarantor of objective truth'. Whether or not this is a fair comment on *Models*, it seems inappropriate here. Quantification guarantees nothing: as Bennett says, albeit in a different context, 'no theory has privileged status, but must instead be supported by its ability to be confirmed by practice and by outcomes'. It may be the case that modellers concentrate on the measurable aspects of phenomena to the neglect of other properties, but that is another issue – one which Cosgrove deals with separately and rather elaborately through his cathedral analogy.

That analogy is nicely chosen but wrongly applied: it involves 'tail-eating grotesques that surround the portals of the gothic cathedral' and 'the soaring columns of its interior, branching into arches and vaults with a clarity that renders the entire edifice a comprehensible and satisfying whole'. Cosgrove identifies the grotesques with the early efforts of model-builders and says that, for him, 'in 1969 the cathedral of geography seemed arrested at the portal'. In my view the analogy works exactly the other way round: the role of formal analysis is, surely, to make comprehensible the structural properties of the cathedral of geography – to show its design and construction. The methods advocated by Cosgrove, with their echoes in the idea of geography as 'informed tourism', seem likely to reveal no more than its decoration.

Cosgrove's analysis belongs to a class of criticisms that appear, in Wilson's words, 'to identify modelling with positivism and a narrow kind of science'. Both Wilson and Haines-Young see this view as naive. It is a view which is reflected both in Harvey's remarks on the restriction of modelling to replicable events and in Sayer's (1984) depiction of the scientific understanding of causality as constant conjunction (which features in the discussion in chapter 7). Flowerdew makes the related point that critics are generally guilty of attacking the mathematical or statistical model 'not in itself but for what it does not do'.

But this is not the end of Cosgrove's critical challenge. Another part of his argument is that 'Moral questions were systematically excluded from geography by the simple act of denying scientific status to alternative value systems' and he claims that 'It is difficult to coordinate [moral] discourse . . . with the imperatives of a mathematical or statistical formulation and their requirements of precision and simple true/false distinctions.' As a long-standing advocate of the development of multi-valued logical approaches to modelling I am pleased (if somewhat surprised) to see Cosgrove suggesting that a fuzzy approach may be useful in this context, but I cannot accept the premise that it is difficult to treat ethical questions in a conventional two-valued logical framework. If that were the case, welfare economics would not exist, nor would the geography that flows from it.

A Counter-criticism

In addition to the criticisms of various aspects and styles of model-building, there is a critical countercurrent flowing through the book. Many contributors see their intellectual expeditions as being chronically undersupported. Geography at large, it seems, is simply not facing up to the challenges of exploring the new and highly taxing intellectual regions that are being opened up. The discipline appears altogether too amateurish in the face of professional scientific competition. Its modelling expeditions are under-funded, ill-equipped and, worst of all, short-handed. Witness the following comments: 'it is remarkable how little climatologists with a geography background have contributed to [the] development and use [of Global Circulation Models]' (Unwin); 'the major developments rest on the shoulders of only a few individuals. Even in the subject at large, there have been remarkably few with the ability to create lasting structures from basic model building activities' (Thornes); 'the numbers of academic geographers involved in this growth in use of geographical data bases have been minuscule compared with those of non-academic geographers and individuals from other disciplines' (Rhind); 'we must recognize this as a "team game"; the solitary academic working in this field – as either an exploiter of data bases, a GIS specialist or a spatial statistician – is at a considerable disadvantage' (Rhind again).

Part of the problem for geography is that many of the regions being opened up transcend traditional subject boundaries and give access to areas that have been thought of (by geographers) as geographical preserves. Thornes speaks of 'a fruitful invasion of physical geography by other scientists anxious and well equipped to do the job' (witness the contribution of the ecologists Harvey and Godfray to this volume); of things changing 'as scientists in related fields realize the potentially rich pickings', and of other scientists stepping in 'where geographers fear to tread'. Openshaw, looking at a different set of issues, comes to essentially the same conclusion: 'if geographers are not prepared to meet . . . emerging needs by re-orientating their modelling work then no doubt other computer sciences will be only too happy to oblige'.

Some contributors clearly feel that geography is in danger of being marginalized. Henderson-Sellers sees the possibility 'that traditional geographers will become entirely excluded from the "climate community" and techniques of geographical analysis, including cartographic truths, will be rediscovered by [the] "new climatologists"'. Kirkby recognizes that 'in many areas of research we are increasingly under threat from scientists who are directing their interests to the physical environment, and we frequently suffer in comparison to them'.

The Culture of the Times

Of course, it is not just the discipline and disciplinary relations that have altered over the past two decades: to use Bennett's phrase, the whole 'culture of the times' has changed. Modelling grew up in a climate of 'consensual agreement on welfarism', under which 'interventionism and social objectives . . . became the dominant concerns' and almost all models, 'became grounded in policy relevance' (Bennett). According to Harvey, 'in the absence of a Marxist option, those who sought remedial action on social problems did so through the medium of an enhanced technical bureaucratic rationality'. It was a time of high hopes and high expectations and, as Bennett says, 'the optimism . . . within geography reflected the general optimism in the 1960s about both technical systems and more general concepts of social improvement'.

Twenty years on 'the social context has changed' and, in Batty's view, 'it is unforgiving in its rejection of all that has gone before. . . . As the boom turned to recession, optimism turned to pessimism, idealism to disillusion.' There was 'a shift from planning growth to managing decline [which] marked not only a disillusion with modelling but with planning itself'. And the irony of all this for (urban) modellers is that they succeeded in developing models 'which could well have produced excellent advice in their day had they been available' just when the culture of the times swung against their deployment.

The landscape of the post-welfare society The new or coming culture of the times is said to be 'post-welfarist'. It is seen as being characterized by self-interest, the cult of information and flexible capital accumulation. Batty says that 'the new landscape is one of a weak public sector and increasingly dominant self-interest . . . if models are demanded at all it is to enhance self-interest. . . . The notion of a collective or public interest and of planners working for it or towards it has more or less disappeared.'

There is also an obsession with acquiring and exploiting data. Information has become tradeable. According to Openshaw, 'the potential importance of models is no longer primarily as a basis for understanding systems behaviour but increasingly as a means of adding value to data'. He sees 'major new opportunities for operational models that can assist both business and services to operate in a more efficient manner'. But the information economy closes opportunities as well as opening them. Ability to participate is determined by access to information systems, hence Rhind's belief that 'acting as "gatekeeper" to knowledge enables one to stay "central to the action"'.

Harvey and Scott see the situation in different terms. For them, what we are witnessing is a transition from a Fordist mass-production system to a regime of flexible accumulation. They argue that 'in contrast with the rigidities of mass production and Keynesian welfare-statism, the new regime is distinguished by a remarkable fluidity of production arrangements, labour markets, financial organization and consumption'. Moreover, it is responsible for 'generating major changes in patterns of geographical development'.

Geography in the post-welfare society Echoing Bennett's observation that modelling grew up in a welfarist context, Harvey and Scott describe the way Marxian social theory developed in the 1960s and 1970s in a Fordist context, focusing on 'the logic of Fordist industrialisation and the Keynesian welfare-statist policies that helped to sustain it'. But by the early eighties 'the Fordist regime of accumulation was being displaced by flexible accumulation as the dominant way of doing capitalist business', so 'an increasing disparity arose between prevailing theory and the actual evolution of most of the advanced capitalist societies'.

By the early 1980s, it seems, the ground on which both the modellers and Marxists had been labouring was being undermined: the welfarist culture which one group worked in and the other worked on was being overtaken (although 'undertaken' might convey the sense of it rather better).

From Harvey and Scott's perspective, 'there were . . . very few arenas of Marxist inquiry that were not called seriously into question'. This questioning heralded 'a decisive retreat from theoretical work' leading to 'a re-assertion of the primacy of empirical research and a fixation on the specificity of the local'. There is a close parallel here with the empiricism favoured by Openshaw. What is more, Openshaw's scepticism about the possibility of reversing the trend away from theory is extended explicitly to the project that Harvey and Scott have in mind for reviving social theory.

That project involves 'a major theoretical effort that transcends the disconnected plethora of approaches and findings that have been generated these last few years in the course of trying to come to grips with the surface appearances of flexible accumulation'. It accepts 'nothing less for its goal than the pursuit of a rigorously formulated and holistic theory of capitalist development in its current phase'.

From his solo contribution it is clear that Harvey must see this project as involving 'the production of objective knowledge from *within* society's confines'. To him, it is simply self-deceiving 'to pretend that we can put ourselves "outside of" or "above" society through the construction of some entirely neutral objective language'. To do so is to be misled into the production of ideology: 'those who stick to the abstract-materialism of natural scientific method and who locate their modelling efforts within that frame are the true ideology producers in contemporary geography'.

Whether or not modellers do have pretences to social neutrality for their work, both Batty and Bennett see modelling as an inescapably social activity: according to Batty 'modelling can never represent a science because its intellectual rationale is entirely determined by a volatile social context'. For his part, Bennett identifies *both* model-building and the construction of radical social theory with the culture of welfarism. He characterizes welfarism as being concerned with 'improving the quality of life and provision of needs through collective and governmental intervention' and he sees its intellectual focus as 'relative deprivation and social, as well as, spatial inequality'. His worry is that 'within the welfarist culture there is a clear ontological shift which is not only dangerous but inherently methodologically circular'.

The ontological shift is said to involve reinterpretation: 'reinterpretation of quantitative and statistical material as purely social constructs' and of the environment as 'merely a social product'; and reinterpretation of geography as a *social* science, where '"social" is defined solely, or primarily, as a concept of class within which relative deprivation is the operational indicator used as evidence of class difference'. The danger in this for Bennett is that 'the ontological shift closes the circle in that a pre-belief (in radical social theory) is confirmed' with the possible consequence that 'the discipline accepts arguments as fact . . . and erects them to the status of *the* theoretical base within which problems must be confronted'. The emerging rapproachment between welfarist modellers and welfarist social theorists (best illustrated here by Wilson's chapter) is taken by Bennett to be adding to this danger.

Remodelling Geography

What are we to make of all this? It would seem that there is an undoubted need to remodel geography to a greater or lesser extent, but to what extent and how should it be done? What view should we take about theories, facts and values, about the subjects we should be addressing, and about the way we should be elaborating and validating our ideas? Where should applied work fit into the picture and how should we change our teaching to ensure our other efforts are not in vain?

Theories, Facts and Values

First, it is important to side with Bennett over the matter of facts. The world *is* real and it *is* knowable. Acquiring social knowledge is peculiarly difficult but it is not impossible. This is not to deny that there is often bias in the way statistics are selected and used by governments, researchers and others, or to deny that facts and figures are sometimes falsified for ideological reasons. Nor is it to deny the subjectivity of observation: Haines-Young puts the point rather well when he says that 'it is precisely because the scientist cannot be objective and value-free that testing and falsifiability are so important'. But it is to deny that there are no such things as facts or, more pointedly, that all statements of 'fact' are ideologically determined: an individual's beliefs may well be predicated on some ideology but facts are not the same things as beliefs. It is also to contradict Wilson's view that 'truth is a consensus'. To my mind, that view is straightforwardly (and dangerously) wrong.

The 'scientific' insistence on the existence of facts should not be taken to imply disinterest in values. Indeed, Cosgrove's suggestion that scientific geography is antipathetic to ethical considerations is not only mistaken in principle, it is inaccurate as a comment on history. To be sure, modellers were slow to take up moral questions explicitly but they have now done so, principally in the form of planning objectives. As Batty observes, 'once the mathematical machinery was in place, [it] could be used to optimize the model . . . in ways . . . that planners and policy-makers might wish. . . . The link between modelling, planning and optimization had finally been worked out.'[2]

[2] In fact, it had not quite been worked out. The objectives in models such as the classic Group Surplus Maximising Model of Huw Williams (see Wilson et al., 1981, chapter 4) or

In practice, the notion that facts exist is generally agreed by contributors (unless I have misunderstood the implications of Harvey and Scott's expressions 'pure empirical form' and 'the gap between . . . theoretical representations and historical geographical events'). So too is the view that we need to construct theories to account for real-world phenomena, although different positions are adopted on the nature of theory[3] and on the best method of theory construction (about which more will be said later). Apart from Cosgrove, the one dissenter on the need for theory is Openshaw, although even he goes part of the way when he acknowledges that 'the need for theory still exists but only to the extent that it will be used if it can be shown to deliver better "working" models or is otherwise useful'. Furthermore, there is wide support for the idea that we should try to unravel the processes that generate spatial patterns (or, if you prefer it, the structures and mechanisms which underlie surface appearances) and this is one of the reasons for the general preference for 'scientific' as opposed to 'technological' modelling. Harvey sums up the general sentiment in favour of theorizing when he says that 'theory construction has to be central to our concerns'.

Just as the emphasis on facts should not be mistaken as a lack of concern for values, so the emphasis on theory should not be assumed to imply a lack of interest in practice. At the moment, as Openshaw observes, 'a great gulf exists between [human geographical] theory and practice' and 'far too many so-called "theories" have never been tested and many more are untestable!'. This charge is as valid when applied to classical location theory as it is in the case of Marxist analyses (both fields rely extensively on ideal types and special conditions). Nor are physical geographical theories immune. The problem we face, then, is one of developing theory in such a way that it can be exposed to empirical test, for as Bennett says, 'it is . . . only by opening theory to challenge that geography can contribute to dialogues on policy or education in a world of changing political, social and economic structures'. Openshaw believes that 'the task of building models that both represent theory and stay within contemporary data supply constraints cannot be

the quadratic programming models of Takayama and Judge (1971) are behavioural propositions, although they are rarely recognized as such. That is, they describe how the agents in the spatial economy seek to behave collectively: the programmes' constraints define the possibilities open to the agents. The models produce socially optimal solutions only to the extent that market equilibria are optimal. A thoroughgoing treatment of planning objectives requires a further objective to be superimposed (producing an hierarchical programming problem), where the variables in the outer objective appear as parameters in the original model.

[3] Compare the different definitions offered in chapters 7 and 16. Rudner's (1966) view quoted in chapter 7 is that 'a theory is a systematically related set of statements, including some lawlike generalisations, that is empirically testable' whereas Harvey and Scott say that they 'use the term "theory" . . . in its usual Marxist sense, to mean the creation of the intellectual preconditions for self-consciousness of the structures of capitalist domination coupled with the construction of coherent representations and analytical tools to facilitate the struggle for human emancipation'.

solved in any satisfactory manner', but we shall see. At the very least we should be striving to reduce the gulf to a gap.

Where do models fit into the quest for theory? According to Haines-Young 'models are not simply calculating devices but re-presentations of our knowledge about the world in a form which is more easily criticized' and they 'remain indispensable tools for testing our understanding of the world, despite claims of the post-modernists to the contrary'. Models are (or should be) expressions of quantitative theories, or quantitative theories together with specific conditions.[4] Thus, the role of models is to express and to facilitate the testing of quantitative theories.

What Should We be Theorizing About?

Given the general acceptance of the view that we should be producing testable theories, is there any agreement on what we should be theorizing about?

Spatial systems From the human geographers a number of propositions emerge. Prominent amongst them are Wilson's suggestions for a new regional geography and for the connection of ideas from modelling to both critical theory and the classics of location theory. The reworking of the classics in the form of optimization models dates back at least to Stevens's (1968) paper, but their casting in a spatial interaction mould is a much more recent development (see Birkin and Wilson, 1986a, b and Wilson and Birkin, 1987). The improvement in the optimization framework for spatial interaction modelling resulting from the efforts of Wilson, Williams and others (see Wilson et al., 1981) provides an excellent opportunity for bringing these two strands of work together and this cross-fertilization should certainly be pursued. The way critical theory might be brought into the picture is much less clear, except that some of the things that critical theorists are concerned with (historical development, capital accumulation, the existence of agents other than firms and households) are noticeable by their absence in nearly all modelling work and many models could be enhanced, even transformed, by their presence.

Regions Wilson's other appeal is for us to 'use all the methods, theories and insights of contemporary geography in all its varieties to write new *regional geographies* at various scales'. This might strike traditional regional geographers as a belated conversion to the one true faith, but it is not as straightforward as that. If I understand it correctly, Wilson's argument about the region is not wholly ontological in the sense that it is not concerned with questions about the existence of regions as distinct,

[4] To this, Haines-Young would insist that we add that models can also be qualitative, although the distinction between model and theory (or model and knowledge base) seems somewhat arbitrary in these circumstances.

'natural' entities. Part of it is epistemological in that it deals with a strategy for acquiring knowledge and testing propositions about the world, where the strategy is the discipline's traditional one of focusing attention on individual regions. One key question then is this: are regions suitable arenas for developing and testing theory? The answer has to be yes, with a qualification. Self-evidently, if theory is to be tested it must be tested somewhere, and as Wilson allows for a variety of scales, 'somewhere' means in some region or regions: But favouring the synthesis of information and ideas about individual regions must not be done at the expense of developing and testing propositions across regions: if, for example, we want to understand famines, we must look at famine processes wherever they occur and test our theories about them accordingly. Wilson's major point, though, is that looking at a region's individual subsystems and doing so from only one geographical perspective is likely to be much less fruitful than examining the various subsystems together and adopting a pluralist perspective.[5] That seems sensible enough and should be welcomed by traditional regional geographers and modellers alike. I await with interest the publication of 'The Geography of West Yorkshire'.

Brave new worlds The agendas of Bennett and Harvey follow directly from the earlier observations on their chapters and on Harvey's joint contribution with Scott. Bennett certainly feels that geography is 'now needing to rethink its overall strategy and priorities' and he sees a major task 'in the field of reappraisal of the welfare state and its alternatives in the light of outcomes'. Harvey wants us 'to search out ways to construct scientifically rigorous theory that can shed light on the often tortuous twists in the historical geography of human occupancy of the earth's surface'. The latter is a grand but important objective and is not inconsistent with anything that has been said so far, although I would want to add the proviso that the testing of such theory should be as rigorous as testing in other contexts, to the extent that historical data constraints will allow. Bennett's proposal involves insisting that theories be tested by comparing the outcomes of policies founded on them with predicted outcomes. This is perfectly sensible in principle, although it may be very difficult in practice. It is certainly right for him to argue that it is foolish to remain attached to theories whose associated policies suffer 'significant and recurrent failures', however 'cherished as idealist visions' the policies may be. However, whilst I would agree both that welfarists[6] have shown a dogged attachment to means to the neglect of ends, and that they have been too quick to regard attacks on the former as attacks on

[5] The pluralist framework Wilson has in mind is not likely to be constructed without considerable effort if Harvey is correct in his assertion that 'the confusion between positivism, mathematics and data analysis has gone too deep to allow any easy disentanglement of that particular knot'.

[6] I have in mind particularly the British Labour movement.

the latter, the idea that the welfarist tradition is in terminal decline seems highly debatable (at least if that tradition is defined in terms of ends not means). There is no doubt that it must adapt and look to new means, little doubt that model-building could play a significant role in evaluating those means,[7] and only a nagging doubt about the eventual demise of the new conservatism. Batty's pessimism about the prospects for modelling in a policy context seem to me to be ill-founded for these reasons.

Battered old worlds As well as pessimism amongst some of the human contributors there is also a degree of parochialism and that is the last charge that geographers should be open to. We should certainly look for our subjects beyond the (geographical) areas implicitly suggested by a number of contributors. I would side with Harvey in saying that 'if the world is beset by all manner of problems – spiralling indebtedness and financial insecurity, militarization, widespread unemployment and not a little social anomie – then it is vital to find a way to represent those phenomena and their geographical dynamics in a critical but objective way'.

Global and regional systems Parochialism is not a charge that can be directed at the physical contributors. As Thornes points out, one of the major trends that has changed the style and subject matter of modelling in physical geography in the 1980s is the development of work at the regional and global scales. Because of the nature of climate and the contributions which geographers could make to understanding it, Henderson-Sellers is particularly keen for this trend to continue. She sees an 'urgent requirement for regionally specific analyses as a basis for improvements in data sets for input into global climate models' and believes that such analyses 'are the responsibility of geographers, since they alone are likely to be trained in the range of disciplines required'. There is a nice, if incomplete, reflection here of Wilson's call for a new regional (human) geography, as there is in Kirkby's assessment of the need for modelling on the grand scale: 'Although there are dangers, as pointed out by Thornes . . . in constructing bigger and more complex models by combining a series of component sub-models, nevertheless there is both scope and need for models which encompass a wide range of mutual interactions. . . . These difficulties are not reasons to avoid large and complex models, but only reasons to build with care and good craftsmanship.'

[7] Bennett is clearly sceptical on this point as evinced by his argument on the dangers of methodological circularity implicit in 'welfarism defined as "relative deprivation"'. This argument has some merit in it but its extension to modelling needs to be treated with caution. He says that 'inasmuch as models help to identify the mechanisms which lead to the development of inequalities, and may even be used to help to reduce some of them, then the model paradigm is also framed and utilized within this welfarist framework'. But how far are models used in this way? In my view, not extensively and certainly not *necessarily*.

Human–environment interactions The physical contributors also appear to have a general interest in the development of evolutionary models and Thornes expresses a specific interest in human–environment interactions. He says that 'despite much arm-waving . . . our knowledge of such interactions is still quite feeble and our ability to model them even weaker', and he adds that 'although physical geographers write extensively about man's impact on the environment . . . very little effort is being made to model these effects either for predictive purposes or in an effort to understand them better'. Yet this would seem to be a vital area if the discipline is to thrive. As Kirkby says, 'physical (and other) geographers must . . . ask what they have to contribute which is not better done in other science and engineering disciplines. The answer to me lies in the willingness to synthesise across conventional boundaries'. The enthusiasm with which we address human–environment interactions must be the key test of that willingness.

How Should we be Theorizing?

Having listed some of the things contributors think we should be theorizing about (and it should be remembered that not all subdisciplinary interests are represented) it is time to turn to the question of method. Cosgrove on the one hand and Harvey and Scott on the other have distinctive views on this matter which I won't reiterate. Batty and Bennett's ideas about the contexts of discovery and application are also relevant but I don't want to return to them just yet. For the remainder of the contributors the question of method brings us back to the issue of modelling styles.

For the most part, the choice of modelling style should be determined by the circumstances. It is a question of using the right tool for the job: screws can be inserted with a hammer but screwdrivers tend to be more effective. Generalized arguments about modelling styles can be somewhat fruitless as a result. However, a general argument in favour of data-based modelling (and explicitly against mathematical modelling) has been advanced by Openshaw. Following Bennett's lead, we might try to assess this argument by asking 'does it work?'.

The inductive approach The Openshaw doctrine, as articulated in the form of a Geographical Analysis Machine, is certainly associated with at least one notable success: the identification of a major (and previously undetected) cluster of acute lymphoblastic leukaemia cases in northern England (see Openshaw et al., 1987). But that success does no more than support the proposition that well-designed empirical work can yield important results – a proposition that was never in doubt. It adds nothing to Openshaw's arguments against theory-driven (as opposed to data-driven) modelling. Interestingly, some light is shed on these arguments by Harvey and Godfray in their paper on island biogeography.

They refer to the production of a set of 'null archipelagoes' for the purposes of comparison with observed species distribution data. The technique resembles that employed by Openshaw et al. to compare real and randomly generated disease clusters. Harvey and Godfray also describe the variety of ways species data can be analysed statistically, and comment that 'it is hardly surprising that different conclusions have been drawn by different investigators'. There are parallels here with the differences that arise between disease cluster analysts as they fall foul of the modifiable areal unit problem and the trap of *post-hoc* hypothesis testing. The difference between the work described by Harvey and Godfray and the method advocated by Openshaw is that the former embraces a deductive as well as an inductive approach (the deductive approach being the analysis of character displacement by means of a modified Lotka–Volterra equation system). The analysis of island biogeography is all the richer for it. Data-based modelling has great potential but Openshaw's argument over-eggs the pudding.

One of the keys to realizing that potential is, of course, the development of better statistical techniques for spatial analysis and the software to operationalize them. As Haining points out 'if every econometrician had to write his own time-series programs this would no doubt have a marked effect on the analysis of such data, and yet that, for the most part, is the situation facing geographers handling spatial data'. Consequently, Haining calls for a major initiative to tackle this problem involving statisticians, computers scientists, geographers and representatives of other disciplines. Many statistical packages are already available, as are various Geographical Information Systems, but the major part of Haining's argument is that 'it will not do to integrate these systems with "standard" statistical packages for purposes of correlation, regression or indeed any other techniques for multivariate spatial analysis'. Our ability to make full use of remote sensing technology will also depend on the development and operationalization of better spatial analytical methods. It is clear that statistical modelling in geography is still immature, but that it is starting to come of age.

The question of predictability The other area of activity previously identified as statistical modelling is stochastic process work, although the distinction between 'statistical' and 'mathematical' in this context is somewhat arbitrary, as noted earlier. The role of stochastic modelling, the competing claims of stochastic and deterministic process descriptions and the associated issue of predictability seem to me to be matters of some importance but they are not pursued at length by any contributor. Unwin touches on them when he argues, in connection with deterministic engineering models of the atmosphere, that 'every realization from such a model should be uniquely determined from the start, yet it is known that these models produce their own stochastic noise. . . . This would not matter if they were stable to small disturbances, but, like the real atmosphere, they are capable of magnifying small differences as they run.' It is not at all clear that this observation supports Unwin's call for 'very careful use of methods from spatial analysis' (which I take to mean statistical methods), since his characterization of the trajectory variations in atmospheric models as 'stochastic noise' is open to challenge. As Unwin knows, Lorenz's classic experiment in 1961 with a primitive atmospheric model led to the formulation of a quite different (and wholly deterministic) account of these variations – an account based on the notion of chaos (see Gleik, 1988). Thus, whilst the conventional approach of ascribing unexplained variations to 'randomness' has many more miles in it yet, we should not ignore the chaotic but deterministic alternative.

The deductive approach The whole realm of non-linear systems theory – of cusps, catastrophes, and bifurcations, of chaos and strange attractors, of master equations and the like – is one of great opportunity. To be sure, we do not want to indulge in another 'orgy of model borrowing' but we should learn the language of nonlinear systems in order to be able to articulate our ideas better and to be in a position to think new things in the first place. We should, but I am not so sure that we will. Convincing people of the merits of mathematical work of any kind is an achievement let alone work of an apparently esoteric kind. Henderson-Sellers does not mince her words on the subject: 'the geographers, and there are many, who refuse to become involved in mathematical manipulation of data and rules because they believe that either a satisfactory mathematical language is not available or, the other side of the same coin, that their particular problem is not mathematically tractable, are fooling no-one but themselves'.

Part of the problem is that it is not generally realized that mathematics, which often appears to students to be static, complete and beyond challenge, has changed dramatically over the past two decades. As Henderson-Sellers points out, 'there is now a wide range of manipulative tools for adventurous geographers which are as similar to conventional mathematics as clouds are to cleavage'. As well as the advances in non-linear systems theory, we have seen fractal geometry, tesseral arithmetic and automata theory come on to the scene, great strides have been made in optimization, and the whole field of fuzzy mathematics has been developed. If we are not to get lost in the scientific land of the 1990s, we are going to have to learn the language.

The role of computing We are also going to have to get to grips with the technology that comes with the territory. Most of us are still adjusting to the idea of using Geographical Information Systems but as Rhind and Openshaw make plain, many more developments are on their way. Sophisticated Decision Support Systems incorporating a variety of geographical models are already being produced (see for example the (1987) work of Fedra et al. at IIASA[8]) and in another twenty years we may have

[8] The existence of work of this kind reinforces the earlier charge of parochialism. Even

the opportunity of using Automated Pattern Analysers, Automated Modelling Systems based on artificial intelligence techniques and, if Openshaw gets his way, so-called auto-theory new knowledge inferencing tools.

In such a dynamic environment, and recognizing the need to use the right tool for the job, can anything general be said about future modelling styles? If another Models conference is held in 2007 there can be little doubt that the subjects under discussion will be computer models, although the adjective will be regarded as superfluous. The separate categories of statistical, mathematical and data-based modelling are already becoming redundant. It is easy to foresee them being superseded by the development of enhanced GIS – Geographical Information Systems with sophisticated spatial analysis and mathematical modelling tool chests. Indeed, this development is already in train. For example, some GIS have basic optimization facilities built into them, and it is a relatively small step either to extend those facilities to a full suite of optimization routines or to tailor a GIS to a particular end use with the aid of certain special routines.[9] Similar steps could be taken in relation to non linear systems theory. Even the exotic world of fuzzy inference is only an interfacing job away,[10] given the existence of systems such as FRIL, and once we get into this world the distinction between qualititative and quantitative breaks down, let alone that between statistical, mathematical and data-based modelling styles.

The reason that such developments are important is that enhanced GIS provide a vehicle for the closure of the gap between theory and practice: theory can be both articulated and tested within them. What is more, with the availability of Decision Support Systems it should be possible to develop applications within the same framework as well. It is easy to let enthusiasm get the better of common sense when contemplating these possibilities. In a way the situation is not unlike the early days of modelling, and it would be as well to learn the lessons of those days. There is certainly no technological fix. The key to theoretical development is careful thought and the exposure of that thought to others and to the arbitration of empirical evidence: Wilson's advocacy of geographical criticism (in the spirit of literary criticism) is to be welcomed in this context. Our guiding principles should be, perhaps, to keep our ambitions and our optimism within bounds, to ensure that our projects remain manageable, to welcome criticism, and (to repeat Kirkby's advice) to build our models with care and good craftsmanship.

though the U.K. and other western countries may be affected by post-modernism and its disregard for the potential value of modelling, not all parts of the world are equally blessed.

[9] Thus, an urban modelling system based on group surplus maximization might be constructed including Newton–Raphson, Fletcher–Reeves, Fletcher–Davidon–Powell and other algorithms.

[10] With that job done, difficulties like the modifiable areal unit problem (which is an artefact of Boolean inference) should become much less acute and a variety of non-spatial classification problems could be greatly eased.

Theories and Applications

Apart from the odd reference to Decision Support Systems, the above argument has been concerned exclusively with what was referred to early in this Afterword as 'scientific' (as opposed to 'technological') modelling. Coupling the words 'scientific' and 'model', though, is by no means uncontentious. According to Batty, 'modelling can never represent a science because its intellectual rationale is entirely determined by a volatile social context'. This suggests that all modelling is essentially technological. Yet Batty also talks about 'the good work accomplished in modelling during the 1970s . . . [which] has never come to be applied in practice'. What makes this work good is the accuracy with which it accounts for certain urban processes. In other words, it is good science. The cultural specificity both of the willingness of bodies to commission and use models and of the social objectives implicit in their application (if they are applied), does not detract from this point.

It is worth comparing Batty's remarks with Openshaw, Rhind and Beaumont's calls for engagement with the institutions in society which now require, or could be persuaded to use, the services of geographers. Disciplinary development may be facilitated, indeed stimulated, by the sort of relationship which once existed with urban planning authorities, and which is now being developed with commercial concerns, health authorities and the like, but it cannot be defined in terms of those relationships. Geography's first purpose has to be to understand the world, and modelling in geography ought to be concerned with the formal articulation and enhancement of that understanding. That is not to say that it ought to be value free – alternative value systems and their implications should never be far from the centre of our concerns – but it is to say that it should, at bottom, be client-free. Academic geographers should not abrogate the responsibility for determining what should be studied and how it should be studied, and they should certainly be prepared to analyse and criticize institutional changes as well as learning to live with them.

Teaching

Only one of the book's major themes remains to be dealt with – the question of teaching. As Haggett and Chorley remind us in their Foreword, teaching was the central theme of *Models in Geography*. In *Remodelling Geography* it has been one of many. Nevertheless, successful teaching is rightly seen by several contributors as the *sine qua non* for the future health of modelling in geography.

Unwin asks 'How do we teach the new models?' Having posed the question, he goes on to say that 'The implication for the future of physical geography as we know it seems obvious: either we become the last generation of geographers able to contribute to modelling research or we pay much more attention to the problems of training the next generation of modellers than we have of late.' Kirkby broadly agrees. He argues that geographical education 'provides students with a minimal level of scientific and mathematical training on which to build. This has been, and continues to be, reflected in our staff and our research, and our future must rest on the development of a more rigorous training.' Bennett is of the same mind: he calls for 'more intensive training in analytical methods including model-based approaches, information systems and elementary analytical skills'.

Work appears to be required in the teaching of mathematics, geographical statistics and basic scientific principles. As Henderson-Sellers says, 'It may be possible to grasp the essence of statistics without mathematical training but modelling is an intrinsically theoretical activity; the tools are mathematical and, more importantly perhaps, computational.' For his part, Haining stresses 'the importance of statistical training for geographers that goes beyond standard statistical methods', while Kirkby argues that physical geographers should have 'an effective training in the physical, chemical and biological principles which underlie their environmental system of interest'. At the same time though, he says that we must not risk 'sacrificing our breadth of understanding or our eye for relationships in the field'.

The Science of the Possible

In the introduction to this Afterword I said that it was generally agreed that modelling has limitations, and that the debate about modelling turns on the nature and extent of those limitations. There is, if you like, a feasible region defined by a set of constraints within which it is sensible to build models. Claims have been made about the existence of both methodological and cultural limitations on human geographical modelling, and the whole field seems to be confronted with technical limitations.

The need to respond to these claims and to the criticisms of modelling that go with them, has been recognized throughout. Haines-Young sums up the sentiment this way: 'To leave such criticisms unanswered is to accept that there are fundamentally different methodologies for the study of human and physical systems. Unfortunately, many of the most interesting and pressing problems which confront mankind are at the human–environment interface. Faced with the loss of what few natural ecosystems remain, the wholesale loss of the genetic resources of the biosphere, and the disruption of many of the earth's essential life support systems, do we simply lapse into the intellectual lethargy of post-modernism and say . . . well, all views are valid?' The question is rhetorical, of course, but answers have been given in full measure.

There are indeed methodological and cultural difficulties, but they are difficulties of practice not of principle. There are also substantial technical difficulties, particularly concerning the confrontation of theory with data. But alternative approaches to modelling have their limitations too. Pluralism is certainly to be encouraged but the test between methods for an empirical discipline like geography has to be 'does it work?' And within a pluralist framework there are questions of balance. The clear message of this book is that the balance has swung too far away from modelling. As Wilson says, 'there is a danger that modelling has been rejected too quickly by too many geographers . . . [it is still] a new and immature subdiscipline'.

The presumption that modelling has passed its apogee is clearly rejected by the modeller–contributors. Indeed, a variety of technical developments suggest that many of the constraints that we are currently bound by will fall away, creating the possibility of significant new developments. According to Openshaw, 'the computer revolution in geography has yet to happen', while Haining argues (by quoting Ripley, 1984), that 'a revolution is needed' in geographical statistics and there is evidence that that revolution is on its way. Couple that with the rapid emergence of new mathematical ideas and it appears that the technical climate for modelling is being transformed.

However, there are other limitations. There are limitations of expertise, as Kirkby makes clear when he says that 'physical geography is in great danger for lack of a consistently rigorous approach to theory, to models and to experimental procedures' and a similar judgement is implicitly made by Macgill when she talks of the 'imperative need for careful assessment of the quality of what is done through different approaches'. There are limitations due to inadequate training, as noted in the previous section. And there are the limitations of working in relative isolation: as Beaumont says, 'the way forward must require much more coordination in the modelling research area'. Perhaps we should explore seriously a suggestion that Alan Wilson has made to create a Nicolas Bourbaki of geographical modelling.[11]

But the greatest limitation is surely of our own making. It is a limitation of vision, purpose and conviction. If we are to remodel geography we will have to recognize and take up the challenges posed by the great geographical issues of our day, we will have to elevate the demands of scholarship above all other considerations, and we will have to exhibit a greater collective faith in our scientific methods. We must stop muddling through but keep modelling through — for another two decades at least.

[11] Nicolas Bourbaki was a collective pseudonym used by an informal working group of French mathematicians.

References

Abrams, P. A. 1986: Character displacement and niche shift analyzed using consumer-resource models of competition. *Theoretical Population Biology*, 29, 107–60.

AC 1988: The proposed standard for digital cartographic data. *American Cartographer*, 15(1), 11–142.

Ackerman, E. A. 1963: Where is the research frontier? *Annals of the Association of American Geographers*, 53, 429–40.

Ahnert, F. 1976: Brief description of a comprehensive three-dimensional process-response model of landform development. *Zeitschrift für Geomorphologie Supplementband*, 25, 29–49.

____ 1987: Approaches to dynamic equilibrium in theoretical simulations of slope development. *Earth Surface Processes and Landforms*, 12(1), 3–16.

Allen, J. R. L. 1974: Reaction, relaxation and lag in natural sediment transport systems; general principles, examples and lessons. *Earth-Science Reviews*, 10, 263–342.

Allen, P. 1983: Planning and decision-making in human systems: modelling self-organisation. In M. Batty and B. Hutchinson (eds), *Systems Analysis in Urban Policy-Making and Planning*, Plenum Press, 491–524.

Anas, A. 1982: *Residential Location Markets and Urban Transportation*, Academic Press.

____ 1986: From physical to economic urban models: the Lowry framework revisited. In B. Hutchinson and M. Batty (eds), *Advances in Urban Systems Modelling*, North-Holland, 163–72.

Anderson, M. G. and Burt, T. 1978: Experimental investigation concerning the topographical control of slope water movement on hillslopes. *Zeitschrift für Geomorphologie, Supplementband*, 29, 52–63.

____ and Howes, S. 1986: Hillslope hydrology models for forecasting in ungauged watersheds. In A. D. Abrahams (ed.), *Hillslope Processes*, Allen & Unwin, 161–86.

Arbia, G. 1986: Problems in the estimation of the spatial autocorrelation function arising from the form of the weights matrix. In R. Haining and D. Griffith (eds), *Transformations through Space and Time*, NATO ASI, Martinus Nijhoff.

____ 1987: Spatial Data configuration in statistical analysis of regional economic and related problems. Unpublished Ph.D., University of Cambridge.

Armstrong, A. C. 1976: A three-dimensional model of slope forms. *Zeitschrift für Geomorphologie, Supplementband*, 25, 20–8.

Ashmore, S. E. 1939: The diurnal range of temperature and its geographical distribution. *Quarterly Journal of the Royal Meteorological Society*, 65, 554–8.

Bagnold, R. A. 1966: An approach to the sediment transport problem from general physics. *U.S. Geological Survey, Professional Papers*, 422–I.

—— 1977: Bed load transport by natural rivers. *Water Resources Research*, 13, 303–12.

Baldwin-Wiseman, W. R. 1934: A cartographic study of drought. *Quarterly Journal of the Royal Meteorological Society*, 60, 523–32.

Barber, K. and Coope, G. R. 1987: Climatic history of the Severn Valley during the last 18000 years. In K. J. Gregory, J. Lewin and J. B. Thornes (eds), *Palaeohydrology in Practice: A river basin analysis*, Wiley, 201–16.

Bardon, K., Elliott, C. and Stothers, N. 1984: *Computer Applications in Local Authority Planning Departments: A Review*, Birmingham, U.K.: Birmingham Polytechnic.

Barringer, T., Robinson, V., Coiner, J. and Bruce, R. 1980: LANDSAT Analysis of Tropical Forest Succession Employing a terrain model. *Proceedings of the 14th International Symposium on Remote Sensing of the Environment*, 1691–700.

Batty, M. 1975: In defence of urban modelling. *The Planner*, 61, 184–7.

—— 1976: *Urban Modelling: Algorithms, Calibrations, Predictions*, Cambridge University Press.

—— 1979: Progress, success, and failure in urban modelling. *Environment and Planning A*, 11, 863–78.

—— 1983: Linear urban models. *Papers of the Regional Science Association*, 53, 5–25.

—— 1985: Formal reasoning in urban planning. In M. Breheny and A. Hooper (eds), *Rationality in Planning*, Pion, 98–119.

—— 1986: Technical issues in urban model development: a review of linear and nonlinear model structures. In B. Hutchinson and M. Batty (eds), *Advances in Urban Systems Modelling*, North-Holland, 133–62.

—— 1987: Models in planning: where do we go from here? *Environment and Planning B*, 14, 119–22.

Baudrillard, J. 1983: *Mirror of Production*, Telos.

Beaumont, J. R. 1986: Modelling should be more relevant: some personal reflections. *Environment and Planning A*, 18, 419–21.

—— 1987: Quantitative methods in the real world: a consultant's view of practice. *Environment and Planning A*, 19, 1441–8.

—— 1988: Britain's future wealth: the need for strategic discussions on research and development. *Environment and Planning A*, 20, 195–201.

Beck, M. B. 1987: Water quality modelling: a review of the analysis of uncertainty. *Water Resources Research*, 1383–442.

Ben Akiva, M. and Lerman, S. 1985: *Discrete Choice Analysis: Theory and Application to Travel Demand*, MIT Press.

Bennett, R. J. 1976: Adaptive adjustment of channel geometry. *Earth Surface Processes*, 1, 136–50.

—— 1979: *Spatial Time Series: Forecasting and Control*, Pion.

—— 1981: Spatial and temporal analysis: spatial time series. In N. Wrigley and R. J. Bennett (eds), *Quantitative Geography*, Routledge & Kegan Paul, 97–103.

—— 1985: Quantification and relevance. In R. J. Johnson (ed.), *The Future of Geography*, Methuen, 211–24.

—— (ed) 1989a: *Decentralisation, Local Government and Markets: Toward a Post-welfare Agenda?*, Oxford University Press.

—— (ed) 1989b: *Territory and Administration in Europe*, Francis Pinter.

—— and Chorley, R. J. 1978: *Environmental Systems: Philosophy, Analysis and*

Control, Methuen.
—— and Haining, R. P. 1985: Spatial structure and spatial interaction: modelling approaches to the statistical analysis of geographical data (with discussion). *Journal of the Royal Statistical Society A*, 148, 1–36.
Berlyand, T. G. and Strokina, L. A. 1980: Global distribution of the total amount of cloudiness. *Hydrometeorological Publication*, Leningrad.
Berry, B. J. L. and Marble, D. F. (eds) 1968: *Spatial Analysis: a reader in statistical geography*, Prentice-Hall.
Berry, F., Bollay, E. and Beers, N. (eds) 1973: *Handbook of Meteorology*, McGraw-Hill.
Besag, J. 1974: Spatial interaction and the statistical analysis of lattice systems. *Journal of the Royal Statistical Society*, 36, Series B, 192–225.
—— 1986: On the statistical analysis of dirty pictures. *Journal of the Royal Statistical Society*, 48, Series B, 259–302.
Beven, K. J. 1977: Hillslope hydrograph analysis by the finite element method. *Earth Surface Processes*, 2, 13–28.
—— 1986: Distributed models. In M. G. Anderson and T. Burt (eds), *Hydrological Forecasting*, Wiley, 405–35.
—— 1987: Towards the use of the catchment geomorphology in flood frequency predictions. *Earth Surface Processes and Landforms*, 12, 69–82.
—— and Kirkby, M. J. 1978: A physically based, variable contributing area model of basin hydrology. *Hydrological Sciences Bulletin*, 24, 43–69.
Birkin, M. and Wilson, A. G. 1986a: Industrial location models 1: a review and an integrating framework. *Environment and Planning A*, 18, 175–205.
—— —— 1986b: Industrial location models 2: Weber, Palander, Hotelling and extensions within a new framework. *Environment and Planning A*, 18, 293–306.
Bivand, R. 1980: A Monte Carlo study of correlation coefficient estimation with spatially auto-correlated observations. *Quaestiones Geographicae*, 6, 5–10.
Bolthausen, E. 1982: On the central limit theorem for stationary mixing random fields. *Annals of Probability*, 10, 1047–50.
Boulton, G. S. 1972: The role of thermal regime in glacial sedimentation. In R. J. Price and D. Sugden (eds), *Polar Geomorphology*, Institute of British Geographers Special Publication 4, 1–19.
—— 1974: Processes and patterns of subglacial sedimentation: A theoretical approach. In A. E. Wright and F. Moseley (eds), *Ice Ages: Ancient and Modern*, Seel House Press, 7–42.
Boyce, D., Day, N. and McDonald, C. 1970: *Metropolitan Plan-Making*, Philadelphia, Pennsylvania: The Regional Science Research Institute.
Brewer, G. D. 1973: *Politicians, Bureaucrats and the Consultant: a critique of urban problem solving*, Basic Books.
Brotchie, J. F., Dickey, J. W. and Sharpe, R. 1980: *TOPAZ – General Planning Technique and its Applications at the Regional, Urban and Facility Planning Levels*, Springer-Verlag.
Brown, W. L., Jr and Wilson, E. O. 1956: Character displacement. *Systematic Zoology*, 5, 49–64.
Brunsden, D. 1985: Geomorphology in the service of society. In R. J. Johnson (ed.), *The Future of Geography*, Methuen, 225–57.
—— and Thornes, J. B. 1979: Landscape sensitivity and change. *Transactions of the Institute of British Geographers*, 4, 4, 463–84.
Budyko, M. I. 1969: The effect of solar radiation variations on the climate of the

Earth. *Tellus*, 21, 611–19.
—— 1974: *Climate and Life*, Academic Press.
Bunge, W. 1966: *Theoretical Geography*, C. W. K. Gleerup.
Burt, T. P., Butcher, D. P., Coles, N. and Thomas, A. D. 1983: The natural history of the Slapton Ley Nature Reserve. XV. Hydrological Processes in the Slapton Wood Catchment. *Field Studies*, 5, 731–52.
Burton, I. 1963: The quantitative revolution and theoretical geography. *The Canadian Geographer*, 7, 141–62.
Burton, M. and Shadbolt, N. 1988: Knowledge engineering. In N. Williams and P. Holt (eds), *Expert Systems for Users*, McGraw-Hill.
Campbell, J. B. 1981: Spatial correlation effects upon accuracy of supervised classification of land cover. *Photogrammetric Engineering and Remote Sensing*, 47, 355–63.
Carslaw, H. S. and Jaeger, J. C. 1959: *Conduction of Heat in Solids*, 2nd edn, Clarendon Press.
Carson, D. J. 1982: Current parameterizations of land-surface processes in atmospheric general circulation models. In P. S. Eagleson (ed.), *Land Surface Processes in Atmospheric General Circulation Models*, Cambridge University Press, 67–108.
Cess, R. D., Briegleb, B. P. and Lian, M. S. 1982: Low-latitude cloudiness and climate feedback: comparative estimates from satellite data. *Journal of Atmospheric Science*, 39, 53–9.
Chalmers, A. F. 1982: '*What is This Thing Called Science?*', 2nd edn, Open University Press.
Chervin, R. M. 1978: The limitations of modelling: the question of statistical significance. In J. Gribbin (ed.), *Climatic Change*, Cambridge University Press, 191–201.
—— and Schneider, S. H. 1976: On determining the statistical significance of climate experiments with General Circulation Models. *Journal of Atmospheric Science*, 33, 405–12.
Chisci, G. and Morgan, R. P. C. 1988: Modelling soil erosion by water: why and how. In R. P. C. Morgan and R. J. Rickson (eds), *Erosion Assessment and Modelling*. Commission of European Communities, Balkema, 23–46.
Chorley, R. J. 1962: Geomorphology and general systems theory. *United States Geological Survey Professional Paper*, 500–B.
—— 1965: A re-evaluation of the geomorphic system of W. M. Davis. In R. J. Chorley and P. Haggett (eds), *Frontiers in Geographical Teaching*, Methuen, 21–38.
—— 1967a: The applications of statistical methods to geomorphology. In G. H. Dury (ed.), *Essays in Geomorphology*, Elsevier, 275–387.
—— 1967b: Models in geomorphology. In R. J. Chorley and P. Haggett (eds), *Models in Geography*, Methuen, 59–96.
—— and Haggett, P. (eds) 1967: *Models in Geography*, Methuen.
—— and Kennedy, B. A. 1971: *Physical Geography: A Systems Approach*, Prentice-Hall.
Chrisman, N. R. 1984: The role of quality information in the long term functioning of a GIS. *Cartographica*, 21, 79–87.
Christaller, W. 1966: *Central Places in southern Germany* (Translated by C. W. Baskin), Prentice Hall.
Clancy, W. J. 1983: The epistemology of a rule-based expert system – a framework for explanation. *Artificial Intelligence*, 20, 215–53.

Clarke, M. and Wilson, A. G. 1987: Towards an applicable human geography: some developments and observations, *Environment and Planning A*, 19(11), 1525–41.

Cliff, A. D. and Ord, J. K. 1973: *Spatial Autocorrelation*, Pion.

____ ____ 1981: *Spatial Processes*, Pion.

Cliff, A. D., Haggett P., Bassett, K., Ord, J. K. and Davies, R. 1975: *Elements of Spatial Structure*, Cambridge University Press.

Clifford, P. and Richardson, S. 1985: Testing the association between two spatial processes. *Statistics and Decisions*, Supplement Issue No. 2, 155–60.

Colwell, R. K. and Winkler, D. 1984: A null model for null models in biogeography. In D. R. Strong, D. Simberloff, L. G. Abele and A. B. Thistle (eds), *Ecological Communities: Conceptual Issues and the Evidence*, Princeton University Press, 344–59.

Copeta, C. (ed) 1986: *E. Dardel, L'uomo e la terra: natura della realta' geografica*, Unicopli.

Coppock, J. T. 1984: National inquiries into digital mapping in the United Kingdom. *Proceedings of the 12th International Conference of the International Cartographic Association*, 81–92, Perth.

Corbett, J. 1979: Topological principles in cartography. *U.S. Bureau of the Census Technical Paper* 48.

Cordova, J. R., Rodriguez-Iturbe, I. and Vaca, P. 1983: On the development of drainage networks. In *Recent Developments in the Explanation and Prediction of Sediment Yield*, International Association for Scientific Hydrology, Publication 137, 239–49.

Cosgrove, D. 1984: *Social Formation and Symbolic Landscape*, Croom Helm.

____ 1985: Prospect, perspective and the evolution of the landscape idea. *Transactions of the Institute of British Geographers*, N.S. 10, 45–62.

____ and Daniels, S. 1988: *The Iconography of Landscape: essays on the symbolic representation, design and use of past environments*, Cambridge University Press.

Couclellis, H. 1986: Artificial intelligence in geography: conclusions on the shape of things to come. *Professional Geographer*, 38, 1–11.

Cripps, E. and Foot, D. 1968: Evaluating alternative strategies, *Official Architecture and Planning*, 31, 928–38.

Crowe, P. R., 1968: Review: *Models in Geography*: the second Madingley Lectures. *Geography*, 57, 423–4.

Cullen, I. 1985: 'Expert systems in planning analysis'. Paper presented to the Annual Meeting of the British Section of the Regional Science Association, Manchester.

Culling, W. E. H. 1957: Equilibrium states in multicyclic streams and the analysis of river terrace profiles. *Journal of Geology*, 65(5), 451–67.

____ 1960: Analytical theory of erosion. *Journal of Geology*, 68, 336–44.

____ 1963: Soil creep and the development of hillside slopes. *Journal of Geology*, 71, 127–61.

____ 1983: Rate process theory of geomorphic soil creep. *Catena Supplement*, 4, 191–214.

____ 1986: Towards a unified theory of particulate flows in a geomorphic setting. *Occasional Papers* 1, Geography Department, Royal Holloway and Bedford New College.

____ 1987: Equifinality: a modern approach to dynamical systems and their potential for geographical thought. *Transactions of the Institute of British*

Geographers, N.S. 12, 57–73.
—— and Datko, M. 1987: The fractal geometry of the soil covered landscape. *Earth Surface Processes and Landforms,* 12(4), 369–85.
Curran, P. 1987: Remote sensing methodologies and geography. *International Journal of Remote Sensing*, 8, 1255–75.
Dangermond, J. and Morehouse, S. 1987: Trends in hardware for GIS. *Proceedings of Auto Carto 8*, 380–5, Baltimore American Congress of Surveying and Mapping, Washington DC.
Daniels, S. J. 1985: Arguments for a humanistic geography. In R. J. Johnson (ed.), *The Future of Geography*, Methuen, 143–57.
—— 1986: The implications of industry: Turner and Leeds. *Turner Studies: his art and epoch 1775–1851*, 6(1), 10–17.
Dardel, E. 1952: *L'homme et la terre: nature de la realite geographique*, Presses Universitaires de France.
David, M. 1977: *Geostatistical Ore Reserve Estimation: developments in geomathematics*, Elsevier.
Davidson, R. S. and Clymer, A. B. 1966: Desirability and applicability of simulating ecosystems. *Annals of the New York Academy of Sciences*, 128, 790–4.
Davis, J. R. and Nannings, P. M. 1985: GEOMYCIN: towards a geographic information system for resource management. *Journal of Environmental Management*, 21, 377–90.
Dear, M. 1986: Postmodernism and planning. *Environment and Planning D, Society and Space*, 4, 367–84.
De Coursey, D. G. 1982: 'ARS small watershed model'. Paper No. 82-2094. American Society of Agricultural Engineers Summer Meeting, Madison.
Degn, H., Holden, A. V. and Olsen, L. F. (eds) 1987: *Chaos in Biological Systems*, Plenum.
Dematteis, G. 1985: *Le metafore della terra: La geografia umuna tra mito e scienza*, G. Feltrinelli, 123.
De Ploey, J. 1982: A stemflow equation for grasses and similar vegetation. *Catena*, 9, 139–52.
De Wit, C. T. 1970: Dynamic concepts in biology. In I. Setlik (ed.), *Prediction and Measurement of Photosynthetic Productivity*, Wageningen. Centre for Agricultural Publications, 17–23.
Department of Education and Science 1987: *Higher education: meeting the challenge*, Cmnd 114, Her Majesty's Stationery Office.
Department of the Environment 1987: *Handling Geographic Information*, Her Majesty's Stationery Office.
—— 1988: *Handling Geographic Information: the government's response to the Committee of Enquiry on the Handling of Geographic Information*, Department of the Environment, London.
Diamond, J. M. 1975: Assembly of species communities. In M. L. Cody, M. and J. M. Diamond (eds), *Ecology and Evolution in Communities*, Harvard University Press, 342–445.
—— 1987: Character replacement: will grebes provide the answer? *Nature*, 325, 16–17.
—— and Gilpin, M. E. 1982: Examination of the 'null' model of Connor and Simberloff for species co-occurrences on islands. *Oecologia*, 52, 64–74.
Diaz, B. M. and Bell, S. B. M. (eds) 1986: *Spatial data processing using Tesseral methods*, Natural Environment Research Council.
Dickinson, R. E. 1984: Modelling evapotranspiration for three-dimensional

global climate models. In J. E. Hansen and T. Takahashi (eds), *Climate Processes and Climate Sensitivity*. American Geophysics Union, Washington, DC, 58–72.
____ 1985: Climate sensitivity. In S. Manabe (ed.), *Issues in Atmospheric and Oceanic Modelling, Part A, Climate Dynamics*, Advances in Geophysics, 28, Academic Press, 99–129.
Dooge, J. C. I. 1972: Mathematical models of hydrological systems. In A. K. Biswas (ed.), *Proceedings of the International Symposium on Modern Technology in Water Systems*, 1, 171–89, Environment Canada.
Douglas, M. 1979: *The World of Goods*, Basic Books.
Dregne, H. E. 1983: *Desertification of Arid Lands*, Harwood.
Dunne, T. and Aubry, B. F. 1986: Evaluation of Horton's theory of sheetwash and rill erosion on the basis of field experiments. In A. D. Abrahams (ed.), *Hillslope Processes*, Allen & Unwin, 31–53.
Durbin, J. 1987: Statistics and statistical science. *Journal of the Statistics Society A*, 150–3.
Dutton, W. H. and Kraemer, K. L. 1985: *Modelling as Negotiating*, Ablex.
Dyckmanm, J. W. 1963: The scientific world of the city planners. *American Behavioral Scientist*, 6, 46–50.
Eagleson, P. S. 1978: Climate, vegetation and soils. *Water Resources Research*, 14(5), 705–76.
____ and Tellers, T. E. 1982: Ecological optimality in water-limited natural soil–vegetation systems. *Water Resources Research*, 18(2), 325–54.
____ and Segarra, F. I. 1985: Water-limited equilibrium of savanna vegetation systems. *Water Resources Research*, 21(10), 1483–93.
Eliade, M. 1978: *The Forge and the Crucible*, University of Chicago Press.
Fairley, J. S. 1981: A north–south cline in the size of the Irish stoat. *Irish Naturalists Journal*, 17, 49–57.
Fedra, K., Li, Z., Wang, Z. and Zhao, C. 1987: Expert systems for integrated development: a case study of Shanxi Province, The People's Republic of China. *IIASA SR-87-001*.
Fenchel, T. 1975: Character displacement and coexistence in mud snails (Hydrobiidae). *Oecologia*, 20, 19–84.
Feyerabend, P. 1975: *Against Method*, Verso.
Flavell, W. S. 1985: Field verification of a stochastic model of soil creep. Unpublished Ph.D. thesis, University of London.
Flowerdew, R. 1986: Three years in British geography. *Area*, 18, 263–4.
Foot, D. 1982: *Operational Urban Models*, Methuen.
Forrester, J. W. 1961: *Industrial Dynamics*, Wright-Allen Press.
____ 1969: *Urban Dynamics*, MIT Press.
Foster, G. R. 1982: Modelling the soil erosion process. In C. T. Haans, H. P. Johnson, and D. L. Brakensiek (eds), *Hydrologic Modelling of Small Watersheds*, American Society of Agricultural Engineering, 297–392.
____, Laflen, J. M. and Alonso, C. V. 1985: A replacement for the Universal Soil Loss Equation. United States Department of Agriculture, Agricultural Research Service, *ARS-30*, 468–72.
Fraedrich, K. 1978: Structural and stochastic analysis of a zero-dimensional climate system. *Quarterly Journal of the Royal Meteorological Society*, 104, 461–74.
____ 1979: Catastrophes and resilience of a zero-dimensional climate system with ice-albedo and green house feedback. *Quarterly Journal of the Royal Meteorological Society*, 105, 147–67.

Frenzen, P. 1955: Westerly flow past an obstacle in a rotating hemispheric shell. *Bulletin of the American Meteorological Society*, 36, 204–10.
Fultz, D. 1961: Development in controlled experiments on larger scale geophysical problems. *Advanced Geophysics*, 7, 1–103.
Funtowicz, S. O. and Ravetz, J. R. 1987: The arithmetic of scientific uncertainty. *Physics Bulletin* 38, 412–14.
____ 1987: Qualified quantities – towards an arithmetic of real experience. In *Measurement, Realism and Objectivity*, forthcoming.
Fye, F. K. 1978: *The AFGWC Automated Cloud Analysis Model*. H.Q. Air Force Global Weather Central, Nebraska, AFGWC TM-78-002.
Galbraith, J. K. 1958: *The Affluent Society*, Hamish Hamilton.
Gandin, L. 1963: *Objective Analysis of Meteorological Fields*, Girometeorologicheskoe Izdatel'stvo, Leningrad.
Garfinkel, D. 1963: Digital computer simulation of ecological systems. *Nature*, 194, 856–7.
Gaschnig, J. 1982: PROSPECTOR: an expert system for mineral exploration. In D. Michie (ed.), *Introductory Readings for Expert Systems*, Gordon and Breach, 47–64.
Ghil, M. 1976: Climate stability for a Sellars-type model. *Journal of Atmospheric Science*, 33, 3–20.
Gilpin, M. E. and Diamond, J. M. 1982: Factors contributing to non-randomness of species co-occurrences on islands. *Oecologia*, 52, 75–84.
Global Atmospheric Research Programme, 1975: The physical basis of climate and climate modelling. *GARP Publication Series No. 16*, WMO/ICSU.
Goddard, J. B. and Armstrong, P. 1986: The 1986 Domesday Project. *Transactions of the Institute of British Geographers*, N. S. 11(3), 290–5.
Goddard, J. and Openshaw, S. 1987: The use and availability of computerised data and the commodification of information. *Environment and Planning A*.
Goodall, D. W. 1975: Ecosystems modelling in the desert biome. In B. C. Patten (ed.), *Systems Analysis and Simulation in Ecology*, Academic Press, 73–94.
Goodchild, M. F. 1982: The fractional Brownian process as a terrain simulation model. In W. G. Vogt and M. H. Mickle (eds), *Modelling and Simulation*, vol. 13, proceedings of the Thirteenth Anniversary Pittsburgh Conference, 1133–6.
Goodenough, D. G., Goldberg, M., Plunkett, G. and Zelek, J. 1987: An expert system for remote sensing. *IEEE Transactions on Geoscience and Remote Sensing*, GE-25, 349–59.
Goodrich, P. 1987: *Legal Discourse: studies in linguistics, rhetoric and legal analysis*, Macmillan.
Goudie, A. S. 1986: The integration of human and physical geography. *Transactions of the Institute of British Geographers*, N.S. 11, 454–8.
Gould, P. 1985a: *The Geographer at Work*, Routledge & Kegan Paul.
____ 1985b: Will geographic self-reflection make you blind? In R. J. Johnson (ed.), *The Future of Geography*, Methuen, 276–90.
Grant, P. R. 1972: Convergent and divergent character displacement. *Biological Journal of the Linnaean Society*, 4, 39–68.
____ 1975: The classical case of character displacement. *Evolutionary Biology*, 8, 237–337.
____ and Abbott, I. 1980: Interspecific competition, island biogeography and null hypotheses. *Evolution*, 34, 332–41.
Graves, G. R. and Gotelli, N. J. 1983: Neotropical land-bridge avifaunas: new approaches to null hypotheses in biogeography. *Oikos*, 41, 322–33.

Gray, F. 1975: Non-explanation in urban geography. *Area*, 7, 228–35.
Green, J. S. A. 1970: Transfer properties of the large scale eddies and the general circulation of the atmosphere. *Quarterly Journal of the Royal Meteorological Society*, 96, 157–85.
Green, N. P. 1986: An assessment of some UK-supported commercially available GIS. Report No. 3 of NERC Remote Sensing Special Topic, *Conceptual Design of a GIS for NERC*, Birkbeck College.
____ 1987: Teach yourself GIS: the design, creation and use of demonstrators and tutors. *International Journal of GIS*, 1(3), 279–90.
____, Finch, S., Rhind, D. W. and Anderson, K. E. 1985: User needs and design constraints. Report No. 1 of NERC Remote Sensing Special Topic, *Conceptual Design of A GIS for NERC*, Birkbeck College.
____, Finch, S., and Wiggins, J. 1985: The 'state of the art' in Geographical Information Systems. *Area*, 17, 295–301.
Greenberger, M., Crenson, M. A. and Crissey, B. L. 1976: *Models in the Policy Process: public decision-making in the computer era*. Russell Sage Foundation.
Gregory, D. 1978: *Ideology, Science and Human Geography*, Methuen.
Gregory, K. J. 1986: *The Nature of Physical Geography*, Arnold.
Griffin, E. 1973: Testing the von Thünen theory in Uruguay. *Geographical Review*, 63, 126–35.
Grossman, R. B. 1985: Some thoughts on the future of EPIC from a soil survey point of view. United States Department of Agriculture, Agricultural Research Service, *ARS-30*, 31–2.
Guelke, L. 1974: An idealist alternative in human geography. *Annals of the Association of American Geographers*, 64, 183–202.
Gyford, J. 1985: *The Politics of Local Socialism*, Allen & Unwin.
Haggett, P. and Chorley, R. J. 1965: Frontier movements and the geographical tradition. In R. J. Chorley and P. Haggett (eds), *Frontiers in Geographical Teaching*, Methuen, 358–78.
____ ____ 1967: Models, paradigms and the new geography. In R. J. Chorley and P. Haggett (eds), *Models in Geography*, Methuen.
Haggett, P., Cliff, A. D. and Frey, A. 1977: *Locational Analysis in Human Geography*, 2nd edn, Arnold.
Haines-Young, R. H. and Petch, J. R. 1980: The challenge of critical rationalism for methodology in physical geography. *Progress in Physical Geography*, 4, 63–78.
____ ____ 1986: *Physical Geography: its nature and methods*, Harper & Row.
Haining, R. P. 1978: A spatial model for High Plains agriculture. *Annals of the Association of American Geographers*, 68, 493–504.
____ 1981: Analysing univariate maps, *Progress in Human Geography*, 5, 58–78.
____ 1987: Trend surface models with regional and local scales of variation with an application to aerial survey data. *Technometrics* 29(4), 461–9.
____ 1988: Estimating spatial means with an application to remotely sensed data. *Communications in Statistics: Theory and Method*, 573–597.
____, Griffith, D. A. and Bennett, R. J. 1984: A statistical approach to the problem of missing spatial data using a first order markov model. *Professional Geographer*, 36, 338–45.
Hall, P. (ed.) 1966: *Von Thunen's Isolated State*, Pergamon Press.
Hansen, J., Johnson, D., Lacis, A., Lebedeff, S., Lee, P., Rind, D. and Russell, G. 1981: Climate impact of increasing carbon dioxide. *Science*, 213, 957–66.

____, Russell, G., Rind, D., Stone, P., Lacis, A., Lebedeff, S., Ruedy, R. and Travis, L. 1983: Efficient three-dimensional global models for climate studies: Models I and II. *Monthly Weather Review*, 111, 609–22.

____, Lacis, A., Rind, D., Russell, G., Stone, P., Fung, I., Ruedy, R. and Lerner, J. 1984: Climate sensitivity: analysis of feedback mechanism. In J. E. Hansen and T. Takahashi (eds), *Climate Processes and Climate Sensitivity*, American Geophysical Union, 130–63.

____, Russell, G., Lacis, A., Fung, I., Rind, D. and Stone, P. 1985: Climate response times: dependence on climate sensitivity and ocean mixing. *Science*, 229, 857–9.

Hare, K. 1953: *The Restless Atmosphere*, Hutchinson.

____ 1962: The Westerlies. *Geographical Review*, 50, 345–67.

____ 1977: Man's world and geographers: a secular sermon. In D. R. Deskins et al. (eds), *Geographic Humanism, Analysis and Social Action: a half century of geography at Michigan*. Michigan Geographical Publication, No. 17, 259–73.

Harris, B. 1965: New tools for planning. *Journal of the American Institute of Planners*, 31, 90–5.

____ 1967: The city of the future: the problems of optimal design. *Papers of the Regional Science Association*, 19, 185–95.

Harris, R. 1981: Remote sensing. In N. Wrigley and R. J. Bennett (eds), *Quantitative Geography*, Routledge & Kegan Paul, 36–45.

____ 1986: Vegetation index models for the assessment of vegetation in marginal areas. In *Proceedings* of the ISLSCP Conference on Parameterization of Land-Surface Characteristics; Use of Satellite Data in Climate Models, and First Results of ISLSCP, Rome, 2–6 December 1985, ESA AP–248.

Harvey, D. 1969: *Explanation in Geography*, Arnold.

____ 1973: *Social Justice in the City*, Blackwell (1988 reissue).

____ 1974a: Population, resources and the ideology of science, *Economic Geography*, 50, 256–77.

____ 1974b: What kind of geography for what kind of public policy?. *Transactions of the Institute of British Geographers*, 63, 18–24.

Harvey, P. H. 1987: On the use of null hypotheses in biogeography. In M. H. Nitecki and A. Hoffman (eds), *Neutral Models in Biology*, Oxford University Press.

____, Colwell, R. K., Silvertown, J. W. and May, R. M. 1983: Null models in ecology. *Annual Reviews of Ecology and Systemics*, 14, 189–211.

Hay, D. A. 1984: An application of von Thunen's model of agricultural location. *Oxford Agrarian Studies*, XVIII, 43–66.

Henderson-Sellers, A. 1986: Increasing cloud in a warmer world. *Climatic Change*, 9, 267–309.

____ 1987: Effects of change in land use on climate in the humid tropics. In Dickenson R. E. (ed), *The Geophysiology of Amazonia*, Wiley, 463–93.

____ and McGuffie, K. 1987: *An Introduction to Climate Modelling*, Wiley.

____ and Robinson P. J. 1986: *Contemporary Climatology*, Longmans.

Henricksen, B. L. and Durkin, J. W. 1986: Growing period and drought early warning in Africa using satellite data. *International Journal of Remote Sensing*, 7, 1583–608.

Hepple, L. 1976: A maximum likelihood model for econometric estimation with spatial series. In I. Masser (ed.), *Theory and Practice in Regional Science*, Pion, 90–104.

Herbert, J. D. and Stevens, B. H. 1960: A model for the distribution of residential

activity in urban areas. *Journal of Regional Science*, 2, 21–36.
Herbertson, A. J. 1905: The natural regions of the world. *Geographical Teacher*, 3, 104–13.
Hewlett, J. D. 1961: Some ideas about storm runoff and baseflow. *U.S. Dept. of Agriculture South East Experimental Station Annual Report*, 62–6.
Hide, R. 1953: Some experiments on thermal convection in a rotating fluid. *Quarterly Journal of the Royal Meteorological Society*, 79, 161.
Hindess, A. 1973: *The Use of Official Statistics in Sociology*, Macmillan.
Hirsch, F. 1977: *The Social Limits to Growth*, Routledge & Kegan Paul.
Hoinville, G. and Smith, T. M. F. 1982: The Rayner review of government statistical services. *Journal of the Royal Statistical Society A*, 145, 195–207.
Holden, A. V. (ed.) 1986: *Chaos*, Manchester University Press.
Holling, C. S. 1964: The analysis of complex population processes. *Canadian Entomologist*, 96, 335–47.
Hotelling, H. 1979: A mathematical theory of migration. *Environment and Planning A*, 10(11), 1233–9.
Huggett, R. J. 1985: *Earth Surface Systems*, Springer Verlag.
Huff, D. L. 1963: A probabilistic analysis of shopping center trade areas. *Land Economics*, 39, 81–90.
Hunt, B. G. 1985: A model study of some aspects of soil hydrology relevant to climatic modelling. *Quarterly Journal of the Royal Meteorological Society*, 111, 1071–85.
Hutchinson, G. E. 1959: Homage to Santa Rosalia, or why are there so many kinds of animals? *American Naturalist*, 93, 145–59.
Ignatieff, M. 1984: *The Needs of Strangers*, Chatto & Windus.
Innis, G. S. 1975: Role of total systems models in the Grasslands Biome study. In B. C. Patten (ed.), *Systems Analysis and Simulation in Ecology*, Academic Press, 13–47.
—— 1978: *Grass Simulation Model*, Ecological Studies, 26, Springer Verlag.
Irvine, J., Miles, I. and Evans, J. (eds) 1979: *Demystifying Social Statistics*, Plato Press.
Jackson, J. B. 1980: Learning about landscapes. In *The Necessity for Ruins and Other Topics*, University of Massachusetts Press.
Jackson, P. 1986: *Introduction to Expert Systems*, Addison Wesley.
Jarvis, P. G. and McNaughton, K. G. 1986: Stomatal control of transpiration: scaling up from leaf to region. *Advanced Ecological Research*, 15, 1–49.
Johnston, R. J. 1983: *Philosophy and Human Geography*, Arnold.
—— 1985: To the ends of the earth. In R. J. Johnston (ed.), *The Future of Geography*, Methuen, 326–8.
—— 1986a: Four fixations and the quest for unity in geography. *Transactions of the Institute of British Geographers*, N.S., 11, 449–453.
—— 1986b: *On Human Geography*, Basil Blackwell.
—— 1986c: *Philosophy and Human Geography*, 2nd edn, Arnold.
Jones, D. K. C. 1983: Environments of concern. *Transactions of the Institute of British Geographers*, 8(4), 429–57.
Journel, A. G. and Huijbregts, J. 1978: *Mining Geostatistics*, Academic Press.
Justice, C. O., Townshend, J. R. G., Holben, B. N. and Tucker, C. J. 1985: Analysis of the phenology of global vegetation using meteorological satellite data. *International Journal of Remote Sensing*, 6, 1271–1318.
Keat, R. and Urry, J. 1975: *Social Theory as Science*, Routledge & Kegan Paul (2nd edition 1982).

Kelley, A. C. and Williamson, J. G. 1984: *'What Drives Third World City Growth'? A dynamic general equilibrium approach'*, Princton University Press.

Kellogg, W. W. and Schware, R. 1981: *Climatic Change and Society: consequences of increasing atmospheric carbon dioxide*, Westview Press.

Kennedy, B. 1979: A naughty world. *Transactions of the Institute of British Geographers*, N.S. 4(4), 550–8.

Kibler, D. F. and Woolhiser, D. A. 1970: The kinematic cascade as a hydrological model. *Colorado State Hydrological Papers*, 39.

Kirk, W. 1963: Problems of geography. *Geography*, 48, 357–71.

Kirkby, M. J. 1971: Hillslope process response models based on the continuity equation. *Institute of British Geographers Special Publication*, 3, 15–30.

____ 1980: The stream-head as a significant geomorphic threshold. In D. R. Coates and J. D. Vitek (eds), *Threshold in Geomorphology*, Allen & Unwin, 53–73.

____ 1984: Modelling cliff development in South Wales: Savigear reviewed. *Zeitschrift für Geomorphologie*, 28(4), 405–26.

____ 1985a: A basis for soil profile modelling in a geomorphic context. *Journal of Soil Science*, 36, 97–121.

____ 1985b: A model for the evolution of regolith mantled slopes. In M. J. Woldenberg (ed.), *Models in Geomorphology*, Allen & Unwin, 213–37.

____ 1986: A two dimensional simulation model for slope and stream evolution. In A. D. Abrahams (ed.), *Hillslope Processes*, Allen & Unwin, 203–22.

____ and Chorley, R. J. 1967: Throughflow, overland flow and erosion. *Bulletin of the International Association of Hydrological Sciences*, 12, 5–21.

____ and Neale, R. H. 1987: A soil erosion model incorporating seasonal factors. In V. Gardiner (ed.), *International Geomorphology*, 1986, Part II, Wiley, 189–210.

____, Burt, T. P., Naden, P. S. and Butcher, D. P. 1987: *Computer Simulation in Physical Geography*, Wiley.

Klemes, V. 1986: Operational testing of hydrological simulation models. *Hydrological Sciences Bulletin*, 31(1), 13–24.

Knevitt, C. 1985: *Space on Earth: Architecture, People and Buildings*, Thames/Methuen.

Krumbein, W. C. and Graybill, F. A. 1965: *An Introduction to Statistical Models in Geology*, McGraw-Hill.

Lai, P. W. 1979: *Transfer Function Modelling*. Concepts and Techniques in Modern Geography, 22, GeoBooks.

Lakshmanan, T. R. and Hansen, W. G. 1965: A retail market potential model. *Journal of the American Institute of Planners*, 31, 134–43.

Langbein, W. B. and Leopold, L. B. 1968: River channel bars and dunes – theory of kinematic waves. *U.S. Geological Survey*, Professional Paper 122L.

Larimore, W. E. 1977: Statistical inference on stationary random fields. *Proceedings of the IEEE*, 65, 196–7.

Laurmann, J. A. and Gates, W. L. 1977: Statistical considerations in the evaluation of climatic experiments with atmospherical general circulation models. *Journal of Atmospheric Science* 34, 1187–99.

Lawton, R. 1983: Space, place and time. *Geography*, 68, 193–207.

Lee, D. B. 1973: Requiem for large-scale models. *Journal of the American Institute of Planners*, 39, 163–78.

Leopold, L. B. and Langbein, W. B. 1962: The concept of entropy in landscape evolution. *U.S. Geological Survey*, Professional Paper 500A.

—— and Maddock, T. 1953: The hydraulic geometry of stream channels and some physiographic implications. *U.S. Geological Survey*, Professional Paper, 252.

Lewis, L. F. 1939: The seasonal and geographical distribution of absolute drought in England. *Quarterly Journal of the Royal Meteorological Society*, 65, 367–82.

Ley, D. 1985: Styles of the time: liberal and neo-conservative landscapes in inner Vancouver 1968–86, *Journal of Historical Geography*, 13, 40–56.

Lillisand, T. M. and Kiefer, R. W. 1979: *Remote Sensing and Image Interpretation*, Wiley.

Lindemann, R. N. 1942: The Trophic-Dynamic Aspect of Ecology. *Ecology* 23, part IV, 399–418.

Linton, D. L. 1968: The assessment of scenery as a natural resource, *Scottish Geographical Magazine*, 84, 219–38.

Livingstone, D. 1985: The history of science and the history of geography: interaction and implications, *History of Science*, 20, 271–302.

London, J. 1957: A study of the atmospheric heat balance, *AFCRC-TR-57–287*, College of Engineering, New York University.

Loomis, R. S. and Williams, W. A. 1964: Maximum crop productivity: an estimate. *Crop Science*, 3, 67–72.

Lösch, A. 1954: Die raumliche Ordnung der Wirtschaft. Translated by W. H. Woglom, and W. F. Stolper, as *Economics of Location*, Yale University Press.

Lough, J. M., Wigley, T. M. L. and Palutikof, J. P. 1983: Climate and climate impact scenarios for Europe in a warmer world. *Journal of Climatology and Applied Meteorology*, 22, 1673–84.

Lovejoy, S. 1982: Area-perimeter relation for rain and cloud areas. *Science*, 216, 185–7.

—— and Schertzer, D. 1986: Scale invariance, symmetries, fractals and stochastic simulations of atmospheric phenomena. *Bulletin of the American Meteorological Society*, 67, 21–32.

——, —— and Ladoy, P. 1986: Fractal characterization of inhomogeneous geophysical measuring networks. *Nature*, 319, 43–4.

Lowenthal, D. 1961: Geography, experience and imagination: towards a geographical epistemology. *Annals Association of American Geographers*, 51, 241–60.

Lowenthal, D. and Prince, H. C. 1964: The English landscape. *Geographical Review*, 54, 309–46.

—— —— 1965: English landscape tastes, *Geographical Review*, 55, 186–222.

Lowry, I. S. 1964: *A Model of Metropolis*, The Rand Corporation.

Lundholm, B. 1984: Approach to evaluation: insights from analysis of a large scale ecosystems project. In F. Di Castri, F. W. G. Baker and M. Hadley (eds), *Ecology in Practice*, Vol. 2, 366–73.

Lynch, K. 1960: *The Image of the City*, MIT Press.

Lyotard, J. F. 1984: *The Post-modern Condition: a report on knowledge*, Manchester University Press.

Macgill, S. M. 1986: Research policy and review 12: Evaluating a heritage of modelling styles. *Environment and Planning A*, 18, 1423–46.

—— and Sheldrick, B. 1987: Monergy: qualifying imperfect measures of need and of performance. *Environment and Planning C: Government and Policy*, 6, 209–23.

Mackenzie, H. G. 1984: *EMYCIN users guide*. Technical Report No. 17, CSIRO Division of Computing Research, Canberra.

Mackinder, H. J. 1887: The scope and methods of geography. *Proceedings of the Royal Geographical Society*, 9, 141–60. Reprinted in Mackinder, 1962: *Democratic Ideals and Reality*, W. W. Norton, 211–40.
—— 1904: The geographical pivot of history. *Geographical Journal*, 22, 421–37.
Manabe, S. 1969: Climate and ocean circulation I: Atmospheric circulation and hydrology of the Earth's surface. *Monthly Weather Review*, 97, 739–74.
—— and Strickler, R. F. 1964: Thermal equilibrium of the atmosphere with a convective adjustment. *Journal of Atmospheric Science*, 21, 361–85.
—— and Wetherald, R. T. 1975: The effects of doubling the CO_2 concentration on the climate of a general circulation model. *Journal of Atmospheric Science*, 32, 3–15.
Mandelbrot, B. B. 1977: *The Fractal Geometry of Nature*, W. H. Freeman & Co.
—— and Wallis, J. R. 1968: Noah, Joseph and operational hydrology. *Water Resources Research*, 4(5), 409–18.
Mankin, J. B., O'Neil, R. V., Schugart, H. H. and Rust, B. W. 1975: The importance of validation in ecosystems analysis. In G. S. Innis (ed.), *New Directions in the Analysis of Ecological Systems*, Simulations Council Proceedings Series V(1), 63–71.
Manley, G. 1947: The geographer's contribution to meteorology. *Quarterly Journal of the Royal Meteorological Society*, 73, 1–10.
Mardia, K. V. and Marshall, R. J. 1984: Maximum likelihood estimation of models for residual covariance in spatial regression. *Biometrika*, 71, 135–46.
Martin, Richard J. 1984: Exact maximum likelihood for incomplete data from a correlated Gaussian process. *Communications in Statistics: Theory and Methods*, 13, A, 1275–88.
Massey, D. and Allen, J. 1984: *Geography Matters*, Cambridge University Press.
Matheron, G. 1971: *The Theory of Regionalized Variables and its Applications*. Les Cahiers du Centre de Morphologie Mathematique de Fontainebleau, No. 5.
MacArthur, R. H. and Wilson, E. O. 1967: *The Theory of Island Biogeography*, Princeton University Press.
McLoughlin, J. B., Nix, C. K. and Foot, D. H. S. 1966: *Regional Shopping Centres: A Planning Report on North West England: Part 2: A Retail Shopping Model*, University of Manchester School of Town and Country Planning.
McNaughton, P. K. G. and Jarvis, G. 1983: Predicting effects of vegetation changes on transpiration and evaporation. *Water Deficits and Plant Growth*, 7, Academic Press, 1–47.
Meehl, G. A. 1987: The Tropics and their role in the global climate system. *Geographical Journal*, 153, 21–46.
Meinig, D. 1983: Geography as art. *Transactions of the Institute of British Geographers*, N.S. 8, 314–38.
Meisner, B. N. and Arkin, P. A. 1987: Spatial and annual variations in the diurnal cycle of large-scale tropical connective cloudiness and precipitation. *Monthly Weather Review*.
Meleshko, V. and Wetherald, R. T. 1981: The effect of a geographical cloud distribution on climate: a numerical experiment with an atmospheric general circulation model. *Journal of Geophysical Research*, 86, 11995–2014.
Mercer, D. 1984: Unmasking technocratic geography. In M. Billinge, D. Gregory and R. Martin (eds), *Recollections of a Revolution: geography as spatial science*, Macmillan, 153–99.
Milligan, B. G. 1985: Evolutionary divergence and character displacement in two

phenotypically-variable competing species. *Evolution*, 39, 1207–22.
Minford, P. 1983: *Rational Expectations and the New Macroeconomics*, Martin Robertson.
——, Davies, D., Pell, M. and Sprague, A. 1983: *Unemployment: Cause and Cure*, Martin Robertson.
Minnis, P. and Harrison, E. F. 1984: Diurnal variability of regional cloud and clear-sky radiative parameters derived from GOES data. Part I: Analysis method. *Journal of Climatology and Applied Meteorology*, 23, 993–1011.
Minogue, K. and Biddiss, M. (eds) 1987: *Thatcherism: personality and politics*, Macmillan.
Minshull, R. 1975: *An Introduction to Models in Geography*, Longman.
Miron, J. 1984: Spatial autocorrelation in regression analysis: a beginner's guide. In G. L. Gaile and C. J. Willmott (eds), *Spatial Statistics and Models*, Reidel.
Mohan, R. 1979: *Urban Economic and Planning Models*, Baltimore, Maryland: Johns Hopkins University Press.
Morgan, M. A. 1967: Hardware models in geography. In R. J. Chorley and P. Haggett (eds), *Models in Geography*, Methuen, 727–74.
Muetzelfeldt, R., Robertson, R., Ushold, M. and Bundy, A. 1987: Computer-aided construction of ecological simulation programs. Department of Artificial Intelligence, University of Edinburgh, Research Paper No. 314.
Murphy, R. E. 1986: Foreword to special issue 'Monitoring the Grasslands of Semi-arid Africa using NOAA AVHRR data'. *International Journal of Remote Sensing*, 7, 1391–4.
Myrup, L. O. 1969: A numerical model of the urban heat island. *Journal of Applied Meteorology*, 8, 908–10.
National Academy of Science (NAS) 1982: *Carbon Dioxide and Climate: a second assessment*, National Academy Press.
Nicolin, B. and Gabler, R. 1987: A knowledge-based system for the analysis of aerial images. *IEEE Transactions on Geoscience and Remote Sensing*, GE-25, 317–29.
North, G. R., Cahalan, R. F. and Coakley, J. A. 1981: Energy balance climate models. *Review of Geophysics and Space Physics*, 19, 91–121.
Nye, J. 1960: The responses of glaciers and ice sheets to seasonal and climatic changes. *Proceedings: Royal Society London*, A256, 559–84.
Nystuen, J. D. 1984: Comment on 'Artificial intelligence and its applicability to geographical problem solving', *Professional Geographer*, 36, 358–9.
Ohring, G. and Clapp, P. F. 1980: The effect of changes in cloud amount on the net radiation at the top of the atmosphere. *Journal of Atmospheric Science*, 37, 447–54.
—— and Gruber, A. 1983: Satellite radiation observations and climate theory. In B. Saltzman (ed.), *Theory of Climate*, Advances in Geophysics, 25, Academic Press.
O'Keeffe, J. H., Danilewitz, B. D. and Bradshaw, J. A. 1987: An 'expert systems' approach to the assessment of the conservation status of rivers. *Biological Conservation*, 40, 69–84.
Olsen, J. S. 1960: Analogue computer models for the movement of nuclides through ecological systems. *Radioecology*, Reinhold, 81–96.
Onstad, C. A. and Foster, G. R. 1975: Erosion modelling on a watershed. *Transactions of the American Society of Agricultural Engineering*, 18(2), 288–92.
Openshaw, S. 1978: *Using Models in Planning: a practical guide*, Corbridge:

Retail Planning Associates.
—— 1983: *The Modifiable Area Unit Problem.* CATMOG 38, GeoBooks.
—— 1985: Fish theory and urban modellers: a Christmas quiz. *Environment and Planning A*, 17, 1546–9.
—— 1986: Modelling relevance. *Environment and Planning A*, 18, 143–7.
—— and Goddard, J. B. 1987: Some implications of the commodification of information and the emerging information economy for applied geographical analysis in the UK. *Environment and Planning A*, 19(11), 1423–39.
—— and Taylor, P. 1981: The modifiable areal unit problem. In N. Wrigley and R. J. Bennett (eds), *Quantitative Geography: a British view*, Routledge & Kegan Paul.
——, Charlton, M. and Craft, A. 1987: Searching for leukaemia clusters using a geographical analysis machine. Paper presented at the European Congress of the Regional Science Association, Athens.
Oram, P. A. 1985: Sensitivity of agricultural production to climatic change. *Climatic Change*, 7, 129–52.
Ord, J. K. 1975: Estimation methods for models of spatial interaction. *Journal of the American Statistical Association*, 70, 120–6.
Outcalt, S. I. 1971: A numerical surface climate simulator. *Geographical Analysis*, 3, 379–92.
Ordnance Survey 1986: *Ordnance Survey Annual Report 1985/86.* Ordnance Survey, Southampton.
—— 1987: *The National Transfer Format.* Ordnance Survey, Southampton.
Padulo, L. and Arbib, M. A. 1974: *System Theory: a unified state–space approach to continuous and discrete time systems.* W. B. Saunders.
Palutikof, J. P., Wigley, T. M. L. and Lough, J. M. 1984: Seasonal climate scenarios for Europe and North America in a high-CO_2, warmer world. *Carbon Dioxide Research Division of Department of Energy, Report TRO12* [available from NTIS, Springfield VA 22161].
Parker, G. 1979: Hydraulic geometry of active gravel rivers. *Journal of Hydraulics Division*, American Society of Civil Engineers, 105, 1185–201.
Patten, B. C. 1975: Ecosystem modelling in the United States International Biological Program. In B. C. Patten (ed.), *Systems Analysis and Simulation in Ecology*, Academic Press, 1–6.
Peitgen, H-O. and Richter, P. H. 1986: *The Beauty of Fractals*, Springer-Verlag.
Pendleton, D. T. 1985: Simulation of Production and Utilisation of Rangeland (SPUR) model. Discussion by SCS, United States Department of Agriculture, Agricultural Research Service, *ARS–30*, 56–8.
Penman, H. L. 1948: Natural evaporation from open water, bare soil and grass. *Proceedings of the Royal Society*, Series A, 193, 120–45.
Persson, T. (ed.) 1980: *Structure and Function of Northern Coniferous Forests – An Ecosystem Study.* Ecological Bulletins, 32, Swedish Natural Sciences Research Council.
Pinker, R. 1979: *The Idea of Welfare*, Heinemann.
Pollard, D. 1979: Simple ice sheet model yields realistic 100kyr glacial cycles. *Nature*, 296, 334–8.
——, Ingersoll, A. P. and Lockwood, J. G. 1980: Response of a zonal climate – systems. In M. Morrisawa (ed.), *Fluvial Geomorphology*, SUNY, Binghamton, 299–310.
—— and Lichty, R. W. 1965: Time, space and causality in geomorphology. *American Journal of Science*, 263, 110–19.

Scruton, R. 1980: *The Meaning of Conservatism*, Macmillan.
ice sheet model to the orbital perturbations during the Quaternary ice ages. *Tellus*, 301–19.

Popper, K. R. 1970: Normal science and its dangers. In I. Lakatos and A. Musgrave (eds), *Criticism and the Growth of Knowledge*, Cambridge University Press, 51–8.

—— 1972: *Conjectures and Refutations*, Oxford University Press.

Potter, G. L. and Gates, W. L. 1984: A preliminary intercomparison of the seasonal response of two atmospheric climate models. *Monthly Weather Review*, 112, 909–17.

——, Ellsaesser, H. W., MacCracken, M. C. and Mitchell, C. S. 1981: Climate change and cloud feedback: the possible radiative effects of latitude redistribution. *Journal of Atmospheric Science* 38, 489–93.

Preisendorfer, R. W. and Barnett, T. P. 1983: Numerical model–reality intercomparison tests using small-sample statistics. *Journal of Atmospheric Science*, 40, 1884–96.

Prigogine, I. and Stengers, I. 1984: *Order Out of Chaos*, Bantam Books.

Putman, S. H. 1983: *Integrated Urban Models*, Pion.

Quenouille, M. H. 1949: Problems in plane sampling. *Annals of Mathematical Statistics*, 20, 355–75.

Ralls, K. and Harvey, P. H. 1985: Geographic variation in size and sexual dimorphism in North American weasels. *Biological Journal of the Linnaean Society*, 25, 119–67.

Ramanathan, V. 1981: The role of ocean–atmosphere in the CO_2 climate problem. *Journal of Atmospheric Science*, 38, 918–30.

——, Pitcher, E. J., Malone, R. C. and Blackmon, M. L. 1983: The response of a spectral general circulation model to refinements in radiative processes. *Journal of Atmospheric Science*, 40, 605–30.

Ravallion, M. 1987: *Markets and Famines*, Clarendon Press.

Ravenstein, E. G. 1885: The laws of migration. *Journal of the Royal Statistical Society*, 48, 167–235.

Rawls, J. 1972: *A Theory of Justice*, Clarendon Press.

Reilly, W. J. 1929: Methods for the study of retail relationships. *Bulletin No. 2944*, Houston, Texas: University of Texas.

Rhind, D. W. 1981: Geographical information systems in Britain. In N. Wrigley and R. J. Bennett (eds), *Quantitative Geography*, Routledge & Kegan Paul, 17–39.

—— 1986: Remote sensing, digital mapping and GIS: the creation of government policy in the UK. *Environment and Planning C*, 91–100.

—— 1987: Recent developments in GIS in the UK. *International Journal of GIS*, 1(3), 229–41.

——, Armstrong, P. and Openshaw, S. 1988: The BBC Domesday system: a nation-wide GIS. *Geographical Journal*, 154, 1, 56–8.

Ripley, B. 1981: *Spatial Statistics*, Wiley.

—— 1984: Present position and potential developments: some personal views. Statistics in the natural sciences. *Journal of the Royal Statistical Society A*, 147, 340–8.

Robbins, S. W. 1985: SCS overview on spatial and temporal variability. United States Department of Agriculture, Agricultural Research Service, *ARS-30*, 287–8.

Rossow, W. B. and Garder, L. 1984: Selection of a map grid for data analysis and

archival. *Journal of Climatology and Applied Meteorology*, 23, 1253–7.
Moscher, F., Kinsella, E., Arking, A., Desbois, M., Harrison, E., Minnis, P., Ruprecht, E., Seze, G., Simmer, C. and Smith, E. 1985: ISCCP cloud algorithm intercomparison. *Journal of Climatology and Applied Meteorology*, 24, 877–903.
Roszak, T. 1986: *The Cult of Information*, Lutterworths.
Rowles, G. 1978: Reflections on experiential field work. In D. Ley and M. S. Samuels (eds), *Humanistic Geography*, Croom Helm, 173–93.
Rudner, R. 1966: *Philosophy of Social Science*, Prentice-Hall.
Rutter, A. J., Kershaw, K. A., Robins, P. C. and Morton, A. J. 1971: A predictive model of rainfall interception over forests. *Agricultural Meteorology*, 9, 367–84.
Sack, R. D. 1980: *Conceptions of Space in Social Thought*, University of Minnesota Press.
Sagan, C., Toon, O. B. and Pollack, J. B. 1979: Anthropogenic albedo changes and the earth's climate. *Science*, 206, 1363–8.
Saltzman, B. 1978: A survey of statistical dynamical models of terrestrial climate. *Advances in Geophysics*, 20, 183–304.
—— and Sutera, A. 1987: The mid-Quaternary climatic transition as the free response of a three variable dynamical model. *Journal of Atmospheric Sciences*, 44(1), 1–6.
Samuels, M. S. 1981: An existential geography. In M. E. Harvey and S. P. Holly, (eds), *Themes in Geographical Thought*, Croom Helm, 115–33.
Sarre, P. 1987: Realism in practice. *Area*, 19, 3–10.
Sauer, R. H. 1978: A simulation model for grassland primary producer phenology and biomass dynamics. In G. S. Innis (ed.), *Grassland Simulation Model*, Ecological Studies, 26, Springer Verlag, 55–87.
Saunders, P. and Williams, A. 1986: The new conservatism: some thoughts on recent and future developments in Urban Studies. *Society and Space: Environment and Planning D*, 4, 393–9.
Savigear, R. A. G. 1952: Some observations on slope development in South Wales. *Transactions of the Institute of British Geographers*, 18, 31–52.
Sayer, A. 1976: A critique of urban modelling. *Progress in Planning*, 6, 187–254.
—— 1977: Gravity models and spatial autocorrelation, or atrophy in urban and regional modelling. *Area*, 9, 183–9.
—— 1979: Understanding urban models versus understanding cities. *Environment and Planning A*, 11, 853–62.
—— 1984: *Method in Social Science: a realist approach*, Hutchinson.
—— 1985: Realism and geography. In R. J. Johnson (ed.), *The Future of Geography*, Methuen, 159–73.
Scheidegger, A. R. 1961: Mathematical models of slope development. *Bulletin of the Geological Society of America*, 72, 37–49.
Schiffer, R. A. and Rossow, W. B. 1985: ISCCP global radiance data set: a new resource for climate research. *Bulletin of the American Meteorological Society*, 66, 1498–505.
Schlager, K. J. 1965: A land use plan design model. *Journal of the American Institute of Planners*, 31, 103–11.
Schluter, D. L., Price, T. D. and Grant, P. R. 1985: Ecological character displacement in Darwin's finches. *Science*, 227, 1056–9.
Schneider, S. H. and Dickinson, R. E. 1974: Climate modelling. *Review of Geophysics and Space Physics*, 12, 447–93.

Schumm, S. A. 1973: Geomorphic thresholds and complex response of drainage
Sellers, P. J. 1985: Canopy reflectance, photosynthesis and transpiration. *International Journal of Remote Sensing*, 6, 1335–72.
—— and Lockwood, J. G. 1981: A computer simulation of the effects of differing crop types on the water balance of small catchments over long time periods. *Quarterly Journal of the Royal Meteorological Society*, 107, 395–414.
——, Mintz, Y., Sud, Y. C. and Dulcher, A. 1986: A simple biosphere model (SiB) for use within general circulation models. *Journal of Atmospheric Science*, 43, 505–31.
Sellers, W. D. 1969: A global climatic model based on the energy balance of the earth–atmosphere system. *Journal of Applied Meteorology*, 8, 392–400.
Shine, K. and Henderson-Sellers, A. 1983: Modelling climate and the nature of climate models: a review. *Journal of Climatology*, 3, 81–94.
Shortliffe, E. H. 1976: *Computer-based Medical Consultation: MYCIN*. American Elsevier.
Shreve, R. L. 1966: Statistical laws of stream numbers. *Journal of Geology*, 74, 17–37.
Shubik, M. 1969: Processing the future. *Science*, 166, 1257–8.
—— 1979: Computers and modelling. In M. L. Dertouzos and J. Moses (eds), *The Computer Age: a twenty year view*, MIT Press, 285–305.
Shugart, H. H. and O'Neil, P. (eds) 1979: Preface, *Systems Ecology*, Hutchinson and Ross, 1–11.
Shuttleworth, W. J., Gash, J. H. C., Lloyd, C. R., Moore, C. J., Roberts, J., Marques Filho, A. de O., Fisch, G., de Paula Silva Filho, V., Molion, L. C. B., de Abreau Sá, L. D., Nobre, J. C. A., Cabral, O. M. A., Patel, S. R. and de Moraes, J. C. 1985: Daily variations of temperature and humidity within and above Amazonian forest. *Weather*, 40, 102–8.
Simmons, A. J. and Bengtsson, L. 1984: Atmospheric general circulation models: their design and use for climate studies. In J. T. Houghton (ed.), *The Global Climate*, Cambridge University Press, 37–56.
Simon, H. A. 1977: *Models of Discovery*, Reidel.
Skilling, H. 1964: An operational view. *American Scientist*, 52, 388A–96A.
Slater, P. N. 1979: A re-examination of the Landsat MSS. *Photogrammetric Engineering and Remote Sensing*, 46, 509–20.
Slatkin, M. 1979: The evolutional response to frequency and density dependent interactions. *American Naturalist*, 114, 384–98.
—— 1980: Ecological character displacement. *Ecology*, 61, 163–77.
Smagorinsky, J. 1983: The beginnings of numerical weather prediction and general circulation modelling: early recollections. In B. Saltzman (ed.), *Theory of Climate*, Academic Press, 3–38.
Smith, E. A. 1986: The structure of the Arabian heat low: Part I: Surface energy budget. *Monthly Weather Review*, 114, 1067–83.
Smith, R. E. and Knisel, W. G. 1985: Summary of methodology for CREASM2 model. United States Department of Agriculture, Agricultural Research Service, *ARS–30*, 33–6.
Smith S. J., 1981: Humanistic method in contemporary social geography. *Area* 13, 293–8.
Smith, T. R. 1984: Artificial intelligence and its applicability to geographic problem solving. *Professional Geographer*, 36, 147–58.
—— and Bretherton, F. P. 1972: Stability and conservation of mass in drainage basin evolution. *Water Resources Research*, 8(96), 1506–29.

—— and Pazner, M. 1984: Knowledge-based control of search and learning in a large-scale GIS. *Proceedings: International Symposium on Spatial Data Handling*, Zurich, August 1984, vol. 2, 498–519.
—— and Peuquet, D. J. 1985: Control of spatial search for complex queries in a knowledge-based geographic information system. *Proceedings: International RSS/CERMA Conference of Advanced Technology for Monitoring and Processing Global Environmental Data*, London, 439–53.
Starfield, A. M. and Bleloch, A. L. 1983: Expert systems: an approach to problem solving in ecological management that are difficult to quantify. *Journal of Environmental Management*, 16, 261–8.
Stephens, G. L. and Webster, P. J. 1979: Sensitivity of radiative forcing to variable cloud and moisture. *Journal of Atmospheric Science*, 36, 1542–56.
Stevens, B. H. 1968: Location theory and programming models: the von Thünen case. *Papers and Proceedings of the Regional Science Association*, 21, 19–34.
Strahler, A. N. 1950: Equilibrium theory of erosional slopes. *American Journal of Science*, 248, 673–96 and 800–14.
—— 1952: Dynamic basis of geomorphology. *Bulletin of the Geological Society of America*, 63, 923–38.
—— 1980: Systems theory in physical geography. *Physical Geography*, 1, 1–27.
Stoddart, D. 1986: *On Geography and its History*, Blackwell.
Strong, D. R., Simberloff, D., Abele, L. G. and Thistle, A. B. (eds) 1984: *Ecological Communities: conceptual issues and the evidence*, Princeton University Press.
Sugden, D. 1978: Glacial erosion by the Laurentide ice sheet. *Journal of Glaciology*, 20, 367–91.
Switzer, P. 1980: Extensions of linear discriminant analysis for statistical classification of remotely sensed satellite imagery. *Mathematical Geology*, 12, 367–76.
—— 1983: Some spatial statistics for the interpretation of satellite data. *Bulletin of the International Statistical Institute*, 50, I, 28, 3.
Takayama, T. and Judge, G. 1971: *Spatial and Temporal Price and Allocation Models*, North Holland.
Taper, M. and Case, T. J. 1985: Quantitative genetic models for the evolution of character displacement. *Ecology*, 66, 355–71.
Theurer, F. D. 1985: Evaluating impacts of soil and water conservation. United States Department of Agriculture, Agricultural Research Service, *ARS–30*, 1–7.
Thom, R. 1975: *Structural Instability and Morphogenesis*, Benjamin.
Thomas, R. W. and Huggett, R. J. 1980: *Modelling in Geography*, Harper & Row.
Thornes, J. B. 1980: Structural instability and ephemeral stream behaviour. *Zeitschrift für Geomorphologie, Supplementband* 36, 233–44.
—— 1985: Ecology of erosion. *Geography*, 70(3), 222–36.
—— 1987: Models for palaeohydrology: in practice. In K. J. Gregory, J. Lewin and J. B. Thornes (eds), *Paleohydrology in Practice*, Wiley, 17–36.
—— 1988: Erosional equilibria under grazing. In J. Bingliff and M. Shackley (eds), *Environmental Archaeology*, Edinburgh University Press.
Thornthwaite, C. W. and Mather, J. R. 1955: The water balance. *Climatology*, VIII, no. 1, Drexel Institute of Technology Laboratory of Climatology, Centerton, New Jersey.
Thurow, L. 1980: *The Zero–Sum Society: distribution and the possibilities for*

economic change, Penguin.
Tobler, W. 1957: Automation and cartography. *Geographical Review*, 49, 526–34.
—— and Kennedy, S. 1983: Geographic interpolation. *Geographical Analysis*, 15, 151–6.
Tomlinson Associates 1987: Review of North American experience of current and potential uses of Geographic Information Systems. Appendix 6 in Department of Environment (1987), *Handling Geographic Information*, Her Majesty's Stationery Office.
Tomlinson, R. F., Calkins, H. W. and Marble, D. F. 1976: Computer handling of geographical data. *Natural Resources Research Services*, No. 13, UNESCO.
Townshend, P. 1984: *Why are the Many Poor?*, Fabian Society.
Triesmann, P. 1974: The radical use of official data. In N. Armistead (ed.), *Reconstructing Social Psychology*, Penguin.
Trieste, D. J. and Gifford, G. F. 1980: Application of the universal soil loss equation to rangelands on a per-storm basis. *Journal of Range Management*, 33, 66–70.
Trovimov, A. M. and Moskovin, V. M. 1984: The dynamic models of geomorphological systems. *Zeitschrift für Geomorphologie*, 28, 77–94.
Tucker, C. J. and Sellers, P. J. 1986: Satellite remote sensing of primary production. *International Journal of Remote Sensing*, 7, 1395–416.
——, Townshend, J. R. G. and Goff, T. E. 1983: Continental land-cover classification using meteorological satellite data. NASA Technical Memorandum 86060.
United States Department of Agriculture, 1982: Proceedings of the Workshop on Estimating Erosion and Sediment Yields on Rangelands, ARS, ARM–W26.
—— 1985: *Proceedings of the Natural Resources Modelling Symposium, ARS-30.*
Unwin, D. J. 1977: Statistical methods in physical geography. *Progress in Physical Geography*, 1, 185–221.
—— 1981a: Climatology. In N. Wrigley and R. J. Bennett (eds), *Quantitative Geography: a British view*, Routledge & Kegan Paul, 261–72.
—— 1981b: Teaching a model-based climatology using energy balance simulation. *Journal of Geography in Higher Education*, 5, 133–8.
Users' Guide to Global Vegetation Index, 1983: Satellite Data Services Division (SDSD), National Climatic Data Center, National Environmental Satellite Data and Information Service, Washington, DC.
Van Dyne, G. M. 1966: *Ecosystems, Systems Ecology and Systems Ecologists*, Oak Ridge Nature Laboratory Report, ORNL 3957.
Van Loon, H., Taljaard, J. J., Sasamori, T., London, J., Hoyt, D. V., Labicze, K. and Newton, C. W. 1972: *Meteorology of the Southern Hemisphere, Meteorology Monogram*, 13, American Meteorological Society.
Voorhees, A. M. 1955: A general theory of traffic movement. *Proceedings of the Institute of Traffic Engineering*, 1, 46–56.
Wardley, N. W. 1984: Vegetation index variability as a function of viewing geometry. *International Journal of Remote Sensing*, 5, 861–70.
Washington, W. M. and Meehl, G. A. 1984: Seasonal cycle experiment on the climate sensitivity due to a doubling of CO_2 with an atmospheric general circulation model coupled to a simple mixed-layer ocean model. *Journal of Geophysical Research*, 89, 9475–503.
—— and Parkinson, C. L. 1986: *An Introduction to Three-dimensional Climate*

Modelling, University Science Books.
Waterman, D. 1986: *Introduction to Expert Systems*, Wiley.
Watt, K. E. F. 1975: Critique and comparison of biome ecosystems modelling. In B. C. Patten (ed.), *Systems Analysis and Simulation in Ecological Modelling*, Academic Press, 139–52.
Webber, M. J. 1987: Profits, crises and industrial change. 1: Theoretical considerations. *Antipode*, 19, 307–28.
Webster, P. J. and Stephens, G. L. 1984: Cloud–radiation interaction and the climate problem. In J. T. Houghton (ed.), *The Global Climate*, Cambridge University Press, 63–78.
Webster, R. 1985: Quantitative spatial analysis of soil in the field. *Advances in Soil Science*, 3, 1–70.
Wetherald, R. T. and Manabe, S. 1986: An investigation of cloud cover change in response to thermal forcing. *Climatic Change*, 8, 5–23.
Whipple, F. J. W. 1942: The variability in India of rainfall and the growth of population. *Quarterly Journal of the Royal Meteorological Society*, 68, 301.
White, M. 1987: Digital map requirements of vehicle navigation. *Proceedings, Auto Carto*, 8, 552–61. American Congress of Surveying and Mapping, Washington, DC.
Whittle, P. 1954: On stationary processes in the plane. *Biometrika*, 41, 434–49.
Wiener, N. 1949 *Extrapolation, Interpolation and Smoothing for Stationary Time Series*, Wiley.
Wiggins, J. C., Hartley, R. P., Higgins, M. J. and Whittaker, R. J. 1987: Computer aspects of a large GIS for the European Community. *International Journal of GIS*, 1(1), 77–88.
Wigley, T. M. L. and Schlesinger, M. E. 1985: Analytical solution for the effect of increasing CO_2 on global mean temperature. *Nature*, 315, 649–52.
——, Jones, P. D. and Kelly, P. M. 1980: Scenario for a warm, high-CO_2 world. *Nature*, 283, 17–21.
Wildavsky, A. 1973: Consumer report. *Science*, 182, 1335–8.
Williams, H. C. W. L. 1977: On the formation of travel demand models and economic evaluation measures of user benefit. *Environment Planning A*, 9, 285–344.
Williams, J. R. 1985: The physical components of the EPCI model. In S. A. El-Swaify, W. C. Modenhauer and A. Lo (eds), *Soil Erosion and Conservation*, Soil Conservation of America, 272–83.
Williamson, M. 1971: *The Analysis of Biological Populations*, Arnold.
Wilson, A. G. 1967: A statistical theory of spatial distribution models. *Transportation Research*, 1, 253–69.
—— 1972: Theoretical geography: some speculations. *Transactions of the Institute of British Geographers*, 57, 31–44.
—— 1974: *Urban and Regional Models in Geography and Planning*, Wiley.
—— 1981a: *Catastrophe Theory and Bifurcation: applications to urban and regional systems*, Croom Helm.
—— 1981b: *Geography and the Environment*, Wiley.
—— 1984: One man's quantitative geography: frameworks, evaluations, uses and prospects. In M. Billinge, D. Gregory and R. Martin (eds), *Recollections of a Revolution*, Macmillan, 200–26.
—— and Bennett, R. J. 1985: *Mathematical Methods in Human Geography and Planning*, Wiley.
—— and Birkin, M. 1987: Dynamic models of agricultural location in a spatial

interaction framework. *Geographical Analysis*, 19(1), 31–56.
——, Coelho, J. D., Macgill, S. M. and Williams, H. C. W. L. 1981: *Optimization in Locational and Transport Analysis*, Wiley.
——, Rees, P. H. and Leigh, C. M. (eds) 1977: *Models of Cities and Regions: theoretical and empirical developments*, Wiley.
Wilson, C. A. and Mitchell, J. F. B. 1986: Diurnal variation and cloud in a general circulation model. *Quarterly Journal of the Royal Meteorological Society*, 112, 347–69.
Wilson, M. F., Henderson-Sellers, A., Dickinson, R. E. and Kennedy, P. J. 1987: Sensitivity of the biosphere/atmosphere transfer scheme (BATS) to the inclusion of variable soil characteristics. *Journal of Climatology and Applied Meteorology*.
Wischmeier, W. H. and Smith, D. D. 1958: Rainfall energy and its relationship to soil loss. *Transactions of the American Geophysical Union*, 39, 285–91.
Wolf, F. L. 1962: *Elements of Probability and Statistics*, McGraw-Hill.
Wolman, M. G. and Miller, J. P. 1960: Magnitude and frequency of forces in geomorphic events. *Journal of Geology*, 68, 54–74.
Wolpert, J. 1964: The decision making process in spatial context. *Annals of the Association of American Geographers*, 54, 337–58.
Wooding, R. A. 1965: A hydraulic model for the catchment–stream problem. I: Kinematic wave theory. *Journal of Hydrology*, 3, 195–223.
Woodmansee, R. H. 1978: Sensitivity analysis of ELM model. In G. S. Innis (ed.), *Grassland Simulation Model*, Ecological Studies, 26, 231–55.
World Climate Programme, 1985: Development of the implementation plan for the International Satellite Land-Surface Climatology project (ISLSCP) – Phase I, *WMO/TD-No. 46, WCP-94*.
World Meteorological Organization, 1986: Report of the International Conference on the Assessment of the Role of Carbon Dioxide and of other Greenhouse Gases in Climate Variations and Associated Impacts, held in Villach, Austria (October 1985), *WMO No. 661*.
Wreford-Watson, J. 1983: The soul of geography. *Transactions of the Institute of British Geographers*, N. S. 385–99.
Wright, J. K. 1947: Terrae incognitae: the place of the imagination in geography. *Annals of the Association of American Geographers*, 37, 1–15.
Wrigley, N. 1985: *Categorical Data Analysis for Geographers and Environmental Scientists*, Longmans.
—— (ed.) 1988: *Store Choice, Store Location and Market Analysis*, Routledge & Kegan Paul.
—— and Matthews, S. 1986: Citation classics and citation levels in geography. *Area*, 18, 185–94.
Wynne, B. 1984: The institutional context of science, models and policy. *The IIASA Energy Study*, Policy Sciences, 17, 227–320.
Yi-Fu Tuan, 1961: Topophilia, or sudden encounter with landscape. *Landscape*, 11, 29–32.
Young, A. 1963: Deductive models of slope evolution. *Nachrichten Akad. Wiss. Gottingen*, Math-Phys Klasse, Jahrgang 1963, 5, 45–66.
Younghusband, F. 1920: Natural beauty and geographical science. *Geographical Journal*, 56, 1–13.
Zelinsky, W. 1973: *The Cultural Geography of the United States*, Prentice-Hall.

Subject Index

Name Index